Progress in Relativity

Edited by Calin Gheorghe Buzea, Maricel Agop and Leo Butler

Published in London, United Kingdom

IntechOpen

Supporting open minds since 2005

Progress in Relativity
http://dx.doi.org/10.5772/intechopen.77603
Edited by Calin Gheorghe Buzea, Maricel Agop and Leo Butler

Contributors
Sikarin Yoo-Kong, Kadiata Ba, Akuro Big-Alabo, Chinwuba Ossia, Giorgio Turchetti, Federico Panichi, Richard Sauerheber, Lawrence Horwitz, Sergey Dmitrievich Prijmenko, Konstantin Lukin, Richard Perry Bocker, B. Roy Frieden, Igor Fomin, Vladimir Gladyshev, Francis Yu, Joás Venâncio, Carlos Batista, Andrzej Radosz, Pawel Gusin, Andy T. Augousti, Aleksander Kaczmarek, Calin Gheorghe Buzea, Decebal Vasnicu, Maricel Agop, Daniel Timofte

First published in London, United Kingdom, 2020 by IntechOpen
IntechOpen is the global imprint of INTECHOPEN LIMITED, registered in England and Wales, registration number: 11086078, 7th floor, 10 Lower Thames Street, London, EC3R 6AF, United Kingdom
Printed in Croatia

British Library Cataloguing-in-Publication Data
A catalogue record for this book is available from the British Library

Additional hard and PDF copies can be obtained from orders@intechopen.com

Progress in Relativity
Edited by Calin Gheorghe Buzea, Maricel Agop and Leo Butler
p. cm.
Print ISBN 978-1-78985-357-5
Online ISBN 978-1-78985-358-2
eBook (PDF) ISBN 978-1-83880-988-1

For EU product safety concerns:
IN TECH d.o.o., Prolaz Marije Krucifikse Kozulić 3, 51000 Rijeka, Croatia, info@intechopen.com or visit our website at intechopen.com.

Meet the editors

Dr. Calin Buzea currently works as a medical physicist at the Regional Institute of Oncology in Iasi, Romania. He is an international researcher with more than 25 years of experience in the field of theoretical, computational, and experimental physics. He is also an expert in medical physics with more than 15 years of clinical experience in hospitals. Today, he is focused on the dissemination and development of precision medicine-based radiomics.

Dr. Maricel Agop currently works as a professor at "Gheorghe Asachi" Technical Univeristy of Iasi, Romania. He is an international researcher with more than 30 years of experience in the field of theoretical physics. His expertise pertains to the topics of non-linearity, chaos theory, quantum mechanics, and fractals. Dr. Agop completed his PhD in Physics in 1983 at University "Al. I. Cuza" Iasi, Romania. He has published more than 200 papers in ISI journals, several books (or book chapters), and participated in national and international conferences and workshops. Dr. Agop is a Corresponding Member of the Romanian Academy of Science and Dr.h.c. of "Vasile Alecsandri" University of Bacau. He is also an editorial board member of several scientific journals.

Dr. Butler is currently an associate professor of Mathematics at the University of Manitoba in Winnipeg, Canada. He completed a degree in economics at the University of Ottawa (1990–1993); worked as a research assistant at the Bank of Canada (1993–1995); and studied mathematics at Queen's University (1995–2000) where he obtained an MSc and PhD. He has held positions at Northwestern University, Queen's University, University of Edinburgh, Central Michigan University, and North Dakota State University. His main research interests include Hamiltonian mechanics. He currently holds an NSERC Discovery Grant on the topic of thermostat dynamics.

Contents

Preface

Some things are absolute and some things are relative. This is a fact of life. If I look at the teapot on my desk, I see that it sits to the left of my cup. If you are sitting opposite me, you will see the teapot to the right of my cup. “Left” and “right” are relative. Whether or not an object is found to the left or to the right of another depends on the observer. While this may be true, if the cup is filled to the top with coffee, all observers should approve this as actual fact, no matter where they sit. That, it’d seem, is an absolute statement, independent of who makes the observation.

The theory of relativity usually incorporates two interconnected theories by Albert Einstein: special relativity and general relativity. Einstein’s special theory of relativity (special relativity) conceived in 1905, available within the paper “On the Electrodynamics of Moving Bodies”, is about what’s relative and what’s absolute about time, space, and motion.

General relativity centers on gravitational, electromagnetic, and velocity fields, as well as functions of space and time, and density distributions that define masses and charges. Space–time is that the arena within which these fields accomplish their combined evolutions. It’s therefore clear that we must first grasp the structure and geometry of space–time. Unluckily, because the velocity of light is so big, routine experience leads us to amass various false impressions about the geometry of space–time. This set of mistaken beliefs is known as Newtonian, or Galilean, space–time. The true (or truer) geometry of space–time was revealed through the improvement of Einstein’s theory of special relativity. The foundation of this theory is the principle of relativity, in line with which the laws of physics are similar in all inertial reference frames. Einstein ran into the current principle by his investigation of Maxwell’s equations.

Special relativity is restricted to things that are moving with regard to inertial frames of reference. That is, during a state of uniform motion with regard to each other, one cannot, by purely mechanical experiments, distinguish one from the opposite. Beginning with the behavior of light (and all other electromagnetic radiation), the theory of special relativity draws conclusions that conflict with daily knowledge, but is fully set by tests that examine subatomic particles at high speeds or measure minor changes between clocks traveling at different velocities. Special relativity discovered that the speed of light is a limit that cannot be reached by any material thing. It is the origin of the famous scientific equation $E = mc^2$, which states that mass and energy are identical physical entities and might be changed one into the other.

Together with quantum mechanics, the theory of relativity is fundamental to modern physics.

This volume deals with extensions of special relativity, general relativity, and their applications in relation to intragalactic and extragalactic dynamics.

In addition, fundamental problems of these extensions are addressed, both classically and quantum mechanically, in Hamiltonian, Lagragian, and matrix formalisms by Richard P. Bocker and B. Roy Frieden in Chapter 2, Sikarin Yoo-Kong in Chapter 5, and Big-Alabo Akuro in Chapter 14, respectively.

Thus, extensions of special relativity are presented extensively by Richard Sauerheber in Chapter 3.

Extensions of general relativity are presented by Francis T.S. Yu in Chapter 6.

The foundations of these extensions are given special attention in the form of Lagrangian formalism by Kadiata Ba in Chapter 7, canonical formalism by Lawrence P. Horwitz in Chapter 9, and variational formalism by Giorgio Turchetti and Federico Panichi in Chapter 10.

In terms of applications, special attention is paid to the nature of light and dark matter, as well as dynamics involving exotic materials such as black holes, wormholes, and other structures involving special topologies. These topics are covered by Joás Venâcio and Carlos Batista in Chapter 1, S. D. Prijmenko and K.A. Lukin in Chapter 4, Maricel Agop et al. in Chapter 8, Giorgio Turchetti and Federico Panichi in Chapter 10, Radosz Andrzej in Chapters 11 and 12, and Fomin Igor in Chapter 13.

Calin Gheorghe Buzea
National Institute of Research and Development for Technical Physics,
Iasi, Romania

Chapter 1

Quasinormal Modes of Dirac Field in Generalized Nariai Spacetimes

Joás Venâncio and Carlos Batista

Abstract

The exact electrically charged solutions to the Dirac equation in higher-dimensional generalized Nariai spacetimes are obtained. Using these solutions, the boundary conditions leading to quasinormal modes of the Dirac field are analyzed, and their correspondent quasinormal frequencies are analytically calculated.

Keywords: quasinormal modes, generalized Nariai spacetimes, Dirac field, boundary conditions

1. Introduction

Quasinormal modes (QNMs) are eigenmodes of dissipative systems. For instance, if a spacetime with an event or cosmological horizon is perturbed from its equilibrium state, QNMs arise as damped oscillations with a spectrum of complex frequencies that do not depend on the details of the excitation. In fact, these frequencies depend just on the charges of the black hole, such as the mass, electric charge, and angular momentum [1, 2]. QNMs have been studied for a long time, and its interest has been renewed by the recent detection of gravitational waves, inasmuch as these are the modes that survive for a longer time when a background is perturbed and, therefore, these are the configurations that are generally measured by experiments [3–29]. Mathematically, this discrete spectrum of QNMs stems from the fact that certain boundary conditions must be imposed to the physical fields propagating in such background [30]. In this chapter, we consider a higher-dimensional generalization of the charged Nariai spacetime [31], namely, $dS_2 \times S^2 \times \ldots \times S^2$, and investigate the dynamics of perturbations of the electrically charged Dirac field (spin 1/2). In such a geometry, the spinorial formalism [32–34] is used to show that the Dirac equation is separable [35] and can be reduced to a Schrödinger-like equation [36] whose potential is contained in the Rosen-Morse class of integrable potentials, which has the so-called Pöschl-Teller potential as a particular case [37, 38]. Finally, the boundary conditions leading to QNMs are analyzed, and the quasinormal frequencies (QNFs) are analytically obtained [5, 39].

2. Presenting the problem

In D dimensions, the dynamics of general relativity in spacetimes with a cosmological constant Λ is described by the Einstein-Hilbert action[1]

$$S = \frac{1}{16\pi}\int d^D x\sqrt{|g|}[\mathcal{R} - (D-2)\Lambda] + S_m, \tag{1}$$

where $\mathcal{R}$ is the Ricci scalar and S_m stands for the action of all matter fields $\{\boldsymbol{\Phi}_i\}$ coupled to gravity appearing in the theory, which can be scalar, spinorial, vectorial, and so on. The least action principle allows to find the equations of motion for the fields $g_{\mu\nu}$ and $\boldsymbol{\Phi}_i$ which are given, respectively, by

$$\mathcal{R}_{\mu\nu} - \frac{1}{2}\mathcal{R}g_{\mu\nu} + \frac{(D-2)}{2}\Lambda g_{\mu\nu} = 8\pi\mathcal{T}_{\mu\nu}, \qquad \frac{\delta S_m}{\delta \boldsymbol{\Phi}_i} = 0, \tag{2}$$

where $\mathcal{T}_{\mu\nu}$ is the symmetric stress-energy tensor associated to $\boldsymbol{\Phi}_i$ defined by the equation

$$\mathcal{T}^{\mu\nu} = \frac{2}{\sqrt{|g|}}\frac{\delta S_m}{\delta g_{\mu\nu}}. \tag{3}$$

Since any symmetry has been imposed, the general solution of the system of Eq. (2) is some metric and fields in the background this metric

$$ds^2 = g_{\mu\nu}(x)dx^\mu dx^\nu, \quad \boldsymbol{\Phi}_i = \boldsymbol{\Phi}_i(x). \tag{4}$$

Now, let the pair $g^{(0)}_{\mu\nu}$ and $\boldsymbol{\Phi}^{(0)}_i$ be a solution for the equations of motion Eq. (2). Then, in order to study the perturbations around this particular solution, we write our fields as a sum of the unperturbed fields $g^{(0)}_{\mu\nu}$ and $\boldsymbol{\Phi}^{(0)}_i$ and the small perturbations $h_{\mu\nu}$ and $\boldsymbol{\Psi}_i$

$$g_{\mu\nu} = g^{(0)}_{\mu\nu} + h_{\mu\nu}, \quad \boldsymbol{\Phi}_i = \boldsymbol{\Phi}^{(0)}_i + \boldsymbol{\Psi}_i, \tag{5}$$

where by "small" we mean that we neglect the quadratic and higher-order powers of the perturbation fields. Inserting the above equation into Eq. (2), we are left with a set of linear equations satisfied by the perturbed fields $h_{\mu\nu}$ and $\boldsymbol{\Psi}_i$. In general, these equations are coupled, namely, $\boldsymbol{\Psi}_i$ is a source for $h_{\mu\nu}$ and vice versa. However, in the special case in which $\boldsymbol{\Phi}^{(0)}_i = 0$, the equations governing the perturbed fields $\boldsymbol{\Psi}_i$ can be decoupled from the metric perturbation $h_{\mu\nu}$ and vice versa. The reason why this happen is that when $\boldsymbol{\Phi}^{(0)}_i = 0$, the stress-energy tensor $\mathcal{T}_{\mu\nu}$ can be set to zero at first order in the perturbation, since $\mathcal{T}_{\mu\nu}$ is typically quadratic or of higher order in the matter fields and, therefore, can be neglected. Therefore, investigating the linear dynamics of generic small perturbations of the matter fields with $\mathcal{T}_{\mu\nu} = 0$ is equivalent to studying the test fields $\boldsymbol{\Psi}_i$ in the background $g^{(0)}_{\mu\nu}$.

In what follows, let us consider a specific matter field $\boldsymbol{\Psi}$ propagating in a generalized version of the Nariai spacetime described in Ref. [31]. Here, $\boldsymbol{\Psi}$ is an

[1] The coefficient of Λ in S can be chosen of several manners. In particular, for any dimension D, in order to insure that the pure dS or pure AdS spacetimes are described by $g_{tt} = 1 - (\Lambda/3)r^2$, as occurs in the case $D = 4$, this coefficient should be $(D-1)(D-2)$.

electrically charged spinorial field of mass m that obeys the Dirac equation minimally coupled to an electromagnetic field in such spacetime. In $D = 2d$, this spacetime is formed from the direct product of the de Sitter space dS_2 with $(d-1)$ copies of the unit spheres S^2 possessing different radii R_j. Thus, the natural line element of the higher-dimensional version of the Nariai spacetime is given by

$$ds^2 = g^0_{\mu\nu} dx^\mu dx^\nu = -f(r)dt^2 + \frac{1}{f(r)} dr^2 + \sum_{j=2}^{d} R_j^2 d\Omega_j^2, \tag{6}$$

where $f(r)$ is a function of the coordinate r and $d\Omega_j^2$ is the line element of the jth unit sphere S^2 as follows

$$f(r) = 1 - \frac{r^2}{R_1^2}, \quad d\Omega_j^2 = d\theta_j^2 + \sin^2\theta_j d\phi_j^2. \tag{7}$$

The radii R_1 and R_j are given by

$$R_1 = \left[\Lambda - \frac{1}{2}Q_1^2 + \frac{Q}{2(D-2)}\right]^{-1/2}, \; R_j = \left[\Lambda + \frac{1}{2}Q_j^2 + \frac{Q}{2(D-2)}\right]^{-1/2}, \tag{8}$$

where Q_1 is an electric charge and Q_j are magnetic charges, while Q is defined by

$$Q = Q_1^2 - \sum_{j=2}^{d} Q_j^2. \tag{9}$$

This spacetime is a locally static solution of Einstein's equation with a cosmological constant Λ and electromagnetic field $\mathcal{F} = d\boldsymbol{\mathcal{A}}$ whose gauge field $\mathcal{A}$ is given by

$$\boldsymbol{\mathcal{A}} = Q_1 r dt + \sum_{j=2}^{d} Q_j R_j^2 \cos\theta_j d\phi_j. \tag{10}$$

The coordinates in the metric are also called static, because they do not depend explicitly on the time coordinate t. One may notice that, in this coordinate system, this background has a local Killing vector ∂_t whose norm vanishes at $r = \pm R_1$. Indeed, $r = \pm R_1$ define closed null surfaces that surround the observer at all times, known as event horizons. The boundary conditions defining QNMs in our spacetime will be posed at these surfaces, as discussed in [39]. For this reason, the dependence of all the components of the field $\boldsymbol{\Psi}$ on the coordinates along the Killing vector ∂_t is assumed to be of the form $e^{-i\omega t}$. Usually, the articles consider that the coordinate r in de Sitter space assume values in the interval $r \in (0, R_1)$ [40–42]. However, this is just justified for de Sitter with $D > 2$, but not for $D = 2$; see [39] for more details. By this reason, our domain of interest will be $r \in (-R_1, R_1)$. In such domain, it is useful to introduce the tortoise coordinate x defined by the equation

$$dx = \frac{1}{f(r)} dr \Rightarrow x = R_1 \operatorname{arctanh}\left(\frac{r}{R_1}\right), \tag{11}$$

in terms of which the line element Eq. (6) becomes

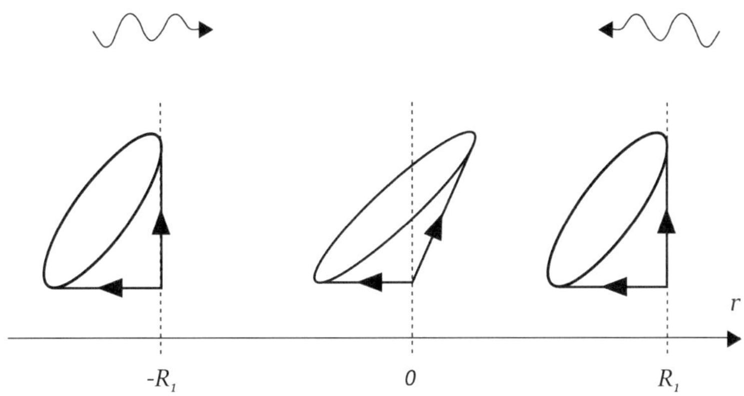

Figure 1.
Illustration of the boundary condition associated to QNMs in our spacetime. The wavy arrows represent the direction of the perturbation field at the boundaries $r = \pm R_1$, while the cones are the local light cones. Mathematically, the wavy arrow pointing to the right represents $e^{-i\omega(t-x)}$, while the wavy arrow pointing to the left represents $e^{-i\omega(t+x)}$. For more details, see Ref. [39].

$$ds^2 = \frac{1}{\cosh^2(x/R_1)}\left(-dt^2 + dx^2\right) + \sum_{j=2}^{d} R_j^2 d\Omega_j^2, \tag{12}$$

and the gauge field can be rewritten as

$$\mathcal{A} = Q_1 R_1 \tanh(x/R_1) dt + \sum_{j=2}^{d} Q_j R_j^2 \cos\theta_j d\phi_j. \tag{13}$$

In particular, note that the tortoise coordinate maps the domain between two horizons, $r \in (-R_1, R_1)$, into the interval $x \in (-\infty, \infty)$.

The QNMs accounting for an important class of fields are associated to Ψ which are solutions to the equations of motion that satisfy specific boundary conditions imposed at the horizons of the spacetime in which the field is propagating; see [5, 6, 43, 44] for more details. In this chapter, we will use the boundary conditions as illustrated in **Figure 1**.

From the mathematical of view, since we are assuming that the time dependence of Ψ is $e^{-i\omega t}$, this boundary condition means that near the horizons $r = \pm R_1$, that is, as $x \to \pm\infty$, the radial component of the field Ψ should behave as $e^{-i\omega(t+x)}$ at $x \to \infty$, while it should go as $e^{-i\omega(t-x)}$ at $x \to -\infty$. The eigenfrequencies of this problem are complex, the reason why they are called QNFs. The real part of the QNFs is associated with the oscillation frequencies of the signal, while the imaginary part is related to its decay in time. This decay in time is closely related to the fact that the event horizon has a dissipative nature.

One interesting feature of this spacetime is that we can compute exactly the QNMs. The exactly solvable systems are usually limits of more realistic systems and allow us to study in detail some properties of a physical process and test some methods which can be used to analyze more complicated systems. Thus they are powerful tools in many research lines. Therefore we expect that the exactly computed QNFs for D-dimensional generalized Nariai spacetime may play an important role in future research [27].

3. Dirac equation in D-dimensional generalized Nariai spacetime

Let us present the construction of a solution to the Dirac equation minimally coupled to the electromagnetic field of D-dimensional generalized Nariai spacetime.

A field of spin 1/2 with electric charge q and mass m propagating in such spacetime is a spinorial field obeying the following version of the Dirac equation:

$$\Gamma^{\alpha}(\nabla_{\alpha} - iq\mathcal{A}_{\alpha})\boldsymbol{\Psi} = m\boldsymbol{\Psi}, \tag{14}$$

where $\mathcal{A}_{\alpha}$ stands for the components of the background gauge field. In $D = 2d$ dimensions, the Dirac matrices Γ^{α} represent faithfully the Clifford algebra by $2^d \times 2^d$ matrices obeying the relation

$$\Gamma_{\alpha}\Gamma_{\beta} + \Gamma_{\beta}\Gamma_{\alpha} = \boldsymbol{g}(\boldsymbol{e}_{\alpha}, \boldsymbol{e}_{\beta})\,\mathbb{I}_d, \tag{15}$$

with $\mathbb{I}_d$ standing for the $2^d \times 2^d$ identity matrix. The index α, β, γ run from 1 to $2d$ and label the vector fields of an orthonormal frame $\{\boldsymbol{e}_{\alpha}\}$. In order to solve the Dirac equation, we must introduce a suitable orthonormal frame of vector fields, which in the case of our background is given by

$$\begin{aligned} \boldsymbol{e}_1 &= -i\cosh(x/R_1)\,\partial_t\ ,\ \ \boldsymbol{e}_j = \frac{1}{R_j \sin\theta_j}\,\partial_{\phi_j}, \\ \boldsymbol{e}_{\tilde{1}} &= \cosh(x/R_1)\,\partial_x\ ,\ \ \boldsymbol{e}_{\tilde{j}} = \frac{1}{R_j}\,\partial_{\theta_j}, \end{aligned} \tag{16}$$

where the index j ranges from 2 to d. In particular, note that

$$\boldsymbol{g}(\boldsymbol{e}_{\alpha}, \boldsymbol{e}_{\beta}) = \delta_{\alpha\beta} \leftrightarrow \begin{cases} \boldsymbol{g}(\boldsymbol{e}_a, \boldsymbol{e}_b) = \delta_{ab}, \\ \boldsymbol{g}(\boldsymbol{e}_a, \boldsymbol{e}_{\tilde{b}}) = 0, \\ \boldsymbol{g}(\boldsymbol{e}_{\tilde{a}}, \boldsymbol{e}_{\tilde{b}}) = \delta_{\tilde{a}\tilde{b}}, \end{cases} \tag{17}$$

where a and $\tilde{a}$ are indices that range from 1 to d. The index a labels the first d vector fields of the orthonormal frame $\{\boldsymbol{e}_a\}$, while the index $\tilde{a}$ labels the remaining d vectors of the frame $\{\boldsymbol{e}_a\}$. The derivatives of the frame vector fields determine the spin connection according to the following relation:

$$\nabla_{\alpha}\boldsymbol{e}_{\beta} = \omega_{\alpha\beta}{}^{\gamma}\,\boldsymbol{e}_{\gamma}. \tag{18}$$

Since the metric g is a covariantly constant tensor, it follows that the coefficients of the spin connection with all low indices $\omega_{\alpha\beta\gamma} = \omega_{\alpha\beta}{}^{\varepsilon}\,\delta_{\varepsilon\gamma}$ are antisymmetric in their two last indices, $\omega_{\alpha\beta\gamma} = -\omega_{\alpha\gamma\beta}$. Note that the indices of the spin connection are raised and lowered with $\delta_{\alpha\beta}$ and $\delta^{\alpha\beta}$, respectively, so that frame indices can be raised and lowered unpunished. In particular, $\omega_{\alpha}^{\beta\gamma} = \omega_{\alpha}^{[\beta\gamma]}$, where indices inside the square brackets are antisymmetrized. The covariant derivative of a spinorial field $\boldsymbol{\Psi}$ is, then, given by

$$\nabla_{\alpha}\boldsymbol{\Psi} = \partial_{\alpha}\boldsymbol{\Psi} - \frac{1}{4}\,\omega_{\alpha}{}^{\beta\gamma}\,\Gamma_{\beta}\Gamma_{\gamma}\boldsymbol{\Psi}, \tag{19}$$

with ∂_{α} denoting the partial derivative along the vector field $\boldsymbol{e}_{\alpha}$.

Our aim is to separate the Dirac Eq. (14). In order to accomplish this, it is necessary to use a suitable representation for the Dirac matrices. We recall that

$$\sigma_1 = \begin{bmatrix} 0 & 1 \\ 1 & 0 \end{bmatrix}, \quad \sigma_2 = \begin{bmatrix} 0 & -i \\ i & 0 \end{bmatrix}, \quad \sigma_3 = \begin{bmatrix} 1 & 0 \\ 0 & -1 \end{bmatrix}, \tag{20}$$

are the Hermitian Pauli matrices and $\mathbb{I}$ denote the 2×2 identity matrix. Using this notation, a convenient representation of the Dirac matrices is the following:

$$\begin{aligned}
\Gamma_a &= \underbrace{\sigma_3 \otimes \ldots \otimes}_{(a-1)\text{ times}} \sigma_1 \underbrace{\otimes \mathbb{I} \otimes \ldots \otimes \mathbb{I}}_{(d-a)\text{ times}}, \\
\Gamma_{\tilde{a}} &= \underbrace{\sigma_3 \otimes \ldots \otimes}_{(a-1)\text{ times}} \sigma_2 \underbrace{\otimes \mathbb{I} \otimes \ldots \otimes \mathbb{I}}_{(d-a)\text{ times}},
\end{aligned} \tag{21}$$

where $\mathbb{I}$ stands for the 2×2 identity matrix. Indeed, we can easily check that the Clifford algebra given in Eq. (15) is properly satisfied by the above matrices.[2] In this case, spinorial fields are represented by the column vectors on which these matrices act. We can introduce a basis of this representation by the direct products of spinors $\boldsymbol{\xi}^s$ given by

$$\boldsymbol{\xi}^+ = \begin{bmatrix} 1 \\ 0 \end{bmatrix}, \quad \boldsymbol{\xi}^- = \begin{bmatrix} 0 \\ 1 \end{bmatrix}, \tag{22}$$

which, under the action of the Pauli matrices, satisfy concisely the relations

$$\sigma_1 \boldsymbol{\xi}^s = \boldsymbol{\xi}^{-s}, \quad \sigma_2 \boldsymbol{\xi}^s = is\boldsymbol{\xi}^{-s}, \quad \sigma_3 \boldsymbol{\xi}^s = s\boldsymbol{\xi}^s. \tag{23}$$

Indeed, in $D = 2d$ dimensions, a general spinor field has 2^d degrees of freedom and can be written as

$$\boldsymbol{\Psi} = \sum_{\{s\}} \boldsymbol{\Psi}^{s_1 s_2 \ldots s_d} \boldsymbol{\xi}^{s_1} \otimes \boldsymbol{\xi}^{s_2} \otimes \ldots \otimes \boldsymbol{\xi}^{s_d}, \tag{24}$$

where each of the indices s_a can take the values "+1" and "−1." Since every s_a can take just two values, it follows that the sum over $\{s\} \equiv \{s_1, s_2, \ldots, s_d\}$ comprises 2^d terms, which is exactly the number of components of a spinorial field in $D = 2d$ dimensions.

In the representation (Eq. (21)), the operator $\Gamma^\alpha \nabla_\alpha$, called Dirac operator, is then represented by

$$\Gamma^\alpha \nabla_\alpha = \sum_{a=1}^{d} (\Gamma_a \nabla_a + \Gamma_{\tilde{a}} \nabla_{\tilde{a}}) = \sum_{a=1}^{d} \underbrace{\sigma_3 \otimes \ldots \otimes}_{(a-1)\text{ times}} D_a \underbrace{\otimes \mathbb{I} \otimes \ldots \otimes \mathbb{I}}_{(d-a)\text{ times}}, \tag{25}$$

where

$$D_a = \sigma_1 \nabla_a + \sigma_2 \nabla_{\tilde{a}}, \tag{26}$$

is the Dirac operator on $\mathbb{R}^2$ with coordinates $\{x^a, y^a\}$. The spinorial basis introduced previously is very convenient, since the action of the Dirac matrices on the spinor fields can be easily computed. Indeed, using Eqs. (21), (23), and (24), we eventually arrive at the following equation

---

[2] In $D = 2d + 1$, besides the $2d$ Dirac matrices Γ_a and $\Gamma_{\tilde{a}}$, we need to add one further matrix, which will be denoted by Γ_{d+1} given by $\Gamma_{d+1} = \underbrace{\sigma_3 \otimes \sigma_3 \ldots \otimes \sigma_3}_{d\text{ times}}$.

$$\Gamma_a \boldsymbol{\Psi} = \sum_{\{s\}} (s_1 s_2 \dots s_{a-1}) \boldsymbol{\Psi}^{s_1 s_2 \dots s_d} \boldsymbol{\xi}^{s_1} \otimes \boldsymbol{\xi}^{s_2} \otimes \dots \otimes \boldsymbol{\xi}^{s_{a-1}} \otimes \boldsymbol{\xi}^{-s_a} \otimes \boldsymbol{\xi}^{s_{a+1}} \otimes$$
$$\dots \otimes \boldsymbol{\xi}^{s_d} = \sum_{\{s\}} (s_1 s_2 \dots s_a) \boldsymbol{\Psi}^{s_1 s_2 \dots s_{a-1} (-s_a) s_{a+1} \dots s_d} \boldsymbol{\xi}^{s_1} \otimes \boldsymbol{\xi}^{s_2} \otimes \tag{27}$$
$$\dots \otimes \boldsymbol{\xi}^{s_{a-1}} \otimes \boldsymbol{\xi}^{s_a} \otimes \boldsymbol{\xi}^{s_{a+1}} \otimes \dots \otimes \boldsymbol{\xi}^{s_d},$$

where from the first to the second line we have changed the index s_a to $-s_a$, which does not change the final result, since we are summing over all values of s_a, which comprise the same list of the values of $-s_a$. Moreover, we have used $(s_a)^2 = 1$. Analogously, we have:

$$\Gamma_{\tilde{a}} \boldsymbol{\Psi} = \sum_{\{s\}} (s_1 s_2 \dots s_{a-1}) (i s_a) \boldsymbol{\Psi}^{s_1 s_2 \dots s_d} \boldsymbol{\xi}^{s_1} \otimes \boldsymbol{\xi}^{s_2} \otimes \dots \otimes \boldsymbol{\xi}^{s_{a-1}} \otimes \boldsymbol{\xi}^{-s_a} \otimes \boldsymbol{\xi}^{s_{a+1}} \otimes$$
$$\dots \otimes \boldsymbol{\xi}^{s_d} = -i \sum_{\{s\}} (s_1 s_2 \dots s_a) s_a \boldsymbol{\Psi}^{s_1 s_2 \dots s_{a-1} (-s_a) s_{a+1} \dots s_d} \boldsymbol{\xi}^{s_1} \otimes \boldsymbol{\xi}^{s_2} \otimes \tag{28}$$
$$\dots \otimes \boldsymbol{\xi}^{s_{a-1}} \otimes \boldsymbol{\xi}^{s_a} \otimes \boldsymbol{\xi}^{s_{a+1}} \otimes \dots \otimes \boldsymbol{\xi}^{s_d}.$$

All that was seen above are necessary tools to attack our initial problem of separating the general Eq. (14). In order to solve such an equation, we need to separate the degrees of freedom of the field, which can be quite challenging in general. Fortunately, the spacetime considered here is the direct product of two-dimensional spaces of constant curvature, which is exactly the class of spaces studied in Ref. [39]. Indeed, in this latter paper, it is shown that the Dirac equation minimally coupled to an electromagnetic field is separable in such backgrounds. In particular, assuming that the components of the spinor field Eq. (24) can be decomposed in the form

$$\Psi^{s_1 s_2 \dots s_d} = \Psi_1^{s_1}(t, x) \Psi_2^{s_2}(\Phi_2, \theta_2) \dots \Psi_d^{s_d}(\Phi_d, \theta_d), \tag{29}$$

where each index s_a can take the values $s_a = \pm 1$, the fields $\Psi_1^{s_1}(t, x)$ satisfy the following differential equation (the reader is invited to demonstrate the equation below or consult more details in [39]):

$$\left[\partial_{\tilde{1}} + \frac{\omega_{1\tilde{1}1}}{2} - iqA_{\tilde{1}} - is_1\left(\partial_1 + \frac{\omega_{\tilde{1}1\tilde{1}}}{2} - iqA_1\right)\right]\Psi_1^{s_1} = (L - is_1 m)\Psi_1^{-s_1}. \tag{30}$$

The separation constant L in the above equation depends on the angular modes. In particular, in the special case of vanishing magnetic charges Q_j, it is determined by the eigenvalues λ_j of the Dirac operator on unit sphere S^2 according to the following relation

$$L = \sqrt{\lambda_2^2 + \lambda_3^2 + \dots + \lambda_d^2} \quad , \quad \lambda_j = \pm 1, \pm 2, \pm 3, \dots, \tag{31}$$

as demonstrated in Appendix A of Ref. [39]. In our frame of vectors, the only components of the spin connection that are potentially nonvanishing are

$$\begin{cases} \omega_{1\tilde{1}1} = -\omega_{11\tilde{1}} = -\dfrac{1}{R_1} \sinh(x/R_1), \\ \omega_{\tilde{j}j\tilde{j}} = -\omega_{\tilde{j}\tilde{j}j} = \dfrac{1}{R_j} \cot\theta_j, \end{cases} \tag{32}$$

and the nonzero components of the gauge field can be written as

$$\mathcal{A}_1 = -iQ_1R_1 \sinh(x/R_1), \quad \mathcal{A}_j = Q_jR_j \cot\theta_j. \tag{33}$$

Now, since the components of the metric are independents of the coordinate t, the vector ∂_t is a Killing vector for this metric. So, it is useful to assume the following time dependence for the field $\Psi_1^{s_1}(t, x)$

$$\Psi_1^{s_1}(t, x) = e^{-i\omega t}\psi^{s_1}(x). \tag{34}$$

Inserting this field along with the gauge field Eq. (33), and taking into account the first relation of the Eq. (32) into the Eq. (30), we end up with the following coupled system of differential equations:

$$\left[\frac{d}{dx} + is_1\omega + \left(isqQ_1R_1 - \frac{1}{2R_1}\right)\tanh(x/R_1)\right]\psi^{s_1} = \frac{(L - is_1m)}{\cosh(x/R_1)}\psi^{-s_1}. \tag{35}$$

In order to solve these equations, we should first decouple the fields ψ^{s_1} and ψ^{-s_1}. Eliminating ψ^{-s_1} we obtain a second-order equation for ψ^{s_1}. Indeed, we can prove that the fields ψ^{s_1} satisfy the following second-order ordinary differential equation

$$\left[\frac{d^2}{dx^2} + \omega^2 - V(x)\right]\psi^{s_1} = 0, \tag{36}$$

which is a Schrödinger-like equation with V being a potential of the form

$$V(x) = A + B\tanh(x/R_1) + \frac{C}{\cosh^2(x/R_1)}, \tag{37}$$

where the parameters A, B, and C are given by

$$\begin{cases} A = \dfrac{1}{4R_1^2} - qQ_1(is_1 + qQ_1R_1^2), \\[2ex] B = -\dfrac{\omega}{R_1}(is_1 + 2qQ_1R_1^2), \\[2ex] C = m^2 + L^2 + \dfrac{1}{4R_1^2} + q^2Q_1^2R_1^2. \end{cases} \tag{38}$$

These are known as potentials of Rosen-Morse type, which are generalizations of the Pöschl-Teller potential [37, 38]. It is straightforward to see that this potential satisfies the following properties:

$$V \to \begin{cases} A + B & \text{at} \quad x \to +\infty, \\ A - B & \text{at} \quad x \to -\infty. \end{cases} \tag{39}$$

In many cases, the potential function V is regular at $r = 0\,(x = 0)$, in particular V can be equal to a constant different from zero. In fact, in our case, we find that

$$V \to A + C \quad \text{at} \quad x \to 0, \tag{40}$$

which clearly is regular. So, we point out that for this potential both limits (Eqs. (39) and (40)) are finite, and thus there is no reason to demand for a regular solution in this point.

Thus, the problem of finding the QNMs is reduced to the searching of the corresponding spectrum of QNFs ω of Eq. (36). Most of the problems concerning the QNMs fall into Schrödinger-like equation with real potentials which vanish at both horizons [5], highlighting the fact that the solutions can be taken to be plane waves. However, clearly this is not the case. Although it is possible to make field redefinitions in order to make the potential real, we shall not do this here. For such procedure we refer the reader to [36]. Once an analytical form for the QNFs of Rosen-Morse type potential is not known, we must find an analytical exact solution of Eq. (36) and impose physically appropriate boundary conditions at the horizons, $x \to \pm\infty$, which define the QNFs in a unique way.

In order to solve Eq. (36), let us make the following change of variable

$$y = \frac{1}{2} + \frac{1}{2}\tanh(x/R_1). \tag{41}$$

In particular, notice that y is defined on the domain $y \in (0, 1)$ with the boundaries $x \to \pm\infty$ being given by $y = 0$ and $y = 1$. In addition to this change of independent variable, if we now set the Ansatz

$$\psi^{s_1}(x) = y^{\alpha}(1-y)^{\beta} H^{s_1}(y), \tag{42}$$

with the parameters α and β being constants conveniently chosen as

$$\alpha = \frac{R_1}{2}\sqrt{A - B - \omega^2} \quad , \quad \beta = -\frac{R_1}{2}\sqrt{A + B - \omega^2}, \tag{43}$$

the functions H^{s_1} must be solutions of the following differential equation

$$y(1-y)\frac{d^2H^{s_1}}{dy^2} + [2\alpha + 1 - (2 + 2\alpha + 2\beta)y]\frac{dH^{s_1}}{dy} - \left[CR_1^2 + (\alpha+\beta)(1+\alpha+\beta)\right]H^{s_1}. \tag{44}$$

This new variable as well as the Ansatz that we have been using are really interesting because in terms of these, it is immediate to see that the functions H^{s_1} satisfy a hypergeometric equation. Indeed, comparing with the standard hypergeometric differential equation

$$y(1-y)\frac{d^2H^{s_1}}{dy^2} + [c - (1 + a + b)y]\frac{dH^{s_1}}{dy} - abH^{s_1} = 0, \tag{45}$$

we find that the constants a, b, and c are given by

$$\begin{cases} a = \frac{1}{2} + \alpha + \beta + \sqrt{\frac{1}{4} - CR_1^2}, \\ b = \frac{1}{2} + \alpha + \beta - \sqrt{\frac{1}{4} - CR_1^2}, \\ c = 2\alpha + 1. \end{cases} \tag{46}$$

Such an equation admits two linearly independent solutions whose linear combination furnishes the following general solution:

$$H^{s_1}(y) = D_2F_1(a,b,c;y) + Ey^{(1-c)}{}_2F_1(1+a+c,1+b+c,2-c;y), \tag{47}$$

where ${}_2F_1$ is the hypergeometric function and D and E are arbitrary integration constants. Given the hypergeometric solution for H^{s_1} is known, one can immediately find the general solution for ψ^{s_1}. Indeed, from Eqs. (42), (46), and (47), we conclude that the solution of Eq. (36), which is regular at the origin, can be written as

$$\begin{aligned}\psi^{s_1} = {}& (1-y)^{\frac{1}{2}(a+b-c)}\big[Dy^{\frac{1}{2}(c-1)}{}_2F_1(a,b,c;y) \\ & + Ey^{-\frac{1}{2}(c-1)}{}_2F_1(1+a-c,1+b-c,2-c;y)\big].\end{aligned} \tag{48}$$

In order to fix the integration constants D and E, we need to apply the appropriate boundary conditions. Inverting the Eq. (41) we find that, near the boundaries $x \to \pm\infty$, the relation between the coordinates x and y assumes the simpler form

$$\begin{aligned} y \simeq e^{+2x/R_1} \quad & \text{at} \quad x \to -\infty, \\ 1-y \simeq e^{-2x/R_1} \quad & \text{at} \quad x \to +\infty. \end{aligned} \tag{49}$$

Thus, taking into account the latter relation and using the fact that at $y = 0\,(x \to -\infty)$ the hypergeometric function ${}_2F_1(a,b,c;0) = 1$, one eventually obtains that near the boundary $x \to -\infty$ the field ψ^{s_1} behaves as

$$\psi^{s_1}\big|_{x\to-\infty} \simeq De^{(c-1)x/R_1} + Ee^{-(c-1)x/R_1}. \tag{50}$$

On the other hand, in order to apply the boundary conditions at $y = 1\,(x \to \infty)$, it is useful to write the hypergeometric functions as functions of $(1-y)$, so that they become united at the boundary. This can be done by rewriting the hypergeometric functions appearing in Eq. (48) by means of the following identity [45]:

$$\begin{aligned} 2F(a,b,c;y) = {}& \frac{\Gamma(c)\Gamma(c-a-b)}{\Gamma(c-a)\Gamma(c-b)_2}F(a,b,a+b-c+1;1-y) \\ & + \frac{\Gamma(c)\Gamma(a+b-c)}{\Gamma(a)\Gamma(b)}(1-y)^{(c-a-b)}{}_2F(c-a,c-b,c-a-b+1;1-y), \end{aligned} \tag{51}$$

where Γ stands for the gamma function. Doing so, and using Eq. (49), we eventually arrive at the following behavior of the solution at $x \to +\infty$:

$$\begin{aligned} \psi^{s_1}\big|_{x\to+\infty} \simeq {}& \left[D\frac{\Gamma(c-a-b)\Gamma(c)}{\Gamma(c-a)\Gamma(c-b)} + E\frac{\Gamma(c-a-b)\Gamma(2-c)}{\Gamma(1-a)\Gamma(1-b)}\right]e^{-(a+b-c)x/R_1} \\ & + \left[D\frac{\Gamma(a+b-c)\Gamma(c)}{\Gamma(a)\Gamma(b)} + E\frac{\Gamma(a+b-c)\Gamma(2-c)}{\Gamma(a-c+1)\Gamma(b-c+1)}\right]e^{(a+b-c)x/R_1}. \end{aligned} \tag{52}$$

Now, from parameters Eqs. (38) and (43), we find that the constants appearing in the hypergeometric equation can be written as

$$
\begin{aligned}
a &= iR_1\sqrt{\mu^2 + q^2Q_1^2R_1^2 + L^2} + (1+s_1)\left(\frac{1}{4} - i\omega\frac{R_1}{2}\right) - i(1-s_1)\frac{qQ_1R_1^2}{2},\\
b &= -iR_1\sqrt{\mu^2 + q^2Q_1^2R_1^2 + L^2} + (1+s_1)\left(\frac{1}{4} - i\omega\frac{R_1}{2}\right) - i(1-s_1)\frac{qQ_1R_1^2}{2},\\
c &= \frac{1}{2} + is_1\left(qQ_1R_1^2 - \omega R_1\right).
\end{aligned}
\tag{53}
$$

In particular, the following relations hold

$$
(c-1)/R_1 = -is_1\omega + is_1 qQ_1R_1 - \frac{1}{2R_1}, \tag{54}
$$

$$
(a+b-c)/R_1 = -i\omega - iqQ_1R_1 + s_1\frac{1}{2R_1}. \tag{55}
$$

Now we are ready to impose the boundary conditions. Obviously, without loss of generality, we can consider that the spin s_1 is already chosen and fixed at $s_1 = +$ or $s_1 = -$ since the QNFs should not depend on the choice of $s_1 = \pm$. Let us impose, for instance, the boundary conditions for the component $s_1 = +$ of the spinorial field. In this case, using the identity Eq. (54) along with the Eq. (34), we eventually arrive at the following behavior of the solution at $x \to \infty$:

$$
\Psi_1^+(t,x)\big|_{x\to-\infty} = De^{-i\omega(t+x)}e^{\left(iqQ_1R_1 - \frac{1}{2R_1}\right)x} + Ee^{-i\omega(t-x)}e^{-\left(iqQ_1R_1 - \frac{1}{2R_1}\right)x}. \tag{56}
$$

Now, **Figure 1** tells us that the field is assumed to move toward higher values of x at the boundary $x \to -\infty$, while at the boundary $x \to -\infty$ it should move toward lower values of x. Then, since the time dependence of the field Ψ_1^+ is of the type $e^{-i\omega t}$, this means that Ψ_1^+ should behave as $e^{-i\omega(t-x)}$ at $x \to -\infty$, while it should go as $e^{-i\omega(t+x)}$ at $x \to +\infty$. Thus, from Eq. (55), we conclude that we must set $D = 0$. In such a case, from Eq. (52), the field Ψ_1^+ becomes

$$
\begin{aligned}
\Psi_1^+\big|_{x\to+\infty} \simeq{}& E\left[\frac{\Gamma(c-a-b)\Gamma(2-c)}{\Gamma(1-a)\Gamma(1-b)}\right]e^{-i\omega(t-x)}e^{\left(iqQ_1R_1 - \frac{1}{2R_1}\right)x}\\
&+E\left[\frac{\Gamma(a+b-c)\Gamma(2-c)}{\Gamma(a-c+1)\Gamma(b-c+1)}\right]e^{-i\omega(t+x)}e^{\left(-iqQ_1R_1 + \frac{1}{2R_1}\right)x}.
\end{aligned}
\tag{57}
$$

Finally, to satisfy the QNM boundary condition near the boundary at $x \to \infty$, we must eliminate the term $e^{-i\omega(t-x)}$ of the above equation. Since E cannot be zero (as otherwise the field would vanish identically), we need the combination of the gamma functions to be zero. Now, once the gamma function has no zeros, the way to achieve this is to let the gamma functions in the denominator diverge, $\Gamma(1-a) = \infty$ or $\Gamma(1-b) = \infty$. Since the gamma functions diverge only at nonpositive integers, we are led to the following constraint:

$$
1-a = -n \quad \text{or} \quad 1-b = -n, \quad \text{where} \quad n \in \{0, 1, 2, \ldots\}. \tag{58}
$$

Using the Eq. (53), we find that these constraints translate to

$$
\omega = \pm\sqrt{m^2 + q^2Q_1^2R_1^2 + L^2} + \frac{i}{R_1}\left(n + \frac{1}{2}\right), \tag{59}
$$

which are the QNFs of the Dirac field propagating in D-dimensional generalized Nariai spacetimes. The real part of a QNF is associated with the oscillation frequency, while the imaginary part is related to its decay rate. At this point, it is worth recalling that L is a separation constant of the Dirac equation that is related to the angular mode of the field.

Likewise, imposing the boundary condition to the component $s_1 = -$ of the spinorial field, we find that we must set $E = 0$ at Eq. (50) and then $c - a = -n$ or $c - b = -n$, with n being a nonnegative integer. This, in its turn, leads to the same spectrum obtained for the component $s_1 = +$ as expected, namely, Eq. (59).

4. Conclusions

In this chapter we have investigated the perturbations on a spinorial field propagating in a generalized version of the charged Nariai spacetime. Besides the separability of the degrees of freedom of these perturbations, one interesting feature of this background is that the perturbations can be analytically integrated. They all obey a Schrödinger-like equation with an integrable potential that is contained in the Rosen-Morse class of integrable potentials. Such an equation admits two linearly independent solutions given in terms of standard hypergeometric functions. This is a valuable property, since even the perturbation potential associated to the humble Schwarzschild background is nonintegrable, despite the fact that it is separable. We have also investigated the QNMs associated to this spinorial field. Analyzing the Eq. (59), namely,

$$\omega_D = \pm\sqrt{m^2 + q^2 Q_1^2 R_1^2 + L^2} + \frac{i}{R_1}\left(n + \frac{1}{2}\right), \tag{60}$$

it is interesting to note that the imaginary parts of the QNFs, which represent the decay rates, do not depend on any details of the perturbation; rather, they only depend on the charges of the gravitational background through the dependence on R_1. On the other hand, the real parts of the QNFs depend on the mass of the field and on the angular mode of the perturbations. Another fact worth pointing out is that the fermionic field always has a real part in its QNFs spectrum, meaning that it always oscillates. This is not reasonable. Indeed, for Klein-Gordon and Maxwell perturbations in the D-dimensional Nariai spacetime, their QNFs are equal to [39].

$$\begin{aligned} \omega_{KG} &= \pm\sqrt{m^2 + \sum_{j=2}^{d} \frac{\ell_j(\ell_j + 1)}{R_j^2} - \frac{1}{4R_1^2}} - \frac{i}{R_1}\left(n + \frac{1}{2}\right), \\ \omega_M &= \pm\sqrt{\sum_{j=2}^{d} \frac{\ell_j(\ell_j + 1)}{R_j^2} - \frac{1}{4R_1^2}} - \frac{i}{R_1}\left(n + \frac{1}{2}\right), \end{aligned} \tag{61}$$

where ℓ_j and m_j are integers, $|m_j| \leq \ell_j$, and $\ell \geq 0$. Due to the negative factor $-1/(4R_1^2)$ inside the square root appearing in the bosonic spectrum, it follows that for small enough R_1, along with small enough mass and angular momentum, the argument of the square root can be negative, so that this term becomes imaginary.

To finish, we believe that a good exercise is to calculate the QNFs of the gravitational field in D-dimensional generalized charged Nariai spacetime. Research on the latter problem is still ongoing and, due to the great number of degrees of freedom in the gravitational field, shall be considered in a future work. The next interesting step is the investigation of superradiance phenomena for the spin 1/2 field. Although bosonic fields like scalar, electromagnetic, and gravitational fields can exhibit superradiant behavior in four-dimensional Kerr spacetime [46], curiously, this is not the case for the Dirac field [36]. Thus, it would be interesting to investigate whether an analogous thing happens in the background considered here [47].

Author details

Joás Venâncio[1]* and Carlos Batista[2]

1 Physics Program, Universidade Federal de Pernambuco, Brazil

2 Department of Physics, Universidade Federal de Pernambuco, Brazil

*Address all correspondence to: joasvenancio@df.ufpe.br

References

[1] Vishveshwara CV. Scattering of gravitational radiation by a Schwarzschild black-hole. Nature. 1970; **227**:936

[2] Regge T, Wheeler JA. Stability of a Schwarzschild singularity. Physical Review D. 1957;**108**:1063

[3] Abbott BP et al. (LIGO scientific and virgo collaborations), observation of gravitational waves from a binary black hole merger. Physical Review Letters. 2016;**116**:061102

[4] Cardoso V. Quasinormal Modes and Gravitational Radiation in Black Hole Spacetimes [doctoral thesis]. Universidade Técnica de Lisboa; 2004. [arXiv:gr-qc/0404093]

[5] Berti E, Cardoso V, Starinets AO. Quasinormal modes of black holes and black branes. Classical and Quantum Gravity. 2009;**26**:163001

[6] Kokkotas KD, Schmidt BG. Quasinormal modes of stars and black holes. Living Reviews in Relativity. 1999;**2**:2

[7] Hod S. Bohr's correspondence principle and the area spectrum of quantum black holes. Physical Review Letters. 1998;**81**:4293

[8] Dreyer O. Quasinormal modes, the area spectrum, and black hole entropy. Physical Review Letters. 2003;**90**: 081301

[9] Maggiore M. The physical interpretation of the spectrum of black hole quasinormal modes. Physical Review Letters. 2008;**100**:141301

[10] Domagala M, Lewandowski J. Black hole entropy from quantum geometry. Classical and Quantum Gravity. 2004; **21**:5233

[11] Konoplya RA, Zhidenko A. Quasinormal modes of black holes: From astrophysics to string theory. Reviews of Modern Physics. 2011;**83**:793

[12] Frolov VP et al. Massive vector fields in Kerr-NUT-(A)dS spacetimes: Separability and quasinormal modes. arXiv: 1804.00030

[13] Zhidenko A. Massive scalar field quasi-normal modes of higher dimensional black holes. Physical Review D. 2006;**74**:064017

[14] Zhidenko A. Linear perturbations of black holes: Stability, quasi-normal modes and tails. [doctoral thesis]. Universidade de São Paulo; 2009. [arXiv:0903.3555]

[15] Liu LH, Wang B. Stability of BTZ black strings. Physical Review D. 2008; **78**:064001

[16] Mukhi S. String theory: A perspective over the last 25 years. Classical and Quantum Gravity. 2011;**28**:153001

[17] Emparan R, Reall HS. Black holes in higher dimensions. Living Reviews in Relativity. 2008;**11**:6

[18] Csáki C. TASI lectures on extra dimensions and branes. In: Shifman M, Vainshtein A, Wheater J, editors. From Fields to Strings: Circumnavigating Theoretical Physics. Vol. 2. Singapore: World Scientific; 2005. p. 967

[19] Maldacena JM. The large-N limit of superconformal field theories and supergravity. International Journal of Theoretical Physics. 1999;**38**:1113

[20] Horowitz GT, Polchinski J. Gauge/ gravity duality. In: Oriti D, editor. Approaches to Quantum Gravity: Toward a New Understanding of Space, Time and Matter. Cambridge, England: Cambridge University Press; 2009. p. 169

[21] Hubeny VE. The AdS/CFT correspondence. Classical and Quantum Gravity. 2015;**32**:124010

[22] Horowitz GT, Hubeny VE. Quasinormal modes of AdS black holes and the approach to thermal equilibrium. Physical Review D. 2000; **62**:024027

[23] Birmingham D, Sachs I, Solodukhin SN. Conformal field theory interpretation of black hole quasinormal modes. Physical Review Letters. 2002; **88**:151301

[24] Nunez A, Starinets AO. AdS/CFT correspondence, quasinormal modes, and thermal correlators in N = 4 SYM. Physical Review D. 2003;**67**:124013

[25] Keranen V, Kleinert P. Thermalization of Wightman functions in AdS/CFT and quasinormal modes. Physical Review D. 2016;**94**:026010

[26] David JR, Khetrapal S. Thermalization of green functions and quasinormal modes. Journal of High Energy Physics. 2015;**07**:041

[27] López-Ortega A. Dirac quasinormal modes of D-dimensional de sitter spacetime. General Relativity and Gravitation. 2007;**39**:1011

[28] Brady PB, Chambers CM. Radiative falloff in Schwarzschild-de sitter spacetime. Physical Review D. 1999;**60**: 064003

[29] Abdalla E et al. Support of dS/CFT correspondence from perturbations of three dimensional spacetime. Physical Review D. 2002;**66**:104018. arXiv:hep-th/0204030

[30] Nollert HP. Quasinormal modes: The characteristic' sound' of black holes and neutron stars. Classical and Quantum Gravity. 1999;**16**:159

[31] Batista C. Generalized charged Nariai solutions in arbitrary even dimensions with multiple magnetic charges. General Relativity and Gravitation. 2016;**48**:160

[32] Venâncio J. The spinorial formalism, with applications in physics [Master dissertation]. Federal University of Pernambuco; 2017. Available from: https://repositorio.ufpe.br/handle/123456789/25303; https://www.researchgate.net/publication/324210081_The_Spinorial_Formalism_with_Applications_in_Physics

[33] Benn I, Tucker R. An Introduction to Spinors and Geometry with Applications in Physics. Adam Hilger; 1987. Available from: http://inspirehep.net/record/256204/; https://www.amazon.com/Introduction-Spinors-Geometry-Applications-Physics/dp/0852741693

[34] Cartan E. The Theory of Spinors. Dover; 1966. Available from: https://store.doverpublications.com/0486640701.html; http://cds.cern.ch/record/104700

[35] Venâncio J, Batista C. Separability of the Dirac equation on backgrounds that are the direct product of bidimensional spaces. Physical Review D. 2017;**95**: 084022

[36] Güven R. Wave mechanics of electrons in Kerr geometry. Physical Review D. 1977;**16**:1706

[37] Dutt R, Khare A, Sukhatme UP. Supersymmetry, shape invariance, and exactly solvable potentials. American Journal of Physics. 1988;**56**:163

[38] Pöschl G, Teller E. Bemerkungen zur Quantenmechanik des anharmonischen Oszillators. Zeitschrift für Physik. 1933;**83**:143

[39] Venâncio J, Batista C. Quasinormal modes in generalized Nariai spacetimes. Physical Review D. 2018;**97**:105025

[40] Hartman T. Lecture notes on classical de Sitter space. 2017. arXiv: 1205.3855 [hep-th]

[41] Anninos D. de Sitter Musings. 2013. arXiv:1205.3855 [hep-th]

[42] Bengtsson I, Sandin P. Anti de sitter space, squashed and stretched. Classical and Quantum Gravity. 2006;**23**:971

[43] López Ortega A. The Dirac equation in D-dimensional spherically symmetric spacetimes. arXiv:0906.2754

[44] Zhidenko A. Linear perturbations of black holes: Stability, quasi-normal modes and tails [doctoral thesis]. Universidade de São Paulo; 2009. ArXiv: 0903.3555

[45] Du ED, Wang B, Su R. Quasinormal modes in pure de sitter spacetimes. Physical Review D. 2004;**70**:064024. arXiv:hep-th/0404047

[46] Abramowitz M, Stegun IA. Handbook of Mathematical Functions with Formulas, Graphs, and Mathematical Tables. New York: Dover; 1972

[47] Rosa JG. Superradiance in the sky. Physical Review D. 2017;**95**:064017

Chapter 2

Eight-by-Eight Spacetime Matrix Operator and Its Applications

Richard P. Bocker and B. Roy Frieden

Abstract

A recent journal article by the authors introduced the eight-by-eight spacetime matrix operator $\hat{M}$ which played a key role in the formulation of Lorentz invariant matrix equations for both the classical electrodynamic Maxwell field equations and the quantum mechanical relativistic Dirac equation for free space. Those new equations we referred to as the Maxwell spacetime matrix and the Dirac spacetime matrix equations. These matrix equations will be briefly reviewed at the beginning of this chapter. Next we will show how the same matrix operator $\hat{M}$ plays a central role in the matrix formulation of other fundamental equations in both electromagnetic and quantum theories. These include the electromagnetic wave and charge continuity equations, the Lorentz conditions and electromagnetic potentials, the electromagnetic potential wave equations, and the quantum mechanical Klein-Gordon equation. In addition, a new generalized spacetime matrix equation, again employing the operator $\hat{M}$, will be described which is a generalization of the Maxwell and Dirac spacetime matrix equations. We will explore time-harmonic plane-wave solutions of this equation as well as the properties of these solutions.

Keywords: special theory of relativity, matrix operators, classical electrodynamics, relativistic quantum mechanics, matter waves, electromagnetic waves, optics, applied mathematics

1. Introduction

The eight-by-eight spacetime matrix operator $\hat{M}$ plays a key role in the matrix formulation of a number of well-known fundamental equations in both the fields of classical electrodynamics and relativistic quantum mechanics (see [1]). The spacetime matrix operator is defined by Eq. (1):

$$\hat{M} \equiv \begin{bmatrix} -\partial_4 & 0 & 0 & 0 & 0 & -\partial_3 & +\partial_2 & -\partial_1 \\ 0 & -\partial_4 & 0 & 0 & +\partial_3 & 0 & -\partial_1 & -\partial_2 \\ 0 & 0 & -\partial_4 & 0 & -\partial_2 & +\partial_1 & 0 & -\partial_3 \\ 0 & 0 & 0 & -\partial_4 & +\partial_1 & +\partial_2 & +\partial_3 & 0 \\ 0 & +\partial_3 & -\partial_2 & +\partial_1 & +\partial_4 & 0 & 0 & 0 \\ -\partial_3 & 0 & +\partial_1 & +\partial_2 & 0 & +\partial_4 & 0 & 0 \\ +\partial_2 & -\partial_1 & 0 & +\partial_3 & 0 & 0 & +\partial_4 & 0 \\ -\partial_1 & -\partial_2 & -\partial_3 & 0 & 0 & 0 & 0 & +\partial_4 \end{bmatrix}. \quad (1)$$

Compact matrix equation	Compact matrix equation description
$\hat{M}\|f\rangle = \|o\rangle$	Maxwell spacetime matrix equation for free space
$\hat{M}\|f\rangle = \|j\rangle$	Maxwell matrix equation with charges and currents
$\hat{M}\hat{M}\|f\rangle = \hat{M}\|j\rangle$	Charge continuity and electromagnetic wave equations
$\hat{M}\|a\rangle = \|f\rangle$	Lorentz conditions and electromagnetic potentials
$\hat{M}\hat{M}\|a\rangle = \|j\rangle$	Electromagnetic potential wave equations
$\hat{M}\|\phi\rangle + \kappa\|\phi\rangle = \|o\rangle$	Dirac spacetime matrix equation for free space
$\hat{M}\hat{M}\|\phi\rangle - \kappa^2\|\phi\rangle = \|o\rangle$	Klein-Gordon spacetime matrix equation for free space
$\hat{M}\|\psi\rangle + \kappa\|\psi\rangle = \|o\rangle$	Generalized spacetime matrix equation for free space

Table 1.
Compact matrix equations where the spacetime matrix operator $\hat{M}$ plays a central role.

The partial derivative symbols are defined by the following:

$$\partial_1 \equiv \frac{\partial}{\partial x} \qquad \partial_2 \equiv \frac{\partial}{\partial y} \qquad \partial_3 \equiv \frac{\partial}{\partial z} \qquad \partial_4 \equiv \frac{1}{ic}\frac{\partial}{\partial t}. \tag{2}$$

The imaginary quantity i represents the square root of minus one, and the physical quantity c corresponds to the speed of light in free space.

Eight compact matrix equations are listed in **Table 1**, each containing the spacetime matrix operator $\hat{M}$. Each of these equations, as well as the ket $|\ \rangle$ vector appearing in these equations, will be discussed in greater detail in the following sections of this chapter. An excellent introduction to bra $\langle\ \ |$ and ket $|\rangle$ vector notation may be found in [2]. The Gaussian system of units (see [3], p. 781) is employed throughout this chapter.

2. Eight-by-eight spacetime matrix operator properties

The spacetime matrix operator $\hat{M}$, defined in Eq. (1), may also be expressed by the following equation:

$$\hat{M} = M_1\partial_1 + M_2\partial_2 + M_3\partial_3 + M_4\partial_4. \tag{3}$$

The four eight-by-eight matrices M_μ, where $\mu = 1, 2, 3, 4$, are simply referred to as the spacetime matrices. These matrices have the following properties:

1. Each matrix M_μ is equal to its own multiplicative inverse

$$M_\mu = M_\mu^{-1}. \tag{4}$$

2. These matrices satisfy the anti-commutation relation

$$M_\mu M_\nu + M_\nu M_\mu = 2\delta_{\mu\nu} I. \tag{5}$$

3. Each matrix M_μ is Hermitian

$$M_\mu = M_\mu^\dagger. \tag{6}$$

4. In addition

$$\hat{M}\hat{M} = \hat{M}^2 = I\,\Box^2. \tag{7}$$

The symbol $\delta_{\mu\nu}$ is the Kronecker delta, and I represents the eight-by-eight identity matrix. The d'Alembertian (see [4], p. 290) and the Laplacian (see [4], p. 15) operators are defined by

$$\Box^2 \equiv \nabla^2 - \frac{1}{c^2}\frac{\partial^2}{\partial t^2} \qquad \text{and} \qquad \nabla^2 \equiv \frac{\partial^2}{\partial x^2} + \frac{\partial^2}{\partial y^2} + \frac{\partial^2}{\partial z^2}. \tag{8}$$

Some authors use the $\Box$ symbol to represent the d'Alembertian operator.

3. Maxwell spacetime matrix equation

The Maxwell field equations play a fundamental role in both classical electrodynamics and physical optics. The propagations of electromagnetic waves through free space (see [4], pp. 514–522), nonconducting media (see [3], pp. 295–309), thin-film optical filters [5], and solid-state crystalline materials [6] are just a few examples where the Maxwell field equations play an important role.

3.1 Maxwell spacetime matrix equation for free space

An earlier eight-by-eight matrix representation of the Maxwell field equations was first introduced by the authors back in 1993 [7]. An improved updated version using the spacetime matrix operator $\hat{M}$ was published recently [1]. For free space, in the absence of charges and currents, this later version is given by

$$\begin{bmatrix} -\partial_4 & 0 & 0 & 0 & 0 & -\partial_3 & +\partial_2 & -\partial_1 \\ 0 & -\partial_4 & 0 & 0 & +\partial_3 & 0 & -\partial_1 & -\partial_2 \\ 0 & 0 & -\partial_4 & 0 & -\partial_2 & +\partial_1 & 0 & -\partial_3 \\ 0 & 0 & 0 & -\partial_4 & +\partial_1 & +\partial_2 & +\partial_3 & 0 \\ 0 & +\partial_3 & -\partial_2 & +\partial_1 & +\partial_4 & 0 & 0 & 0 \\ -\partial_3 & 0 & +\partial_1 & +\partial_2 & 0 & +\partial_4 & 0 & 0 \\ +\partial_2 & -\partial_1 & 0 & +\partial_3 & 0 & 0 & +\partial_4 & 0 \\ -\partial_1 & -\partial_2 & -\partial_3 & 0 & 0 & 0 & 0 & +\partial_4 \end{bmatrix} \begin{bmatrix} iE_1 \\ iE_2 \\ iE_3 \\ 0 \\ B_1 \\ B_2 \\ B_3 \\ 0 \end{bmatrix} = \begin{bmatrix} 0 \\ 0 \\ 0 \\ 0 \\ 0 \\ 0 \\ 0 \\ 0 \end{bmatrix}. \tag{9}$$

The compact matrix form of Eq. (9) is given by

$$\hat{M}|f\rangle = |o\rangle. \tag{10}$$

The wave function $|f\rangle$ is an eight-by-one ket vector containing, in general, six nonzero scalar components associated with the electric field vector $\mathbf{E} = (E_1\, E_2\, E_3)$ and the magnetic induction vector $\mathbf{B} = (B_1\, B_2\, B_3)$. The elements (4,1) and (8,1) in $|f\rangle$ have purposely been set equal to zero. The case when these two elements are nonzero will be considered when the generalized spacetime matrix equation for free space is discussed. The ket vector $|o\rangle$ represents the eight-by-one null vector.

The Maxwell spacetime matrix equation (9) when expanded is equivalent to two divergences and two curl equations, namely,

$$\nabla \cdot \mathbf{E} = 0 \qquad \text{and} \qquad \nabla \cdot \mathbf{B} = 0 \tag{11}$$

$$\nabla \times \mathbf{E} + \frac{1}{c}\frac{\partial}{\partial t}\mathbf{B} = 0 \qquad \text{and} \qquad \nabla \times \mathbf{B} - \frac{1}{c}\frac{\partial}{\partial t}\mathbf{E} = 0. \tag{12}$$

We recognize these four equations as the traditional Maxwell field equations (Gaussian units) for free space in the absence of charges, currents, and ordinary matter terms (see [8], pp. 362–368).

For electromagnetic waves, time-harmonic plane-wave solutions of the form

$$\mathbf{E}(\mathbf{r}, t) = \mathbf{E_o} \exp\{i(\mathbf{k}\cdot\mathbf{r} - \omega t)\} \qquad \text{and} \qquad \mathbf{B}(\mathbf{r}, t) = \mathbf{B_o} \exp\{i(\mathbf{k}\cdot\mathbf{r} - \omega t)\} \tag{13}$$

will next be substituted back into the previous four vector equations. This yields the following set of equations:

$$\mathbf{k}\cdot\mathbf{E_o} = 0 \qquad \text{and} \qquad \mathbf{k}\cdot\mathbf{B_o} = 0 \tag{14}$$

$$\mathbf{k}\times\mathbf{E_o} = +\frac{\omega}{c}\mathbf{B_o} \qquad \text{and} \qquad \mathbf{k}\times\mathbf{B_o} = -\frac{\omega}{c}\mathbf{E_o}. \tag{15}$$

The quantities $\mathbf{k}$ and ω correspond to the wave vector and the angular frequency associated with the electromagnetic wave; $\mathbf{r}$ and t represent the position vector and the instantaneous time. From the preceding equations, we find the vectors $\mathbf{E}_o$, $\mathbf{B}_o$, and $\mathbf{k}$ are mutually perpendicular. That is,

$$\mathbf{k}\perp\mathbf{E_o} \qquad \mathbf{E_o}\perp\mathbf{B_o} \qquad \mathbf{k}\perp\mathbf{B_o}. \tag{16}$$

These properties represent transverse electromagnetic waves. We also obtain the important results

$$E_o = B_o \tag{17}$$

and

$$\omega = kc \qquad \lambda f = c \qquad \text{where} \qquad \omega = 2\pi f \qquad k = 2\pi/\lambda. \tag{18}$$

The quantities k, f, and λ represent the wave number, the frequency, and the wavelength, respectively, associated with the electromagnetic wave. So for free space, the magnitudes of the electromagnetic field vectors $\mathbf{E_o}$ and $\mathbf{B_o}$ are equal, a well-known result in electromagnetic wave propagation. Recall we are using Gaussian units.

3.2 Maxwell spacetime matrix equation with charges and currents

The Maxwell spacetime matrix equation, with the addition of charge and current terms [1], is given by

$$\begin{bmatrix} -\partial_4 & 0 & 0 & 0 & 0 & -\partial_3 & +\partial_2 & -\partial_1 \\ 0 & -\partial_4 & 0 & 0 & +\partial_3 & 0 & -\partial_1 & -\partial_2 \\ 0 & 0 & -\partial_4 & 0 & -\partial_2 & +\partial_1 & 0 & -\partial_3 \\ 0 & 0 & 0 & -\partial_4 & +\partial_1 & +\partial_2 & +\partial_3 & 0 \\ 0 & +\partial_3 & -\partial_2 & +\partial_1 & +\partial_4 & 0 & 0 & 0 \\ -\partial_3 & 0 & +\partial_1 & +\partial_2 & 0 & +\partial_4 & 0 & 0 \\ +\partial_2 & -\partial_1 & 0 & +\partial_3 & 0 & 0 & +\partial_4 & 0 \\ -\partial_1 & -\partial_2 & -\partial_3 & 0 & 0 & 0 & 0 & +\partial_4 \end{bmatrix} \begin{bmatrix} iE_1 \\ iE_2 \\ iE_3 \\ 0 \\ B_1 \\ B_2 \\ B_3 \\ 0 \end{bmatrix} = \frac{4\pi}{c} \begin{bmatrix} J_{e1} \\ J_{e2} \\ J_{e3} \\ c\rho_m \\ iJ_{m1} \\ iJ_{m2} \\ iJ_{m3} \\ -ic\rho_e \end{bmatrix}. \tag{19}$$

The compact matrix form of the Maxwell spacetime matrix equation is given by

$$\hat{M}|f\rangle = |j\rangle. \tag{20}$$

Eq. (19), when expanded, is equivalent to two divergences and two curl equations. The resulting four vector equations are referred to as the microscopic Maxwell field equations (see [8], pp. 283–290). They are given by

$$\nabla\cdot\mathbf{E} = +4\pi\rho_e \quad \text{and} \quad \nabla\cdot\mathbf{B} = +4\pi\rho_m \tag{21}$$

$$\nabla\times\mathbf{E} + \frac{1}{c}\frac{\partial}{\partial t}\mathbf{B} = -\frac{4\pi}{c}\mathbf{J}_m \quad \text{and} \quad \nabla\times\mathbf{B} - \frac{1}{c}\frac{\partial}{\partial t}\mathbf{E} = +\frac{4\pi}{c}\mathbf{J}_e. \tag{22}$$

The various scalar and vector quantities appearing in the microscopic Maxwell vector equations are the electric field vector $\mathbf{E} = (E_1\, E_2\, E_3)$, the magnetic induction vector $\mathbf{B} = (B_1\, B_2\, B_3)$, the electric current density vector $\mathbf{J}_e = (J_{e1}\, J_{e2}\, J_{e3})$, the magnetic current density vector $\mathbf{J}_m = (J_{m1}\, J_{m2}\, J_{m3})$, the electric charge density ρ_e, the magnetic charge density ρ_m, and the speed of light c in free space. Both magnetic charge and magnetic current density (see [8], pp. 283–290) have been included in the Maxwell vector equations for purposes of completeness. They, of course, may be set equal to zero since hypothetical magnetic monopoles have not been discovered in nature. The ket vector $|f\rangle$ represents the eight-by-one column vector on the left-hand side of Eq. (19). The ket vector $|j\rangle$ corresponds to the eight-by-one column vector on the right-hand side of Eq. (19) multiplied by the factor $4\pi/\mathrm{c}$.

3.3 Charge continuity and electromagnetic wave equations

Charge continuity equations for electric (see [8], p. 15) and magnetic charges as well as the electromagnetic wave equations involving electric and magnetic charges and currents may be easily obtained by simply multiplying both sides of the Maxwell spacetime matrix equation in compact form (20) by the spacetime matrix operator $\hat{M}$. That is,

$$\hat{M}\hat{M}|f\rangle - \hat{M}|j\rangle. \tag{23}$$

Expanding this single matrix equation yields the charge continuity and electromagnetic wave equations:

$$\nabla\cdot\mathbf{J}_e + \frac{\partial}{\partial t}\rho_e = 0 \quad \text{and} \quad \nabla\cdot\mathbf{J}_m + \frac{\partial}{\partial t}\rho_m = 0 \tag{24}$$

$$\Box^2\mathbf{E} = \frac{4\pi}{c^2}\frac{\partial}{\partial t}\mathbf{J}_e + 4\pi\nabla\rho_e + \frac{4\pi}{c}\nabla\times\mathbf{J}_m \quad \text{and} \quad \Box^2\mathbf{B} = \frac{4\pi}{c^2}\frac{\partial}{\partial t}\mathbf{J}_m + 4\pi\nabla\rho_m - \frac{4\pi}{c}\nabla\times\mathbf{J}_e. \tag{25}$$

3.4 Lorentz conditions and electromagnetic potentials

By using the spacetime matrix operator $\hat{M}$, we can determine the relationship between electromagnetic fields and vector-scalar potentials as well as determine expressions for the Lorentz conditions (see [9], pp. 179–181) in a single matrix equation. The following matrix equation provides the desired relation:

$$\begin{bmatrix} -\partial_4 & 0 & 0 & 0 & 0 & -\partial_3 & +\partial_2 & -\partial_1 \\ 0 & -\partial_4 & 0 & 0 & +\partial_3 & 0 & -\partial_1 & -\partial_2 \\ 0 & 0 & -\partial_4 & 0 & -\partial_2 & +\partial_1 & 0 & -\partial_3 \\ 0 & 0 & 0 & -\partial_4 & +\partial_1 & +\partial_2 & +\partial_3 & 0 \\ 0 & +\partial_3 & -\partial_2 & +\partial_1 & +\partial_4 & 0 & 0 & 0 \\ -\partial_3 & 0 & +\partial_1 & +\partial_2 & 0 & +\partial_4 & 0 & 0 \\ +\partial_2 & -\partial_1 & 0 & +\partial_3 & 0 & 0 & +\partial_4 & 0 \\ -\partial_1 & -\partial_2 & -\partial_3 & 0 & 0 & 0 & 0 & +\partial_4 \end{bmatrix} \begin{bmatrix} -A_{e1} \\ -A_{e2} \\ -A_{e3} \\ -\phi_m \\ -iA_{m1} \\ -iA_{m2} \\ -iA_{m3} \\ i\phi_e \end{bmatrix} = \begin{bmatrix} iE_1 \\ iE_2 \\ iE_3 \\ 0 \\ B_1 \\ B_2 \\ B_3 \\ 0 \end{bmatrix}. \tag{26}$$

The compact matrix form of Eq. (26) is given by

$$\hat{M}|a\rangle = |f\rangle. \tag{27}$$

The ket vector $|a\rangle$ corresponds to the eight-by-one column vector on the left-hand side of Eq. (26). Equation (26), when expanded, yields the Lorentz conditions and the relationship between electromagnetic fields and potentials:

$$\nabla \cdot \mathbf{A}_e + \frac{1}{c}\frac{\partial}{\partial t}\phi_e = 0 \quad \text{and} \quad \nabla \cdot \mathbf{A}_m + \frac{1}{c}\frac{\partial}{\partial t}\phi_m = 0 \tag{28}$$

$$\mathbf{E} = -\nabla\phi_e - \frac{1}{c}\frac{\partial}{\partial t}\mathbf{A}_e - \nabla \times \mathbf{A}_m \quad \text{and} \quad \mathbf{B} = -\nabla\phi_m - \frac{1}{c}\frac{\partial}{\partial t}\mathbf{A}_m + \nabla \times \mathbf{A}_e. \tag{29}$$

The new scalar and vector quantities appearing in the above equations are the electric vector potential $\mathbf{A}_e = (A_{e1}\, A_{e2}\, A_{e3})$, the magnetic vector potential $\mathbf{A}_m = (A_{m1}\, A_{m2}\, A_{m3})$, the electric scalar potential ϕ_e, and the magnetic scalar potential ϕ_m. So again we see how the eight-by-eight spacetime matrix operator $\hat{M}$ plays a central role in tying together important electromagnetic relations.

3.5 Electromagnetic potential wave equations

It is well-known that the electromagnetic vector and scalar potentials satisfy wave equations (see [9], pp. 179–181). This can be easily shown by multiplying both sides of Eq. (27) by the spacetime matrix operator $\hat{M}$. This gives

$$\hat{M}\hat{M}|a\rangle = \hat{M}|f\rangle. \tag{30}$$

Next replace the term $\hat{M}|f\rangle$ by the ket vector $|j\rangle$ using Eq. (20). This yields

$$\hat{M}\hat{M}|a\rangle = |j\rangle. \tag{31}$$

Expanding this single matrix equation yields eight partial differential equations which can be easily combined to form the following four potential wave equations:

$$\Box^2\phi_e = -4\pi\rho_e \quad \text{and} \quad \Box^2\phi_m = -4\pi\rho_m \tag{32}$$

$$\Box^2\mathbf{A}_e = -\frac{4\pi}{c}\mathbf{J}_e \quad \text{and} \quad \Box^2\mathbf{A}_m = -\frac{4\pi}{c}\mathbf{J}_m. \tag{33}$$

The single compact matrix (Eq. (31)) is therefore equivalent to these four potential wave equations.

4. Dirac spacetime matrix equation

The nonrelativistic Schrödinger wave equation (see [10], pp. 143–146) plays a fundamental role in quantum mechanical phenomena where the spin property of nonrelativistic particles may be ignored. This equation is usually first met in modern physics textbooks. However, when a particle with half-integer spin and/or moving at relativistic speeds is involved, the relativistic Dirac equation [11] comes into play.

4.1 Dirac spacetime matrix equation for free space

Using the spacetime matrix operator $\hat{M}$, the authors introduced in their most recent publication [1] a modified version of the traditional Dirac equation, referred to as the Dirac spacetime matrix equation. In the absence of electromagnetic potentials [11], the Dirac spacetime matrix equation for free space is given by

$$\begin{bmatrix} -\partial_4 & 0 & 0 & 0 & 0 & -\partial_3 & +\partial_2 & -\partial_1 \\ 0 & -\partial_4 & 0 & 0 & +\partial_3 & 0 & -\partial_1 & -\partial_2 \\ 0 & 0 & -\partial_4 & 0 & -\partial_2 & +\partial_1 & 0 & -\partial_3 \\ 0 & 0 & 0 & -\partial_4 & +\partial_1 & +\partial_2 & +\partial_3 & 0 \\ 0 & +\partial_3 & -\partial_2 & +\partial_1 & +\partial_4 & 0 & 0 & 0 \\ -\partial_3 & 0 & +\partial_1 & +\partial_2 & 0 & +\partial_4 & 0 & 0 \\ +\partial_2 & -\partial_1 & 0 & +\partial_3 & 0 & 0 & +\partial_4 & 0 \\ -\partial_1 & -\partial_2 & -\partial_3 & 0 & 0 & 0 & 0 & +\partial_4 \end{bmatrix} \begin{bmatrix} iU_1 \\ iU_2 \\ iU_3 \\ 0 \\ L_1 \\ L_2 \\ L_3 \\ 0 \end{bmatrix} + \kappa \begin{bmatrix} iU_1 \\ iU_2 \\ iU_3 \\ 0 \\ L_1 \\ L_2 \\ L_3 \\ 0 \end{bmatrix} = \begin{bmatrix} 0 \\ 0 \\ 0 \\ 0 \\ 0 \\ 0 \\ 0 \\ 0 \end{bmatrix}. \quad (34)$$

The compact matrix form of Eq. (34) is given by

$$\hat{M}|\phi\rangle + \kappa|\phi\rangle = |o\rangle. \quad (35)$$

The wave function $|\phi\rangle$ is an eight-by-one ket vector containing, in general, six nonzero scalar components associated with two vector quantities $\mathbf{U} = (U_1\ U_2\ U_3)$ and $\mathbf{L} = (L_1\ L_2\ L_3)$. The elements (4,1) and (8,1) in $|\phi\rangle$ have purposely been set equal to zero. The case when these two elements are nonzero will also be considered when the generalized spacetime matrix equation for free space is discussed later in this chapter. The ket vector $|o\rangle$ represents the eight-by-one null vector. The constant κ is defined by

$$\kappa \equiv m_o c/\hbar. \quad (36)$$

Here m_o represents the rest mass of the matter-wave particle under consideration, c again is the speed of light in free space, and $\hbar$ is equal to the Planck constant h divided by 2π.

The Dirac spacetime matrix equation (34) when expanded is equivalent to eight partial differential equations. These eight equations can be rewritten as two divergence and two curl equations [1], namely,

$$\nabla \cdot \mathbf{U} = 0 \quad \text{and} \quad \nabla \cdot \mathbf{L} = 0 \quad (37)$$

$$\nabla \times \mathbf{U} = -\frac{1}{c}\frac{\partial}{\partial t}\mathbf{L} - i\kappa\mathbf{L} \quad \text{and} \quad \nabla \times \mathbf{L} = +\frac{1}{c}\frac{\partial}{\partial t}\mathbf{U} - i\kappa\mathbf{U}. \quad (38)$$

We refer to these equations as the Dirac spacetime vector equations for free space. It is noted that these equations resemble the four Maxwell field equations for free space in the absence of charge, current, and ordinary matter terms.

The simplest solutions of these vector equations are time-harmonic plane-wave solutions of the form

$$\mathbf{U}(\mathbf{r},t) = \mathbf{U_o} \exp\{i(\mathbf{p}\cdot\mathbf{r} - Et)/\hbar\} \quad \text{and} \quad \mathbf{L}(\mathbf{r},t) = \mathbf{L_o} \exp\{i(\mathbf{p}\cdot\mathbf{r} - Et)/\hbar\}. \quad (39)$$

The quantities $\mathbf{p}$ and E correspond to the linear momentum and the total energy of the associated matter-wave particle; $\mathbf{r}$ and t represent the position vector and the instantaneous time. For particles with nonzero rest mass m_o, the following special theory of relativity equations (see [10], pp. 21–25) may also be useful:

$$E = \gamma m_o c^2 \qquad p = \gamma m_o v \qquad \text{where} \qquad \gamma = 1/\sqrt{1-\beta^2} \qquad \beta = v/c. \quad (40)$$

The quantities γ and β are known as the Lorentz factor and the speed parameter, respectively. The symbol v represents the relativistic speed of the matter-wave particle. Substitution of the preceding time-harmonic plane-wave solutions back into the Dirac spacetime vector equations yield the following set of vector equations for matter waves:

$$\mathbf{p}c \cdot \mathbf{U_o} = 0 \qquad \text{and} \qquad \mathbf{p}c \cdot \mathbf{L_o} = 0 \quad (41)$$

$$\mathbf{p}c \times \mathbf{U_o} = +E\,\frac{(\gamma-1)}{\gamma}\,\mathbf{L_o} \qquad \text{and} \qquad \mathbf{p}c \times \mathbf{L_o} = -E\,\frac{(\gamma+1)}{\gamma}\,\mathbf{U_o}. \quad (42)$$

From the previous equations we find the three vectors $\mathbf{U}_o$, $\mathbf{L}_o$, and $\mathbf{p}c$ are mutually perpendicular. That is,

$$\mathbf{p}c \perp \mathbf{U_o} \qquad \mathbf{U_o} \perp \mathbf{L_o} \qquad \mathbf{p}c \perp \mathbf{L_o} \quad (43)$$

These properties represent transverse waves. In addition, we also obtain the important result:

$$(\gamma+1)\,U_o^2 = (\gamma-1)\,L_o^2. \quad (44)$$

The magnitudes of the vectors $\mathbf{U}_o$ and $\mathbf{L}_o$ are related through the Lorentz factor γ, which depends on the speed parameter β, which ultimately depends on the speed v of the nonzero rest-mass particle. Note, for γ much greater than unity, characteristic of a relativistic particle, the magnitudes of the vectors $\mathbf{U}_o$ and $\mathbf{L}_o$ are nearly equal. On the other hand, for γ close to unity, characteristic of a nonrelativistic particle, the magnitude of the vector $\mathbf{L}_o$ is much greater than the magnitude of the vector $\mathbf{U}_o$. One other important result is

$$E^2 = p^2c^2 + m_o^2c^4 \qquad \text{which implies} \qquad E = \pm\sqrt{p^2c^2 + m_o^2c^4}. \quad (45)$$

The $\pm$ sign is associated with the quantum mechanical energy E of a matter-wave particle, like a half-integer spin electron. This was first interpreted by Paul A. M. Dirac. He recognized the negative energy levels predicted by his relativistic equation could not be ignored. This led to his concept of a hole theory of positrons. For a detailed discussion on negative energy states (see [11]).

4.2 Klein-Gordon spacetime matrix equation

The Klein-Gordon equation (see [12], pp. 118–129) is yet another quantum mechanical relativistic equation which is the field equation of the quanta associated with spin-less (spin-0) particles. An example of a spin-less particle is the recently discovered Higgs boson.

A version of the Klein-Gordon equation can be easily derived by simply starting with the compact matrix form of the Dirac spacetime matrix equation for free space, namely, Eq. (35). Multiply both sides by the spacetime matrix operator $\hat{M}$. This gives

$$\hat{M}\hat{M}|\phi\rangle + \kappa\hat{M}|\phi\rangle = |o\rangle. \tag{46}$$

Next replace the term $\hat{M}|\phi\rangle$ with $-\kappa|\phi\rangle$ using Eq. (35). We obtain

$$\hat{M}\hat{M}|\phi\rangle - \kappa^2|\phi\rangle = |o\rangle. \tag{47}$$

We refer to this equation as the Klein-Gordon spacetime matrix equation for free space. Using the fourth property of the spacetime matrix operator $\hat{M}$, it can be easily shown that Eq. (47) is equivalent to the following two equations involving the vectors **U** and **L**:

$$\Box^2\mathbf{U} - \kappa^2\mathbf{U} = 0 \qquad \text{and} \qquad \Box^2\mathbf{L} - \kappa^2\mathbf{L} = 0. \tag{48}$$

Therefore, the vectors **U** and **L** also satisfy Klein-Gordon type equations.

5. Generalized spacetime matrix equation

In this section, we will introduce for the first time a new matrix equation where again the spacetime operator $\hat{M}$ plays a central role. We will refer to this equation as the generalized spacetime matrix equation for free space.

5.1 Big unanswered questions and mysteries in physics and astronomy

The number of unanswered questions and mysteries regarding the universe from the smallest to the largest, in the fields of physics and astronomy, is unimaginable. There are many references, too numerous to list here, which address this topic. However, an excellent comprehensive list of unsolved problems in physics appears in [13] for various broad areas of physics. These areas include general physics, quantum physics, cosmology, general relativity, quantum gravity, high-energy physics, particle physics, astronomy, astrophysics, nuclear physics, atomic physics, molecular physics, optical physics, classical mechanics, condensed matter physics, plasma physics, and biophysics. The following is a partial list of some of the most important questions and mysteries being addressed today by physicists and astronomers around the globe:

How did the universe begin and what is the ultimate fate of the universe?
Is the universe infinite or just very big?
Why is there more matter than antimatter in the universe?
What came before the big bang?
Why are the galaxies distributed in clumps and filaments?
Are there additional dimensions?
Is spacetime fundamentally continuous or discrete?
How can we create a quantum theory of gravity?
What is dark energy and dark matter?
Do dark gravity, dark charge, and dark antimatter exist?

What happens inside a black hole and do naked singularities exist?
Why does time seem to flow only in one direction?
Is time travel really possible?
Is string theory or M-theory a viable theory of everything?
What kind of physics underlies the standard model?
Are there really just three generations of leptons and quarks?
Do gravitons exist?
Are protons unstable?
Do magnetic monopoles exist?
What are the masses of neutrinos?
Do the quarks or leptons have any substructure?
Do tachyons exist and can information travel faster than light?
Why do the particles have the precise masses they do?
Do fundamental physical constants vary over time?
Why are the strengths of the fundamental forces what they are?
Do parallel universes exist and is there a multiverse?
Was our spatially 3-D universe formed out of a vacuum by a 2-D hologram?
Was the hologram formed by a flow of information? If so, what form?
Does pair production formed, spontaneously, out of a vacuum?
Are they likewise formed out of a flow of information?
Do life processes, such as ion flows through cell membranes, form likewise as flows of information?

As we can see, even with all of the discoveries made over the past several hundred years, there is so much we do not understand and so much yet to be discovered about our universe and possibly beyond.

So far we have described the first seven compact matrix equations listed in **Table 1** where the spacetime matrix operator $\hat{M}$ plays a fundamental role. We found that each of these seven equations correspond to a variety of fundamental equations, in both classical electrodynamics and relativistic quantum mechanics. In the next subsection, we will discuss in detail the eighth compact matrix equation listed in **Table 1**. This eighth equation is associated with a new matrix equation which we will refer to as the generalized spacetime matrix equation for free space. As we will see, there are several theoretical implications resulting from our study of the generalized spacetime matrix equation which perhaps may be added as unanswered questions or mysteries to the preceding list.

5.2 Generalized spacetime matrix equation for free space

We define the generalized spacetime matrix equation for free space by the following equation:

$$\begin{bmatrix} -\partial_4 & 0 & 0 & 0 & 0 & -\partial_3 & +\partial_2 & -\partial_1 \\ 0 & -\partial_4 & 0 & 0 & +\partial_3 & 0 & -\partial_1 & -\partial_2 \\ 0 & 0 & -\partial_4 & 0 & -\partial_2 & +\partial_1 & 0 & -\partial_3 \\ 0 & 0 & 0 & -\partial_4 & +\partial_1 & +\partial_2 & +\partial_3 & 0 \\ 0 & +\partial_3 & -\partial_2 & +\partial_1 & +\partial_4 & 0 & 0 & 0 \\ -\partial_3 & 0 & +\partial_1 & +\partial_2 & 0 & +\partial_4 & 0 & 0 \\ +\partial_2 & -\partial_1 & 0 & +\partial_3 & 0 & 0 & +\partial_4 & 0 \\ -\partial_1 & -\partial_2 & -\partial_3 & 0 & 0 & 0 & 0 & +\partial_4 \end{bmatrix} \begin{bmatrix} \Delta_1 \\ \Delta_2 \\ \Delta_3 \\ \Delta_4 \\ \Omega_1 \\ \Omega_2 \\ \Omega_3 \\ \Omega_4 \end{bmatrix} + \kappa \begin{bmatrix} \Delta_1 \\ \Delta_2 \\ \Delta_3 \\ \Delta_4 \\ \Omega_1 \\ \Omega_2 \\ \Omega_3 \\ \Omega_4 \end{bmatrix} = \begin{bmatrix} 0 \\ 0 \\ 0 \\ 0 \\ 0 \\ 0 \\ 0 \\ 0 \end{bmatrix}. \quad (49)$$

The compact matrix form of Eq. (49) is given by

$$\hat{M}|\psi\rangle + \kappa|\psi\rangle = |o\rangle. \tag{50}$$

This is the eighth compact matrix equation in **Table 1**. Note the similarity between the generalized spacetime matrix equation for free space and the Dirac spacetime matrix equation for free space (34) when $\kappa = m_o c/\hbar$ and the Maxwell spacetime matrix equation for free space (9) when $\kappa = 0$. In those equations we purposely set the (4,1) and (8,1) elements in the ket vectors identically equal to zero. Doing so allowed us to convert those matrix equations to vector equations (involving three-dimensional vectors only) which are described in greater detail in [1].

In Eq. (49), we no longer restrict elements (4,1) and (8,1) to be equal to zero. The wave function $|\psi\rangle$ can be thought of as being composed of two four-dimensional vectors $\Delta = (\Delta_1\ \Delta_2\ \Delta_3\ \Delta_4)$ and $\Omega = (\Omega_1\ \Omega_2\ \Omega_3\ \Omega_4)$. The implications by avoiding the earlier restrictions on elements (4,1) and (8,1) will be investigated shortly. We will find some new predictions and surprises by removing these restrictions.

5.3 Eigenvalue spacetime matrix equations

Our primary goal now is to determine the properties of time-harmonic plane-wave solutions satisfying the generalized spacetime matrix (Eq. (49)) for free space. The approach we will take is to cast Eq. (49) into an eigenvalue equation and use the methods of linear algebra to determine the set of orthonormal eigenvectors and corresponding eigenvalues satisfying this eigenvalue equation. (For an excellent book on linear algebra and the solution of eigenvalue equations; see [14], pp. 189–190.) For now let $\kappa = m_o c/\hbar$, the same constant in the Dirac spacetime matrix equation. Later on we will look at the special case when $\kappa = 0$.

We first multiply Eq. (49) by the factor $\hbar c M_4$. The matrix M_4 is the fourth of the spacetime matrices first introduced in Eq. (3). After doing so, with minor algebraic manipulation, we obtain the following matrix equation:

$$\hbar c\begin{bmatrix} -\kappa & 0 & 0 & 0 & 0 & +\partial_3 & -\partial_2 & +\partial_1 \\ 0 & -\kappa & 0 & 0 & -\partial_3 & 0 & +\partial_1 & +\partial_2 \\ 0 & 0 & -\kappa & 0 & +\partial_2 & -\partial_1 & 0 & +\partial_3 \\ 0 & 0 & 0 & -\kappa & -\partial_1 & -\partial_2 & -\partial_3 & 0 \\ 0 & +\partial_3 & -\partial_2 & +\partial_1 & +\kappa & 0 & 0 & 0 \\ -\partial_3 & 0 & +\partial_1 & +\partial_2 & 0 & +\kappa & 0 & 0 \\ +\partial_2 & -\partial_1 & 0 & +\partial_3 & 0 & 0 & +\kappa & 0 \\ -\partial_1 & -\partial_2 & -\partial_3 & 0 & 0 & 0 & 0 & +\kappa \end{bmatrix}\begin{bmatrix} \Delta_1 \\ \Delta_2 \\ \Delta_3 \\ \Delta_4 \\ \Omega_1 \\ \Omega_2 \\ \Omega_3 \\ \Omega_4 \end{bmatrix} = i\hbar\frac{\partial}{\partial t}\begin{bmatrix} \Delta_1 \\ \Delta_2 \\ \Delta_3 \\ \Delta_4 \\ \Omega_1 \\ \Omega_2 \\ \Omega_3 \\ \Omega_4 \end{bmatrix}. \tag{51}$$

The compact matrix form of this equation is given by

$$\hat{H}|\psi\rangle = i\hbar\frac{\partial}{\partial t}|\psi\rangle. \tag{52}$$

This equation has the same identical form as the nonrelativistic Schrödinger equation (see [12], pp. 118–129). However, the Hamiltonian matrix operator $\hat{H}$ is entirely different. This equation represents the canonical form of the generalized spacetime matrix (Eq. (49)).

For time-harmonic plane-wave solutions, the ket vector $|\psi\rangle$ may be expressed as

$$|\psi\rangle = |\psi_o\rangle \exp\left[+i(\mathbf{p}\cdot\mathbf{r} - Et)/\hbar\right]. \tag{53}$$

Again the quantities $\mathbf{p}$ and E correspond to the linear momentum vector and the total energy; $\mathbf{r}$ and t represent the position vector and the instantaneous time. After substituting the eight-by-one ket vector $|\psi\rangle$ back into Eq. (51), we obtain the following eigenvalue equation:

$$pc\begin{bmatrix} -\mu & 0 & 0 & 0 & 0 & +i\alpha_3 & -i\alpha_2 & +i\alpha_1 \\ 0 & -\mu & 0 & 0 & -i\alpha_3 & 0 & +i\alpha_1 & +i\alpha_2 \\ 0 & 0 & -\mu & 0 & +i\alpha_2 & -i\alpha_1 & 0 & +i\alpha_3 \\ 0 & 0 & 0 & -\mu & -i\alpha_1 & -i\alpha_2 & -i\alpha_3 & 0 \\ 0 & +i\alpha_3 & -i\alpha_2 & +i\alpha_1 & +\mu & 0 & 0 & 0 \\ -i\alpha_3 & 0 & +i\alpha_1 & +i\alpha_2 & 0 & +\mu & 0 & 0 \\ +i\alpha_2 & -i\alpha_1 & 0 & +i\alpha_3 & 0 & 0 & +\mu & 0 \\ -i\alpha_1 & -i\alpha_2 & -i\alpha_3 & 0 & 0 & 0 & 0 & +\mu \end{bmatrix}\begin{bmatrix} \Delta_1 \\ \Delta_2 \\ \Delta_3 \\ \Delta_4 \\ \Omega_1 \\ \Omega_2 \\ \Omega_3 \\ \Omega_4 \end{bmatrix} = E\begin{bmatrix} \Delta_1 \\ \Delta_2 \\ \Delta_3 \\ \Delta_4 \\ \Omega_1 \\ \Omega_2 \\ \Omega_3 \\ \Omega_4 \end{bmatrix}. \tag{54}$$

We will refer to Eq. (54) as the eigenvalue spacetime matrix equation. The compact matrix form of Eq. (54) is represented by

$$H|\psi\rangle = E|\psi\rangle. \tag{55}$$

The eight-by-eight matrix H is Hermitian which implies the eigenvalues E are real (see [14], p. 222). The following equations define various quantities appearing in Eq. (54):

$$\mu \equiv m_o c^2/pc \qquad \text{and} \qquad \mathbf{p} \equiv p\,(\alpha_1\ \alpha_2\ \alpha_3). \tag{56}$$

The quantity p is the magnitude of the linear momentum vector $\mathbf{p}$, and α_1, α_2, α_3 represent the direction cosines, associated with the direction of the linear momentum vector $\mathbf{p}$.

5.4 Wave propagation along the +z direction for $\kappa = m_o c/\hbar$

Without loss of generality, let us consider matter-wave propagation along the +z direction, that is,

$$\mathbf{p} = p\,(0\ 0\ 1). \tag{57}$$

Eq. (54) reduces to the following simplified form:

$$\begin{bmatrix} -E_o & 0 & 0 & 0 & 0 & +ipc & 0 & 0 \\ 0 & -E_o & 0 & 0 & -ipc & 0 & 0 & 0 \\ 0 & 0 & -E_o & 0 & 0 & 0 & 0 & +ipc \\ 0 & 0 & 0 & -E_o & 0 & 0 & -ipc & 0 \\ 0 & +ipc & 0 & 0 & +E_o & 0 & 0 & 0 \\ -ipc & 0 & 0 & 0 & 0 & +E_o & 0 & 0 \\ 0 & 0 & 0 & +ipc & 0 & 0 & +E_o & 0 \\ 0 & 0 & -ipc & 0 & 0 & 0 & 0 & +E_o \end{bmatrix}\begin{bmatrix} \Delta_1 \\ \Delta_2 \\ \Delta_3 \\ \Delta_4 \\ \Omega_1 \\ \Omega_2 \\ \Omega_3 \\ \Omega_4 \end{bmatrix} = E\begin{bmatrix} \Delta_1 \\ \Delta_2 \\ \Delta_3 \\ \Delta_4 \\ \Omega_1 \\ \Omega_2 \\ \Omega_3 \\ \Omega_4 \end{bmatrix} \tag{58}$$

where

$$E_o \equiv m_o c^2. \tag{59}$$

The matrix in Eq. (58) is an eight-by-eight square matrix. A compact matrix version of Eq. (58) may be expressed as follows:

$$H|\psi_n\rangle = E_n|\psi_n\rangle \qquad n = 1, 2, 3, \ldots 8. \tag{60}$$

At this point we are now in a position to determine eight eigenvectors $|\psi_n\rangle$ and the corresponding eigenvalues E_n satisfying the eigenvalue (Eq. (58)). We chose to use the matrix software program MATLAB [15] for determining the eigenvalues and eigenvectors. As it turns out, there are only two unique eigenvalues given by

$$E_+ = +E \quad \text{and} \quad E_- = -E \qquad \text{where} \qquad E = \sqrt{E_o^2 + p^2c^2}. \tag{61}$$

From the special theory of relativity (see [10], pp. 21–25), the following relations may also be of use:

$$E = \gamma E_o \qquad p = \gamma m_o v \qquad pc = \gamma\beta E_o \qquad \gamma = 1/\sqrt{1-\beta^2} \qquad \beta = v/c. \tag{62}$$

As before, γ and β are referred to as the Lorentz factor and speed parameter, respectively. For each of the two eigenvalues, there are four unique eigenvectors. The eight eigenvectors $|\psi_n\rangle$ form an orthonormal set, that is,

$$\langle\psi_m|\psi_n\rangle = \delta_{mn}. \tag{63}$$

The symbol δ_{mn} represents the Kronecker delta. In **Table 2** is a summary of the eigenvalues and orthonormal eigenvectors satisfying the eigenvalue spacetime matrix (Eq. (58)).

The constants a and b appearing in **Table 2** are defined by

$$a = \frac{\sqrt{2}}{2}\sqrt{\frac{\gamma+1}{\gamma}} \qquad a^2 + b^2 = 1 \qquad b = \frac{\sqrt{2}}{2}\sqrt{\frac{\gamma-1}{\gamma}} \tag{64}$$

Inspection of the contents of **Table 2** reveals the following important results:

1. $|\psi_1\rangle$ and $|\psi_2\rangle$ represent transverse waves with positive energy $+\gamma E_o$.

2. $|\psi_3\rangle$ and $|\psi_4\rangle$ represent transverse waves with negative energy $-\gamma E_o$.

3. $|\psi_5\rangle$ and $|\psi_6\rangle$ represent non-transverse waves with positive energy $+\gamma E_o$.

4. $|\psi_7\rangle$ and $|\psi_8\rangle$ represent non-transverse waves with negative energy $-\gamma E_o$.

For wave propagation in the +z direction, the transverse waves have eigenvector solutions $|\psi\rangle$ where elements (3,1), (4,1), (7,1), and (8,1) are identically equal to zero. In other words, $\Delta = (\Delta_1\ \Delta_2\ \ 0\ \ 0)$ and $\Omega = (\Omega_1\ \Omega_2\ \ 0\ \ 0)$. For this case, Δ_1, Δ_2 and Ω_1, Ω_2 correspond to the x and y components. Thus, for wave propagation in the +z direction, the transverse wave solutions only have x and y vector components characteristic of a transverse wave in three dimensions.

E_n	E_1	E_2	E_3	E_4	E_5	E_6	E_7	E_8
E	$+\gamma E_o$	$+\gamma E_o$	$-\gamma E_o$	$-\gamma E_o$	$+\gamma E_o$	$+\gamma E_o$	$-\gamma E_o$	$-\gamma E_o$
$\vert\psi_n\rangle$	$\vert\psi_1\rangle$	$\vert\psi_2\rangle$	$\vert\psi_3\rangle$	$\vert\psi_4\rangle$	$\vert\psi_5\rangle$	$\vert\psi_6\rangle$	$\vert\psi_7\rangle$	$\vert\psi_8\rangle$
Δ_1	0	$+b$	0	$+a$	0	0	0	0
Δ_2	$+ib$	0	$+ia$	0	0	0	0	0
Δ_3	0	0	0	0	$+b$	0	$+a$	0
Δ_4	0	0	0	0	0	$+ib$	0	$+ia$
Ω_1	$-a$	0	$+b$	0	0	0	0	0
Ω_2	0	$-ia$	0	$+ib$	0	0	0	0
Ω_3	0	0	0	0	0	$-a$	0	$+b$
Ω_4	0	0	0	0	$-ia$	0	$+ib$	0

Table 2.
Eigenvalues and orthonormal eigenvectors associated with the generalized spacetime matrix equation for wave propagation in the +z direction when $\kappa = m_o c/\hbar$.

On the other hand, for wave propagation in the $+z$ direction, the non-transverse waves have eigenvector solutions $|\psi\rangle$ where elements (1,1), (2,1), (5,1), and (6,1) are identically equal to zero. That is to say, $\Delta = (0\ \ 0\ \ \Delta_3\ \Delta_4)$ and $\Omega = (0\ \ 0\ \ \Omega_3\ \Omega_4)$. This implies, Δ_3 and Ω_3 represent z-components. Δ_4 and Ω_4 represent the fourth components (unknown origin) in a four-dimensional space. Thus, for wave propagation in the $+z$ direction, the non-transverse wave solutions have a z vector component (longitudinal in nature) and a fourth vector component (neither transverse nor longitudinal in nature, perhaps a "time" component) of a non-transverse wave in four dimensions.

5.5 Traditional Dirac equation

The authors, in their most recent publication [1], indicated solutions of their Dirac spacetime matrix equation for free space could be mapped into solutions satisfying the traditional Dirac matrix equation. We wish to explore this in greater detail. The traditional Dirac equation, in the absence of electromagnetic potential terms, is given by

$$\begin{bmatrix} +\partial_4 & 0 & -i\partial_3 & -\partial_2 - i\partial_1 \\ 0 & +\partial_4 & +\partial_2 - i\partial_1 & +i\partial_3 \\ +i\partial_3 & +\partial_2 + i\partial_1 & -\partial_4 & 0 \\ -\partial_2 + i\partial_1 & -i\partial_3 & 0 & -\partial_4 \end{bmatrix} \begin{bmatrix} \Sigma_1 \\ \Sigma_2 \\ \Sigma_3 \\ \Sigma_4 \end{bmatrix} + \kappa \begin{bmatrix} \Sigma_1 \\ \Sigma_2 \\ \Sigma_3 \\ \Sigma_4 \end{bmatrix} = \begin{bmatrix} 0 \\ 0 \\ 0 \\ 0 \end{bmatrix} \tag{65}$$

This equation corresponds to the special case employing the Dirac representation (see [12], pp. 694–706) for details. The compact matrix form of Eq. (65) is given by

$$\hat{D}|\sigma\rangle + \kappa|\sigma\rangle = |o\rangle. \tag{66}$$

The Dirac matrix operator $\hat{D}$ represents the four-by-four matrix operator on the left-hand side of Eq. (65), $|\sigma\rangle$ is the four-by-one ket vector appearing twice on the left-hand side, and $|o\rangle$ is the four-by-one null ket vector appearing on the right-hand side. For time-harmonic plane-wave solutions, the ket vector $|\sigma\rangle$ may be expressed as

$$|\sigma\rangle = |\sigma_o\rangle \exp\left[+i(\mathbf{p}\cdot\mathbf{r} - Et)/\hbar\right]. \tag{67}$$

Substituting this time-harmonic plane-wave solution back into the traditional Dirac equation (65) ultimately leads to the corresponding eigenvalue equation:

$$pc\begin{bmatrix} +\mu & 0 & +\alpha_3 & -i\alpha_2+\alpha_1 \\ 0 & +\mu & +i\alpha_2+\alpha_1 & -\alpha_3 \\ +\alpha_3 & -i\alpha_2+\alpha_1 & -\mu & 0 \\ +i\alpha_2+\alpha_1 & -\alpha_3 & 0 & -\mu \end{bmatrix}\begin{bmatrix}\Sigma_1\\ \Sigma_2\\ \Sigma_3\\ \Sigma_4\end{bmatrix} = E\begin{bmatrix}\Sigma_1\\ \Sigma_2\\ \Sigma_3\\ \Sigma_4\end{bmatrix}. \tag{68}$$

For the special case of wave propagation in the +z direction, the preceding eigenvalue equation reduces to the following simplified form:

$$\begin{bmatrix} +E_o & 0 & +pc & 0 \\ 0 & +E_o & 0 & -pc \\ +pc & 0 & -E_o & 0 \\ 0 & -pc & 0 & -E_o \end{bmatrix}\begin{bmatrix}\Sigma_1\\ \Sigma_2\\ \Sigma_3\\ \Sigma_4\end{bmatrix} = E\begin{bmatrix}\Sigma_1\\ \Sigma_2\\ \Sigma_3\\ \Sigma_4\end{bmatrix}. \tag{69}$$

Again using the matrix software MATLAB, the four orthonormal eigenvectors and corresponding eigenvalues satisfying Eq. (69) are listed in the **Table 3**.

The quantities a and b appearing in **Table 3** are defined by

$$a \equiv \frac{\sqrt{2}}{2}\sqrt{\frac{\gamma+1}{\gamma}} \qquad a^2+b^2=1 \qquad b \equiv \frac{\sqrt{2}}{2}\sqrt{\frac{\gamma-1}{\gamma}}. \tag{70}$$

Note, the quantities a and b appearing in the traditional Dirac equation eigenvectors listed in **Table 3** are the same a and b quantities appearing in the generalized spacetime matrix equation eigenvectors listed in **Table 2** for $\kappa = m_o c/\hbar$.

E_n	E_1	E_2	E_3	E_4
E	$+\gamma E_o$	$+\gamma E_o$	$-\gamma E_o$	$-\gamma E_o$
$\|\sigma_n\rangle$	$\|\sigma_1\rangle$	$\|\sigma_2\rangle$	$\|\sigma_3\rangle$	$\|\sigma_4\rangle$
Σ_1	$-\frac{\sqrt{2}}{2}a$	$-\frac{\sqrt{2}}{2}a$	$+\frac{\sqrt{2}}{2}b$	$+\frac{\sqrt{2}}{2}b$
Σ_2	$-\frac{\sqrt{2}}{2}a$	$+\frac{\sqrt{2}}{2}a$	$+\frac{\sqrt{2}}{2}b$	$-\frac{\sqrt{2}}{2}b$
Σ_3	$-\frac{\sqrt{2}}{2}b$	$-\frac{\sqrt{2}}{2}b$	$-\frac{\sqrt{2}}{2}a$	$-\frac{\sqrt{2}}{2}a$
Σ_4	$+\frac{\sqrt{2}}{2}b$	$-\frac{\sqrt{2}}{2}b$	$+\frac{\sqrt{2}}{2}a$	$-\frac{\sqrt{2}}{2}a$

Table 3.
Eigenvalues and orthonormal eigenvectors associated with the traditional Dirac equation for wave propagation in the +z direction when $\kappa = m_o c/\hbar$.

5.6 Linear transformation equation

For the special case of a matter wave traveling through free space in the $+z$ direction, we found the orthonormal set of eigenvectors and corresponding eigenvalues, for both the generalized spacetime matrix (Eq. (49)) and the traditional Dirac equation (65), when $\kappa = m_o c/\hbar$. These two sets of orthonormal eigenvectors are related [1] through the following linear transformation matrix equation:

$$\begin{bmatrix} \Sigma_1 \\ \Sigma_2 \\ \Sigma_3 \\ \Sigma_4 \end{bmatrix} = \frac{\sqrt{2}}{2} \begin{bmatrix} 0 & 0 & 0 & 0 & +1 & -i & +1 & -i \\ 0 & 0 & 0 & 0 & +1 & +i & -1 & -i \\ -1 & +i & -1 & +i & 0 & 0 & 0 & 0 \\ -1 & -i & +1 & +i & 0 & 0 & 0 & 0 \end{bmatrix} \begin{bmatrix} \Delta_1 \\ \Delta_2 \\ \Delta_3 \\ \Delta_4 \\ \Omega_1 \\ \Omega_2 \\ \Omega_3 \\ \Omega_4 \end{bmatrix}. \tag{71}$$

The compact matrix form of Eq. (71) is given by

$$|\sigma\rangle = Z|\psi\rangle. \tag{72}$$

When we substitute each the eight eigenvectors $|\psi_n\rangle$ from **Table 2** back into Eq. (71), we obtain the following results:

1. The four transverse eigenvectors in **Table 2** map into the four eigenvectors in **Table 3**:

$$|\sigma_1\rangle = Z|\psi_1\rangle \qquad |\sigma_2\rangle = Z|\psi_2\rangle \qquad |\sigma_3\rangle = Z|\psi_3\rangle \qquad |\sigma_4\rangle = Z|\psi_4\rangle. \tag{73}$$

2. The four non-transverse eigenvectors in **Table 2** map into the same four eigenvectors in **Table 3**:

$$|\sigma_1\rangle = Z|\psi_5\rangle \qquad |\sigma_2\rangle = Z|\psi_6\rangle \qquad |\sigma_3\rangle = Z|\psi_7\rangle \qquad |\sigma_4\rangle = Z|\psi_8\rangle. \tag{74}$$

Therefore, whether we use the four transverse eigenvector solutions or the four non-transverse eigenvector solutions satisfying the generalized spacetime matrix (Eq. (49)), the same four eigenvector solutions satisfying the traditional Dirac equation (65) are obtained using Eq. (71). It is noted the four transverse eigenvector solutions could have been obtained from the four Dirac vector equations (37) and (38).

5.7 Wave propagation along the +z direction for $\kappa = 0$

For the special case of wave propagation in the $+z$ direction, when $\kappa = 0$, time-harmonic plane-wave solutions satisfying the generalized spacetime matrix equation for free space (49) yield the set of eigenvectors and eigenvalues presented in **Table 4**. The eight eigenvectors $|\psi_n\rangle$ also form an orthonormal set, that is,

$$\langle\psi_m|\psi_n\rangle = \delta_{mn}. \tag{75}$$

E_n	E_1	E_2	E_3	E_4	E_5	E_6	E_7	E_8
$p_n c$	$+pc$	$+pc$	$-pc$	$-pc$	$+pc$	$+pc$	$-pc$	$-pc$
$\|\psi_n\rangle$	$\|\psi_1\rangle$	$\|\psi_2\rangle$	$\|\psi_3\rangle$	$\|\psi_4\rangle$	$\|\psi_5\rangle$	$\|\psi_6\rangle$	$\|\psi_7\rangle$	$\|\psi_8\rangle$
Δ_1	0	$+b$	0	$+a$	0	0	0	0
Δ_2	$+ib$	0	$+ia$	0	0	0	0	0
Δ_3	0	0	0	0	$+b$	0	$+a$	0
Δ_4	0	0	0	0	0	$+ib$	0	$+ia$
Ω_1	$-a$	0	$+b$	0	0	0	0	0
Ω_2	0	$-ia$	0	$+ib$	0	0	0	0
Ω_3	0	0	0	0	0	$-a$	0	$+b$
Ω_4	0	0	0	0	$-ia$	0	$+ib$	0
v_n	v_1	v_2	v_3	v_4	v_5	v_6	v_7	v_8
	$+c$	$+c$	$-c$	$-c$	$+c$	$+c$	$-c$	$-c$

Table 4.
Eigenvalues and orthonormal eigenvectors associated with the generalized spacetime matrix equation for wave propagation in the +z direction when $\kappa = 0$.

The constants a and b appearing in **Table 4** are now defined by

$$a \equiv \frac{\sqrt{2}}{2} \qquad a^2 + b^2 = 1 \qquad b \equiv \frac{\sqrt{2}}{2} \tag{76}$$

Inspection of the contents of **Table 4** reveals the following important results:

1. $|\psi_1\rangle$ and $|\psi_2\rangle$ represent transverse waves moving with speed $+$c.

2. $|\psi_3\rangle$ and $|\psi_4\rangle$ represent transverse waves moving with speed $-$c.

3. $|\psi_5\rangle$ and $|\psi_6\rangle$ represent non-transverse waves moving with speed $+$c.

4. $|\psi_7\rangle$ and $|\psi_8\rangle$ represent non-transverse waves moving with speed $-$c.

For wave propagation in the +z direction, the transverse waves have eigenvector solutions $|\psi\rangle$ where elements (3,1), (4,1), (7,1), and (8,1) are identically equal to zero. In other words, $\Delta = (\Delta_1\ \Delta_2\ \ 0\ \ 0)$ and $\Omega = (\Omega_1\ \Omega_2\ \ 0\ \ 0)$. For this case, Δ_1, Δ_2 and Ω_1, Ω_2 correspond to the x and y components. Thus, for wave propagation in the +z direction, the transverse wave solutions only have x and y vector components, characteristic of a transverse wave in three dimensions. Only those waves propagating at a speed in free space of +c represent real electromagnetic waves.

On the other hand, for wave propagation in the +z direction, the non-transverse waves have eigenvector solutions $|\psi\rangle$ where elements (1,1), (2,1), (5,1), and (6,1) are identically equal to zero. That is to say, $\Delta = (0\ \ 0\ \ \Delta_3\ \Delta_4)$ and $\Omega = (0\ \ 0\ \ \Omega_3\ \Omega_4)$. This implies, Δ_3 and Ω_3 represent z-components. Δ_4 and Ω_4 represent the fourth components in a four-dimensional space. Thus, for wave propagation in the +z direction, the non-transverse wave solutions have a z vector component (longitudinal in nature) and a fourth vector component (neither transverse nor longitudinal in nature) of a non-transverse wave in four dimensions. Perhaps there is new physics regarding these additional solutions.

5.8 Unresolved issues regarding the generalized spacetime matrix equation

The eigenvectors and eigenvalues associated with the generalized spacetime matrix equation, for the special case of a time-harmonic plane-wave propagating in free space in the $+z$ direction, have been determined for both $\kappa = m_o c/\hbar$ and $\kappa = 0$. The following are the key points found in this analysis:

1. For the case when $\kappa = m_o c/\hbar$, we found there were four orthonormal eigenvectors (two having positive energy eigenvalues $+\gamma E_o$ and two having negative energy eigenvalues $-\gamma E_o$) describing waves having transverse properties. From **Table 2**, each of these four eigenvectors have components $\Delta_3 = \Delta_4 = \Omega_3 = \Omega_4 = 0$. Using the linear transformation equation (71), these eigenvectors map nicely into four orthonormal eigenvectors satisfying the traditional Dirac equation.

2. For the case when $\kappa = m_o c/\hbar$, we found there were also four orthonormal eigenvectors (again two having positive energy eigenvalues $+\gamma E_o$ and two having negative energy eigenvalues $-\gamma E_o$) describing waves having non-transverse properties. From **Table 2**, each of these four eigenvectors have components $\Delta_1 = \Delta_2 = \Omega_1 = \Omega_2 = 0$. Again, using the linear transformation equation (71), these four eigenvectors map nicely into the same four orthonormal eigenvectors satisfying the traditional Dirac equation as mentioned in Case 1.

3. Therefore, for the case when $\kappa = m_o c/\hbar$, the generalized spacetime matrix equation (49) for free space provides eight orthonormal eigenvector solutions (both transverse and non-transverse) which map into four orthonormal eigenvector solutions satisfying the traditional Dirac equation (65).

4. For the case when $\kappa = 0$, we found there were four orthonormal eigenvectors (two associated with waves propagating in free space with speed $+c$ and two associated with waves propagating in free space with speed $-c$) describing waves having transverse properties. From **Table 4**, each of these four eigenvectors have components $\Delta_3 = \Delta_4 = \Omega_3 = \Omega_4 = 0$. For the case of transverse waves propagating with $+c$, these eigenvectors are associated with real electromagnetic waves predicted by the traditional Maxwell equations.

5. For the case when $\kappa = 0$, we found there were also four orthonormal eigenvectors (two associated with waves propagating in free space with speed $+c$ and two associated with waves propagating in free space with speed $-c$) describing waves having non-transverse properties. From **Table 4**, each of these four eigenvectors has components $\Delta_1 = \Delta_2 = \Omega_1 = \Omega_2 = 0$.

6. The generalized spacetime matrix equation for $\kappa = 0$ when $\Delta_4 \equiv 0$ and $\Omega_4 \equiv 0$ is simply the Maxwell spacetime matrix equation for free space. The generalized spacetime matrix equation for $\kappa = m_o c/\hbar$ when $\Delta_4 \equiv 0$ and $\Omega_4 \equiv 0$ is simply the Dirac spacetime matrix equation for free space. In addition, the Dirac spacetime matrix equation for free space is equivalent to the four Dirac spacetime vector equations (37) and (38) for free space resembling the four Maxwell vector equations (11) and (12) for free space.

In the de Broglie-Bohm picture of quantum mechanics, Hardy [16] and Bell [17] suggest empty waves represented by wave functions propagating in spacetime, but not carrying energy or momentum, can exist. This same concept was called ghost waves or ghost fields by Albert Einstein (see [18]). The controversy as to whether matter waves correspond to real waves or ghost waves has been and is still a subject of debate and controversy.

In Section 5.1, we mentioned that the number of unanswered questions and mysteries regarding the universe from the smallest to the largest, in the fields of physics and astronomy, is unimaginable. Allowing the elements Δ_4 and Ω_4 to have nonzero values in the generalized spacetime matrix equation certainly raises a number of unanswered questions. The following is the author's list of 12 unanswered questions and mysteries regarding our analysis of the generalized spacetime matrix equation for free space:

For relativistic quantum mechanics—matter waves:
What class of particles do the transverse eigenvectors represent?
Do the transverse eigenvectors represent real or ghost waves?
What class of particles do the non-transverse eigenvectors represent?
Do non-transverse eigenvectors represent real or ghost waves?
Are the transverse and non-transverse eigenvectors equivalent in some way?

For classical electrodynamics—electromagnetic waves:
What can be said about those waves propagating with speed $-c$?
Do these represent a new type of electromagnetic wave?
What can be said about those waves having a longitudinal component?
What can be said about those waves having a fourth component?
Could these be associated with undiscovered electromagnetic waves?

And two last questions:
Why do the *Dirac* and *Maxwell* vector equations resemble each other?
Does the spacetime matrix operator $\hat{M}$ have more surprises in store?

6. Conclusions

1. The four classical electromagnetic microscopic Maxwell field equations have been rewritten as a single matrix equation, referred to as the Maxwell spacetime matrix equation, using the spacetime matrix operator $\hat{M}$. The Maxwell spacetime matrix equation is relativistic invariant under a Lorenz transformation.

2. The square eight-by-eight matrix operator $\hat{M}$ has several benefits as summarized next. Other fundamental equations of electromagnetic theory have also been expressed as single matrix equations using the spacetime matrix operator $\hat{M}$, namely, the electromagnetic wave and charge continuity equations, the Lorentz conditions and electromagnetic potentials, and the electromagnetic potential wave equations.

3. The traditional relativistic Dirac equation for free space has been expressed as a new matrix equation, referred to as the Dirac spacetime matrix equation for free space, using the same spacetime matrix operator $\hat{M}$. The Dirac spacetime matrix equation is also relativistic invariant under a Lorenz transformation.

4. Solutions of the new Dirac spacetime matrix equation can be easily transformed into solutions satisfying the traditional relativistic Dirac equation using the linear transformation matrix Z.

5. The Dirac spacetime matrix equation is equivalent to four new relativistic quantum mechanical vector equations. We referred to these equations as the Dirac spacetime vector equations. In the absence of electromagnetic potentials, these vector equations resemble the four classical electromagnetic microscopic Maxwell field vector equations in the absence of charge and current densities.

6. Multiplication of the Dirac spacetime matrix equation by the spacetime matrix operator $\hat{M}$ leads to the relativistic Klein-Gordon spacetime matrix equation.

7. Four transverse orthonormal eigenvectors as well as the four non-transverse orthonormal eigenvectors satisfying the Dirac spacetime matrix equation map, via the linear transformation matrix Z, into the same set of four orthonormal eigenvectors satisfying the traditional Dirac equation.

8. A new generalized spacetime matrix equation employing the operator $\hat{M}$ was introduced. This equation is a generalization of the Maxwell and Dirac spacetime matrix equations for free space. We explored time-harmonic plane-wave solutions of this equation as well as their properties. Some of results obtained may suggest new physics.

Acknowledgements

We are most appreciative of the help by Ms. Trin Riojas of the Optical Sciences Center in coordinating computer station inputs/outputs between authors and publishers. The past informal discussions with Dr. Arvind S. Marathay of the Optical Sciences Center are also greatly appreciated. This research did not receive any specific grant from funding agencies in the public, commercial, or not-for-profit sectors.

Author details

Richard P. Bocker[1*†] and B. Roy Frieden[2†]

1 San Diego State University, San Diego, California, United States of America

2 University of Arizona, Tucson, Arizona, United States of America

*Address all correspondence to: rp44bocker@gmail.com

† These authors contributed equally.

References

[1] Bocker R, Frieden B. A new matrix formulation of the Maxwell and Dirac equations. Heliyon. 2018;**4**(12):e01033. DOI: 10.1016/j.heliyon.2018.e01033

[2] Messiah A. Quantum Mechanics. New York: Dover; 2014. pp. 245-250. ISBN: 13:9780486784557

[3] Jackson J. Classical Electrodynamics. 3rd ed. New York: Wiley; 1999. DOI: 10.1119/1.19136

[4] Lorrain P, Corson D, Lorrain F. Electromagnetic Fields and Waves. 3rd ed. New York: Freeman; 1988. ISBN: 10: 0716718235

[5] Macleod H. Thin-Film Optical Filters. 2nd ed. New York: McGraw-Hill; 1989. pp. 1-312. ISBN: 0-07-044694-6

[6] Fowles G. Introduction to Modern Optics. New York: Holt, Rinehart and Winston; 1968. pp. 168-183. DOI: 10.1119/1.1975142

[7] Bocker R, Frieden B. Solution of the Maxwell field equations in vacuum for arbitrary charge and current distributions using the methods of matrix algebra. IEEE Transactions on Education. 1993;**36**:350-356. DOI: 10.1109/13.241610

[8] Ohanian H. Classical Electrodynamics. Boston: Allyn and Bacon; 1988. ISBN-10:0205105289

[9] Jackson J. Classical Electrodynamics. New York: Wiley; 1962. Library of Congress, Catalog Card Number 62-8774

[10] Serway R, Moses C, Moyer C. Modern Physics. Philadelphia: Saunders; 1989. ISBN: 0-03-004844-3

[11] Schiff L. Quantum Mechanics. 3rd ed. New York: McGraw-Hill; 1968. pp. 472-488. ISBN: 13: 978-0070552876

[12] Roman P. Advanced Quantum Theory. Palo Alto: Addison-Wesley; 1965. ISBN-13: 9780201064957

[13] Wikipedia. List of Unsolved Problems in Physics [Internet]. Available from: https://en.wikipedia.org/wiki/List_of_unsolved_problems_in_physics [Accessed: 29 April 2019]

[14] Strang G. Linear Algebra and its Applications. 2nd ed. New York: Academic Press; 1976. ISBN: 0-12-673660-X

[15] Gilat A. Matlab: An Introduction with Applications. 5th ed. New York: Wiley; 2014. ISBN: 9781118629864

[16] Hardy L. On the existence of empty waves in quantum theory. Physics Letters A. 1992;**167**:11-16. DOI: 10.1016/0375-9601(92)90618-V

[17] Bell J. Six possible worlds of quantum mechanics. Foundations of Physics. 1992;**22**:1201-1215. DOI: 10.1007/BF01889711

[18] Selleri F, Van der Merwe A. Quantum Paradoxes and Physical Reality. Dordrecht: Kluwer; 1990. ISBN: 978-94-009-1862-7

Chapter 3

Clarifying Special Relativity

Richard Sauerheber

Abstract

Special relativity for light requires substantial correction. The notion that time dilates for observers in motion has been disproven theoretically, experimentally, and mathematically. Absolute time is not altered by the motion of objects or human activity. The original concept used distance and light velocity improperly to compute time. When the displacement of objects in relation to the traveling direction of a photon of light is considered properly, both stationary and moving observers compute time for any particular event that is equal. Light photons travel at intrinsic speed c in the propagation direction but have component velocities less than c. Although light velocity c cannot be altered by motion of its source in the propagation direction, photons from a lateral moving source experience a lateral velocity component and angle travel from the source at speed c in that direction. Due to motion of the earth in its orbit, objects that are seen are images from a former location when the light departed. More or less time is required for light to traverse objects in motion than when stationary. This is not due to dilation of absolute time. Fixed light speed ensures that differing distances require differing times for light to travel.

Keywords: light, time, relative measurements, nature of light

1. Introduction to relativity

Most all measurements in the physical world are subject to relativity. Any object viewed from a distance appears smaller than its size seen at a closer distance. Its actual size however is the same, independent of the distance from which it is viewed. Likewise, time can feel very long when one is bored but very short when one is entertained, when the actual time is the same independent of such feelings. Relativity is this fact that perceptions for a particular object or event can differ for different observers and can depend on one's vantage point.

It is true that different observers watching a given event will describe that event differently. In this way relativity can be of particular value. The different disciples' accounts of Jesus reflect features differently that present a more complete picture for those reading the gospels. In other cases, relativity may be a hindrance that must be adjusted for, such as when determining the actual true value of a scientific measurement. Relativity can cause a measurement to be made incorrectly. Even though the absolute time for an event itself does not respond to physical changes of matter and is independent of whether matter exists or not, a long-held notion called time dilation is widely taught as fact in many Physics texts and must be explained. Time dilation stemmed from thoughts regarding the use of light and physical objects to attempt to measure time. Some examples use a lateral moving "light box" containing a light source that is represented to measure time, and other instances use a linear moving rod or train car that must be traversed by light. In all cases, observers moving with the device perceive a different distance of travel than that for observers who are stationary who notice the motion of the device. In reality, light must travel farther to reach a receding target or less to reach an approaching target, whether a box, train, or rod that moves during the photon travel time, all while the time required to travel that distance is a single correct value. The true time for any event is not affected by one's position to observe the event. Umpires not in a position to observe well a play in a sports event often make incorrect decisions. Likewise measurement of time using fixed-speed light that interacts with matter depends on the relative motion of that matter. If it is observed easily, the correct time may be determined. If not, and not properly corrected, a meaningless time will result.

All scientific instruments must be calibrated for variables that affect readings, or the readings will be incorrect, and this includes the measurement of time using light. A watch with a lead weight placed on its hands ticks more slowly and reports a wrong time for an event. And this does not change the actual time required for the event to occur! Time cannot be measured correctly with a moving light box, or train or rod UNLESS the direction and magnitude of motion of the box or rod or train in relation to the propagating photon are known, and used to determine actual displacement of the light in the direction at which its velocity is known. Any light clock velocity unequal to zero, or moisture in the air that slows light speed, causes light clocks to report a time that is not the correct time. Real time for an event is not subject to motion of a device attempting to measure it. Real time is determined by the event itself, independent of whether an observer runs away, runs toward, or remains still with respect to the event. Twins are the same age, whether one runs differently than the other or travels in spacecraft.

2. Light is massless and propagates at fixed speed c

A photon of light is electromagnetic energy that can only exist while traveling at a fixed fast speed in a given medium and that propagates in perpetuity if uninterrupted by physical matter. James Clerk Maxwell (1865) successfully derived mathematically the speed with which light must propagate in a given medium from the point in space at which it is produced, where $c = \mathbf{E}/\mathbf{B} = 1/(\varepsilon\upsilon)^{1/2}$. **E** and **B** are the amplitudes of the electrical and magnetic field orthogonal components of light, and ε and υ are the electrical permittivity and magnetic permeability of the particular translucent medium in which light propagates. In vacuum, the speed of light is approximately $c = 2.99792 \times 10^8$ m/s. The Nobel prize-winning American physicist from Poland, Albert Michelson, directly measured the speed of light experimentally in the San Gabriel mountains of California in 1926. The round-trip for light to travel from Mount Wilson near Pasadena to Lookout mountain at Mt. Baldy (Mt. San Antonio) near Alta Loma is a ground distance of 44 miles. Knowing the rotation speed of a rotating slotted set of mirrors and thus the time between successive slots through which light passed, the speed of light was computed to 6 digit accuracy at 2.99792×10^8 m/s, confirming experimentally the correctness of the Maxwell theoretic derivation. We now know that because the earth orbits the sun and light travels in fixed straight paths at a speed that does not add to that of the earth, that the true time to travel this 44 mile round-trip is slightly different because the total travel distance is larger due to the earth's motion during the photon travel time.

3. Time for light to traverse a moving object is relative: one-dimensional case

The distance between two trees along the ground is **D** km (**Figure 1**). Because the speed of light in its propagation direction is c from the spatial coordinate at which

Figure 1.
The ground distance traveled by a photon from one tree to another tree differs from the actual distance the photon travels because the trees move along with the orbiting earth while the speed of light is nevertheless constant.

it departs its source (that is, a stationary point in space) then the time required for a light photon to travel from the pine to the oak here would be $t = \mathbf{D}/c$ if the earth were stationary. However the earth orbits the sun at 30 km/s and, ignoring any contribution from any translational or rotational motion of the solar system, the more accurate distance the photon actually travels between the trees is directed distance **D** plus the distance the earth moves parallel to **D** (in the direction the photon travels) during the photon travel time between the trees, where $t = (\mathbf{D} + 30\ t)/c$ so that $t = \mathbf{D}/(c - \mathbf{v}) = \mathbf{D}/(c - 30)$. Thus the true time for an event involving light interacting with material mass must be computed with proper vector algebra where the actual distance of travel must be known. Moreover, a stationary observer on earth easily computes an incorrect time as **D**/c because it is simple to assume the actual distance traveled is only **D** when it is not. The total distance is greater than **D** when the earth moves in the direction the photon travels and less than **D** when the earth moves opposite to the photon. Note that c in this case is the magnitude of the velocity of the photon in its travel direction and thus c and **D** are both vector quantities in this case.

The original notion of time dilation derived from thoughts regarding a forward moving rod traversed by light. It was assumed that an observer on the rod would compute the time for light to traverse the rod forward and then backward again as 2 **L**/c where L is the length of the rod. A stationary observer watching the moving rod would notice the distance traveled by the rod while light was traversing it and would compute a different time than 2 **L**/c. On closer inspection it is clear that the actual time required for light to make such a round trip on a moving rod depends on the velocity of the rod (1, 3) and when computed properly by the moving observer matches the time computed by the stationary observer. If the rod were to move at very fast sub-light speeds, the light would not reach the end of the rod for a long time interval but using a clock instead of the length of the rod by the moving observer would report that prolonged time correctly. It was mistakenly assumed that the shorter time for light to return to the rod, where the relative velocity would be $c + \mathbf{v}$ when the rod approaches the light, would cancel the longer time the light requires to reach the end of the receding rod on the forward trip. This is not the case, where the total travel round-trip time is given by $t = \mathbf{D}/(c - \mathbf{v}) + \mathbf{D}/(c + \mathbf{v}) = 2\mathbf{D}/(c - \mathbf{v}^2/c)$. Notice that if $\mathbf{v} = c$, t would be infinite since the light would not catch the rod end. When **v** is zero then $t = 2\mathbf{D}/c$, the time for a round-trip travel for light on a stationary rod. The faster the **v** of the rod, the longer is the time required to round-trip the rod. There is no dilation of absolute time, simply a longer time is required for a longer trip to be completed.

This system involves both classical and special relativity to understand. Classically distances traveled by any object moving at a particular speed toward a target that is also moving always depend on the relative motion of the target. The actual distance traveled may be greater or less than the original distance to the target at the time the light departs its source. The actual distance traveled by a photon to the target is relative to the distance moved by the target. One may perceive the photon only traveled the distance of separation between the source and target, when in actuality the photon travels a different path since the target moved during the time of travel for the light, while light speed is fixed.

Also in the light trees example here, since light speed is fixed at c with respect to a stationary coordinate in space, this special behavior of light requires one only use this speed or its proper component in relation to the travel directions of the light and the moving target. The speed of light is fixed in a given medium, unlike physical objects which pick up additional speed and energy $1/2\ mv^2$ from moving sources. Both sound waves and light waves also travel at a fixed speed even from moving sources. The frequency and wavelength of the sound or light are changed, but not the speed which is the product of frequency f times wavelength λ and for light we write $c = f\lambda$. Although the frequency of light and its intrinsic energy $E = hf$ where h is Planck's constant are increased by a source moving in its propagation direction, it is not possible to increase its speed which is fixed at c. A rifle bullet travels between the two trees on the moving earth at a combined speed of muzzle velocity plus earth orbit velocity, so the time to reach the target tree is simply **D** divided by the muzzle velocity because the extra distance moved by the trees during the bullet travel time is matched and overcome by the extra velocity the bullet has from the moving earth. This is not the case for light which must travel at fixed speed c independent of motion of its source or the target toward which it speeds. This is the key aspect of special relativity. Light, but not true for physical objects, emanating from moving sources requires different times to travel to a target in motion than when stationary.

Another special property of light is that it is massless and its propagation speed c is not exceeded by any object having mass. However before proceeding to the two dimensional light box case, it must be emphasized that light velocity is simple to exceed, because light is only velocity c specifically in its travel direction [1]. Velocity components for a light ray are less than c and may be easily exceeded by physical objects. Merely point a laser light North and you walk East and you will reach an Eastern target that the light does not, because you exceeded the Eastward component velocity of the light ray.

From the above relativity considerations, to compute time for an event that involves using light interacting with physical objects, it is necessary to match distance and velocity vectors, or the computed time will simply be incorrect. The original concept of time dilation unfortunately did not consider vector algebra when computing time for theoretic light timing devices in either the linear or lateral motion cases and these have now been corrected. The concept of dilation of time, presumed to occur when light is used as a time piece, has been disproven, theoretically, mathematically, and experimentally (1–3).

3.1 Two-dimensional case: a light box

A light photon in a *stationary* light box travels the height of the box **d** in time **d**/c (**Figure 2**).

However, in a *lateral moving* light box, a photon must have a horizontal component velocity equal to that of the box **v** in order to hit the moving box top spot. A stationary observer sees the true path **r** of the photon (**Figure 3**) and calculates the correct time **r**/c.

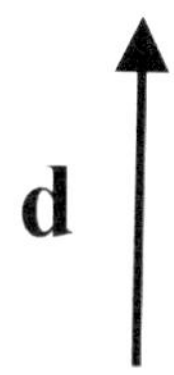

Figure 2.
*The arrow represents the path a photon could take from its source to the top of a box with height **d** that is imagined to be stationary.*

Figure 3.
*The arrow represents the path of a photon emitted from its source to the top of the box of height **d** when the box is imagined to be moving rapidly laterally. The path distance to reach the to of the box is now length **r**.*

The vector **r** does not represent a beam of light, but rather the path traced by a single photon. This is because the photon that arrives at the top of the box left the source when it was at its leftward position earlier as shown. The source at the time of arrival is vertically directly under the arriving photon in the box, and photons leaving the source at that time begin their angled path from this new source location and will arrive at the target after it shifts to another location. So photons all angle travel with this bearing and arrive at the target later at another shifted position. Several photons produced in succession cause the illusion that all traveled vertically to the target, when actually all angle travel distance **r** to arrive. An observer inside the box who only notices the vertical component of the photons could falsely compute time as if it were **d**/c. But light photons in the moving box did not follow the path along vector **d**. Each actually follows a vector parallel to **r** to reach the moving spot on the box top. **d**/c is false because it is a vector mismatch. Correct displacements for light must be determined not by appearance, but by truth, before time can be calculated correctly. Just like a virtual image is not a real image, the appearance that light followed **d** for the moving box is a mirage, not the real displacement path **r**. Notice that a stationary observer far to the right might also assume the photons only moved upward distance **d** and could compute time incorrectly, so the incorrect computation has nothing to do with motion (or not) of an observer.

It is improper to claim that time "slows down" for some event simply because an observer moves during that event. Light cannot sense that an observer is in motion, to adjust its time required to travel a particular distance. Stated simply, a longer distance requires a longer travel time for light at a fixed velocity than a shorter distance, regardless of the state of motion of any observer.

To avoid a vector algebra blunder, it is always mathematically necessary to couple the correct light velocity component with the vector component actually traveled with that velocity. In this way, time calculated for any particular event is the same for any observer, regardless of their state of motion. Note that in time **d**/c, the photon above travels distance **d** (since velocity c times time t equals distance: $\mathbf{d} = c(\mathbf{d}/c) = \mathbf{d}$). This means that the photon traveling along vector **r** travels a distance **d** in time **d**/c but of course has not yet reached the box top at **r**.

Experiments conducted at Palomar Community College with a laser light [1–3] demonstrate that light photons that propagate at speed c pick up lateral velocity when produced by a lateral moving source and thus have a component velocity

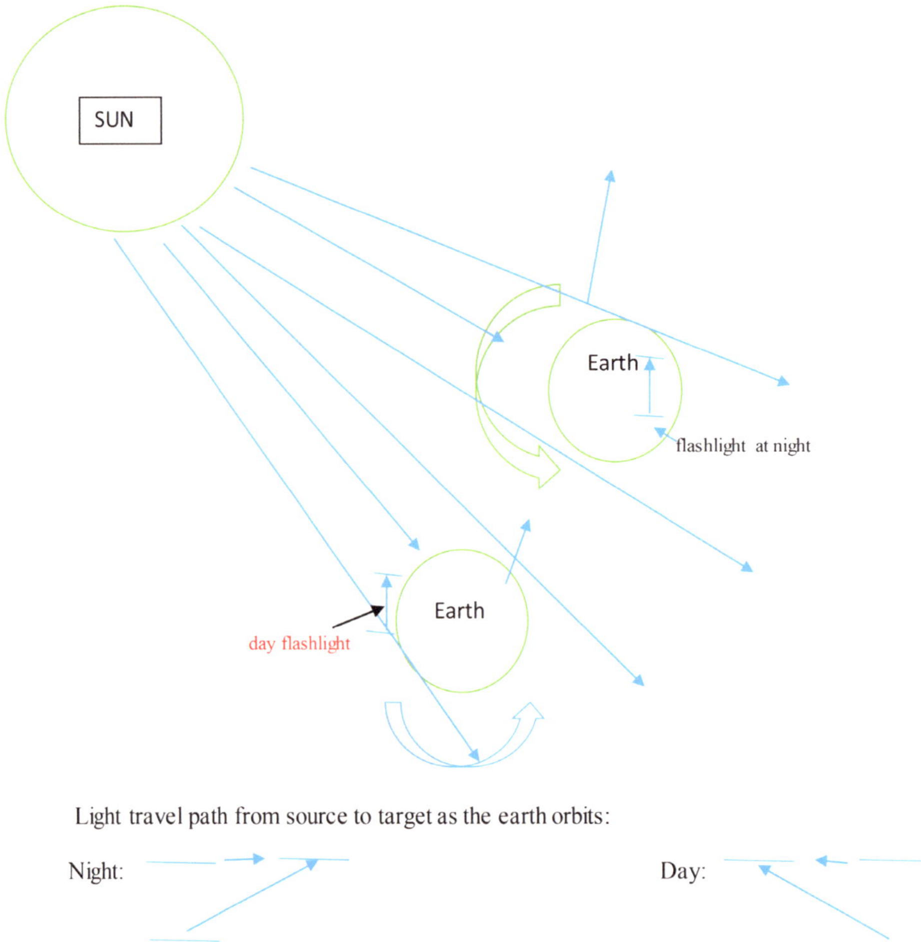

Figure 4.
Diagram depicting the experiment conducted with a laser light continuously illuminating a target 30 m distant while on the rotating and orbiting earth. Light photons travel north to the target, while earth and target orbit 65,000 mph laterally around the Sun. This experiment proves that light photons angle-travel in a straight line and catch the target on earth which shifts laterally 1.3 mm during the time it takes for the photon to arrive. The photons do not simply travel 30 m north and miss the center of the target. Because the earth spins like a twirling figure skater also circling a rink, the 3 mm shift is east of its original position when the photon left the source at midnight but 3 mm west of its original position when a photon leaves at noon. The light continuously on for 24 h periods nevertheless always lands on the center of the target. The photon travel path is thus always larger than the 30 m distance along the ground to the target, because the earth never stops orbiting. Thus photons take longer to reach the target because the earth target is always shifting away from the light. This extension of travel time is not a "dilation" of absolute time due to the motion, but is simply due to the longer distance traveled compared to the time to travel 30 m if the earth were not moving. Anyone who computes time as 30/c rather than (30 + d)/c is sincere, but wrong. Time does not slow down or dilate when objects move. Motion has nothing to do with the fact that absolute time marches on. The special theory of relativity is here modified to indicate that time dilation does not exist, while light remains special in propagating at fixed forward speed independent of motion of the source.

equal to that of the source (**Figure 4**). While a photon travels at speed c 30 m to a target, an observer on earth, orbiting with the source and target, would be incorrect to computer time as 30/c because the photon actually travels farther than the ground distance to the target. Since the earth orbits at 30 km/s, the target and source move laterally 3 mm West at noon (or 3 mm East at midnight) during the time required for a photon to propagate 30 m North. Thus the true travel time to arrive at the new target position is slightly longer than if the earth were stationary. A stationary observer in outer space could see such an angled travel path, while a moving observer on earth would not and the photon travel distance would then need to be calculated. This distance must be used by either stationary or moving observers in order to compute a correct time interval when using light to measure time. Light boxes are inferior timepieces because physical objects on the orbiting rotating earth are always in motion.

Light speed in fog is slower than in dry air. So the number of light box ticks (photon round-trips) in fog is always less than for dry air, for any specific event being timed by a 'light clock". Twins 1 year from now will not be different in age simply because one lives in fog and measures time with a foggy light box. This foggy clock reports a different time for an event compared to what a dry clock reports, not because the actual time is different, but because the clock is affected. Light "clocks" must be calibrated for humidity, or else the reported time is incorrect.

As for any scientific instrument, all variables affecting its operation must be calibrated. Light boxes are affected by lateral velocity from the point at which the light leaves its source. The equation for this dependence is correctly derived in Physics textbooks as: $t = \mathbf{d}/(c^2 - \mathbf{v}^2)^{1/2}$, where **v** is the lateral velocity of the box with respect to a stationary point. At $\mathbf{v} = 0$, the box is stationary, and the time reported by the clock is $\mathbf{d}/c$ for the event where light travels **d**. But the clock in motion reports a smaller number of ticks (or round-trips for a light photon inside the box) for that same given event being timed. The observer inside the box who uses $t = \mathbf{d}/c$ for the moving box has wrongly placed 0 for **v** into the formula. The formula must be followed to obtain a correct time, and **v** is not 0 for the moving box. $\mathbf{d}/c$ is a nonsensical computation for a moving box because more time is required for a tick at the longer distance required by light to travel. Time doesn't slow, it is simply that it takes longer for light to arrive.

Only when velocity of the box (and humidity of the air inside) is calibrated can a correct time interval be reported. Humidity is needed to know the value of c, and **v** is needed to know the displacement distance that photons travel before the clock registers a tick.

The typical Physics textbook conclusion, that since a moving light box ticks more slowly then absolute time itself "slows" [4] does not appreciate that the clock operation is altered by its own motion, similar to being slowed when operating in fog. The explanations of the Hafele-Keating experiment with atomic clocks in airplanes, environmental muon lifespans, the perihelion progression of the planet Mercury, or the actual meaning of the Michelson-Morley split light beam interferometer data have all been presented earlier without need to invent the notion of time dilation [2].

An additional proper way to calculate t for the moving observer inside a light box, moving lateral with velocity $\mathbf{v}_x$, is to match the vertical net displacement **d** with the velocity component for the photon in that vertical direction, which is $\mathbf{v}_y = c\sin\theta$ (where θ is the angle made by the vector **r** from the horizontal). Here $\sin\theta$ is $\mathbf{d}/\mathbf{r}$. Time then becomes $t = \mathbf{d}/c\sin\theta = \mathbf{d}/(c\mathbf{d}/\mathbf{r}) = r/c$, the same time as properly computed by the stationary observer. Although there are several other possible incorrect ways to compute time for this event, these are not further discussed here.

Note that if the light box moves in the direction of the long axis of the box, the box top recedes from the propagating photon, and the equation for time is $t = \mathbf{d}/(c - \mathbf{v})$ (from Einstein, 1905) [5]. This is because the relative, net velocity of the photon toward the top of the box, $c - \mathbf{v}$, depends directly on the receding velocity **v** of the box. The equation becomes very complicated if the box velocity is neither perpendicular nor parallel to the orientation of the box. If the box were to remain aligned with the Y axis, then the time for a photon to traverse the moving box with horizontal velocity $\mathbf{v}_x$ and $\mathbf{v}_y$ is $t = [2\mathbf{v}_y \pm (4\mathbf{v}_y{}^2 + 4\mathbf{d}^2 + c^2 = \mathbf{v}_x{}^2 = \mathbf{v}_y{}^2))^{1/2}]/c$. If the box velocity involves three dimensions, the equation becomes even more complex, which proves that a light box is an improper device for measuring the time for an event. (A light box is however a good motion detector since light arrives at a position other than a target spot when only slight motion of the source with respect to the detector occurs).

4. Simultaneity is not relative

A similar problem affects textbook examples attempting to prove that *simultaneity* is somehow dependent on motion of an observer [6]. Either two events occurred at the same time t, or they occurred at two different times. For example, two light waves or sound waves of fixed speed, produced at the same instant, arrive at an observer at different times if the observer is in motion and shifts from the midpoint. This is due to different distances for each wave to reach him. This does not mean that the waves were themselves produced at different times, but that they traveled different distances to reach him. The stationary observer directly at the midpoint between the two sources would of course conclude correctly that the waves were produced simultaneously. A moving observer must adjust the distance to the source origins by the amount he shifted from the midpoint during the wave travel time, to know whether they were simultaneously produced.

5. Special relativity

Relativity for light is indeed special because light is special. Unlike for objects with mass, light speed from its origin in space in the propagation direction is fixed in a given medium. Firing a bullet inside a lateral moving box similar to the light box example does not change the time required for the bullet to reach the top of the box. This is because for physical objects that have no fixed speed, the intrinsic muzzle speed provided by the source adds to the additional velocity provided by the moving box so that the total speed is greater. Thus the time to arrive at the top of the box, traveling the additional distance due to motion of the box, is the same as that required to reach the top of the box when stationary, $\mathbf{d/v}$ where $\mathbf{v}$ is the intrinsic velocity in the vertical direction for the bullet. Whether the box is moving or not, $\mathbf{v}$ in the vertical direction remains the same. The horizontal component is in addition to its intrinsic component in the vertical direction, so the bullet has more kinetic energy due to the added lateral velocity. Light however must travel at fixed intrinsic speed $c = \mathbf{E/B}$ which is always a constant in a given medium in the direction it propagates, regardless of motion of its source. Indeed, sunlight from the edge of the spinning sun that recedes from view is redshifted compared to light from the edge spinning toward the earth which is blue shifted, while the various colors of light all travel at the same intrinsic speed c. A forward moving light source produces light with greater energy but it is not kinetic energy and is rather intrinsic electromagnetic energy. Light reflected or scattered from physical objects, such as the well-known Compton scattering, loses some energy and departs with a lower energy and lower frequency but travels with a longer wavelength, again at required speed c. These properties of light mean that a light box would be a useful device for measuring relative motions of objects, such as during earthquakes or the ground motion associated with tidal drift, but would not be useful to measure time because light speed does not add to source speed.

6. General relativity

General relativity usually centers on the notion that force fields can be indistinguishable in some experiments and attempts to explain the nature of gravity. But it is mistaken to extrapolate that force fields, whether due to gravity or due to contact forces, are identical and indistinguishable. For example, gravity requires no physical contact with the object being accelerated, as Newton wrote gravity generates forces from great distances. Contact forces that replicate that magnitude of acceleration do require physical contact. If a box were accelerated laterally by either gravity or a physically applied force, then differences would be simple to notice. A weight hanging on a string from the roof for example would not accelerate together with the box if an applied force were responsible for accelerating of the box, while the ball and string would accelerate along with the box if gravity were the responsible agent. There is no solid evidence for general relativity that is non-classical.

The notion is also mistaken that light has mass and is subject to gravity. The expression $E = mc^2$ reflects the fact that mass contains latent energy. This was proven directly with the atomic bomb where annihilated mass in a nuclear reaction causes the release of vast amounts of energy per gram of matter. Radiant energy from the sun likewise is produced from the annihilation of mass due to nuclear fusion reactions. And in reverse, the formation of mass when the universe of matter was Created must have been from the conversion of massive amounts of energy. However the formula is not a statement of congruence, and rather is a relation expressing equal magnitudes but not necessarily quality of energy. For example, the radiant energy from the sun's mass becomes fast traveling light photons, and light is not mass. And mass is not light. So the relation is being misused when one assumes that $m = E/c^2$ somehow proves that "light has mass." Likewise mass is not light, even though the relation was used properly by deBroglie to help prove that electrons with mass oscillate in orbitals around nuclei with wavelengths much like light has. Although the calculated deBroglie wavelengths for an electron in a hydrogen atom match the circumferences of orbitals in the hydrogen atom, mass is not light, just as light is not mass and is not subject to gravity as masses are. Light of course can be bent or refracted by the sun's corona matter, but gravity alone cannot act on light because light has no mass.

Recent photographs of a structure in deep space referred to as a black hole is not a hole. The belief is that material inside it is so dense that light is prevented from escaping it, but if matter is in its center then it is not a black hole, but rather a black body. And there is no proof that light is absent in it since light is invisible unless it is detected by either an eye or a camera. The object appears to block light behind it much like an eclipse, and there is no proof that light is swallowed into it, rather than either being extinguished by absorption or blocked forming a shadow. Again, gravity acts on masses, not light which is massless.

7. Nature of gravity

Earlier work [7] discussed the uniqueness of gravity that is distinct from electric or magnetic fields and from light. Gravity emanates from all objects with mass. Important characteristics of gravity are what it is not. Gravity is not energy, does not require loss of mass or loss of energy to exist around an object [8], is not a wave or a force, and is not spatially reducible. Gravity is an accelerating region in which masses produce a force. Gravity is not diffracted or reflected like light and is not attenuated or diminished by objects in its presence like electric fields are. Even a miniscule electron has gravity emanating around it because electrons have mass. Two neutrons separated by distance $\mathbf{r}$ have a gravitational force between them of $\mathbf{F} = Gm^2/\mathbf{r}^2$ where m is the mass of a neutron and G is the universal gravitation constant. There is no region in space around one neutron where the other neutron is able to escape the gravitation from the other. This is found by experiment and indicates that gravity should not be considered a force field characterized with field lines since this could imply that at some distance $\mathbf{r}$ from the mass that there could be a spatial position at that distance where that gravity might be absent.

Gravity intensity at a given distance from a mass is spatially irreducible. This is because even a miniscule electron senses its presence at any position at which it is located at a distance **r**. Pluto is a huge 6 billion km from the sun and nevertheless is smoothly turned by gravity at every position in its orbit, not only preventing its escape into space but also causing the tracing of an orbit that follows an elliptic mathematical function. Gravity is not reducible at any spatial position along the orbit. All planets in the solar system fall endlessly in perpetuity along elliptic paths in a dynamic equilibrium that is always striving to increase entropy while minimizing orbital energy [9]. Orbiting bodies around the sun instantly detect changes in gravity from the wobbling travel pattern of the sun even at great distances, which causes the bodies to change speed to travel in a smooth elliptic orbit. Galaxies in the universe may also behave in such a way, where each are gravitationally attracted to maintain order in the universe of matter where rotating galaxies maintain relative positions possibly in a dynamic equilibrium steady state.

8. Position is relative

The question, where are you?, requires context and relativity to answer. If the position of a person is desired in relation to the longitude and latitude coordinates on earth's surface, or in relation to a street address or city on earth, then an answer can be given because the spatial coordinate is provided in relation to a particular described position. However since the earth and all objects in the solar system are in constant motion, the true spatial coordinate of where one is located is not actually known with respect to a theoretic stationary 3-D (x, y, z) coordinate in space that one might refer to as an origin from which other coordinates may be measured and stated. And even if the entire universe of matter (as a whole unit) were not rotating or undergoing translational motion so that an origin point in space could be defined, the answer would also depend on time. Due to motion of the particular galaxy and solar system on which one might be located, the position one provides is technically only true at the particular time when the answer was given. The spatial coordinate is quickly changing while one provides the answer. Finally, the definition of you is also relative, where the position in space of the head is different than the feet or the body's center of mass, all at different elevations in 3-D space.

Figure 5 shows the position of the moon in perigee (at its closest approach to the earth) in relation to a tree on earth as a function of time. The moon shifts toward the tree about 5° of angular rotation in 20 min. This observation cannot distinguish whether the shift is caused by the orbiting motion of the moon (that does not spin), being accelerated while continuously changing its direction, or rather is due to either the spinning or to the orbiting motions of the earth. However, with extra data it is indeed possible to determine the major contributor. The moon orbits the earth and returns to its full moon position again in 28.3 days, at an angular velocity of 9.4 h

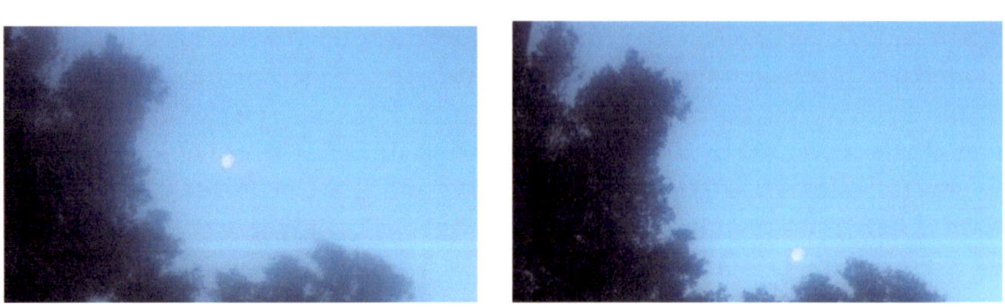

Figure 5.
Photographs of the full moon Feb 23, 2019 traversing behind a pine tree as time proceeds.

to shift 5°. The earth spins or rotates on its own axis 360° in 24 h, or 20 min for 5°, and thus is mostly responsible for the observed shift. In reality of course the actual observed shift is also affected by the fact that the moon and earth co-orbit around a common barycenter. In any event, the specific causes of effects that involve general relativity can indeed be determined by collection of additional information.

9. Nature of light

Unlike gravity that contains no energy itself, light is composed of individual photons of electromagnetic energy hf, formed from electrons that drop to lower energy levels in a source such as the sun, or a tungsten filament in a bulb, or a radio antenna. Light consists of orthogonal electric and magnetic field components that self-induce and self-annihilate rhythmically in perpetuity when uninterrupted. Thomas Young (1801) first demonstrated the wave nature of light which thus can undergo diffraction and interference and can be reflected, scattered, refracted, and absorbed by various media. Light in the visible frequency range is actually not visible to the naked eye. For example most light from the sun emanates into outer space and is not seen. Only light that reaches one's eyes is sensed. This means that anything assumed to be seen is actually a sensed image made by light reflected from the object at an earlier time. Since physical objects in the universe and on earth are always in a state of motion, objects are actually in a different spatial position at the time their light image is sensed. For example it takes 7.5 min for sunlight to reach the earth, so sunrise and sunset actually occurred 7.5 min before these events are actually sensed or "seen".

Photographs of light reflected from a candle prove that light emanates in all three dimensional space even though that light cannot be directly seen (**Figure 6**). The mirror reflects light directly toward the camera for detection from any position, reflecting light that was produced by the candle on the right while the light that exists on the left is invisible. The light on the left is made visible upon reflection by the mirror relocated on the left, while the original light that still exists on the right remains invisible.

Similarly, because light from sources such as stars propagates in all dimensional space even though it cannot be seen, distances to stars can be directly and conclusively determined by parallax. A star is at a particular time of night from an earth location positioned among background stars in shifted locations depending on the location of earth in its orbit. At summertime, the earth shown on the right in **Figure 7** detects light emanated from a star, while light from that star of course exists on the left but is invisible, being not reflected to the eye. In the winter when the earth is positioned on the left, the light is detected from the star while light on the right still exists but remains invisible. From the shifted relative position of the star between summer and winter, the distance to that star can be properly computed

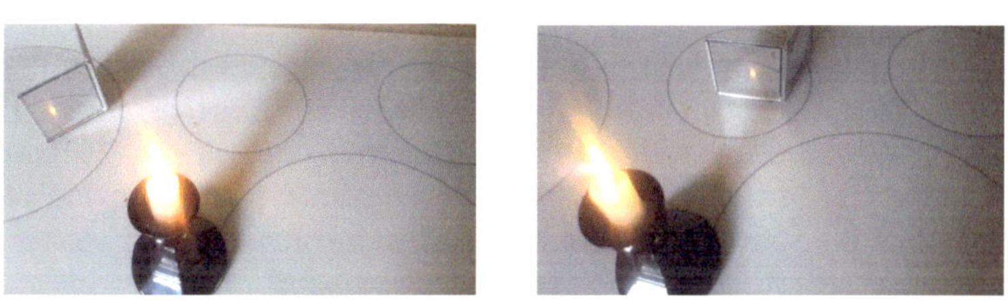

Figure 6.
Light from a source propagates in all directions in space but is invisible. It is sensed by either directly entering the eye (as seen here directly from the candle) or after reflecting the invisible light from the candle to the eye.

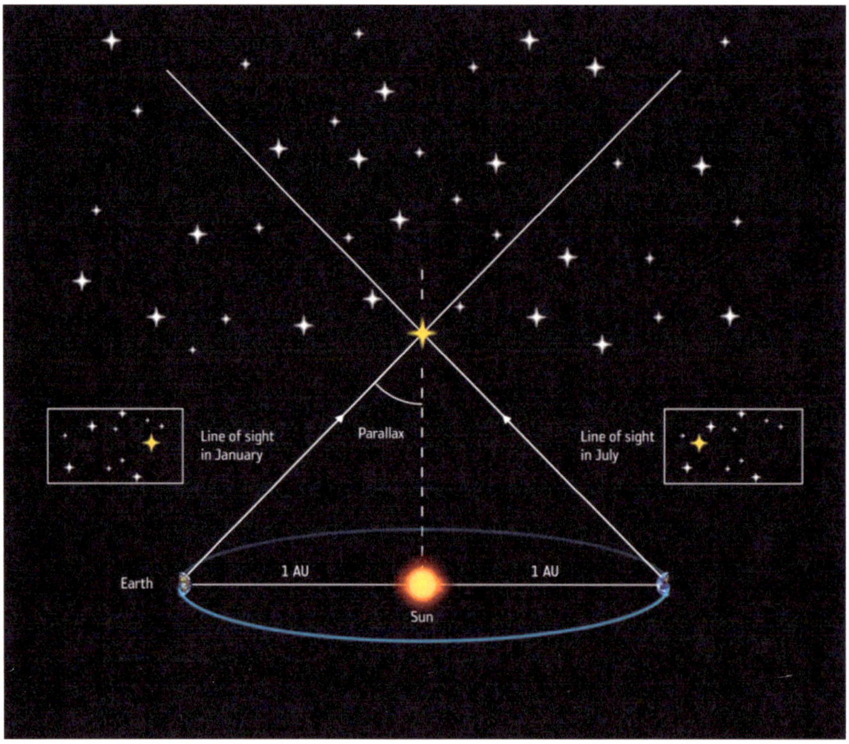

Figure 7.
Earth view of star in relation to distant background stars at summer and winter.

by triangulation. For example the secant of the elevation angle is **r**/(1 AU) where **r** is the distance from the earth to the star and 1 AU is 93 million miles. The picture is distorted intentionally for clarity, where the nearest star to the sun is the Alpha Centauri group at 4.37 light years away (25.6 trillion miles or 266,000 AUs) so that its elevation angle is actually greater than 89.9°. The farthest stars capable of being triangulated with modern space telescopes have a parallax angle so close to 0° that the distance is over 6000 light years. This means the these stars, where light is now arriving here on earth from them, must be at least as old as 6000 years and are at a distance of about 35 quadrillion miles from earth [(6000 years) (365 days/year) (24 h/day) (3600 s/h) (186,000 miles/s)].

Newton first proved in 1665 at Woolsthorpe Manor in England that light beams are actually composed of individual units he called corpuscles which we now call photons. Light has no mass since each photon must propagate in a given medium at a fixed speed c = E/B determined by properties of that medium. Photons speed up upon entering a more favorable medium and slow and retain that lesser fixed speed c in a less favorable medium. Photons follow one another in succession along a fixed bearing in cases where the light source is either stationary (which does not exist in nature since all galaxies rotate, and perhaps undergo translational motion and may vibrate with respect to each other over time, etc.) or moving in the direction photons propagate. Most light from either natural or artificial sources is actually composed of photons that are traveling in directions determined by the lateral motion of its source. This is because the first photon is emitted when the source is at location (x, y, z) but the next photon is produced when the source is at a slightly shifted location due to star or galactic motion. A laser light beam directed to a target while the source and target are in lateral motion consists of photons that traveled different paths to arrive at the

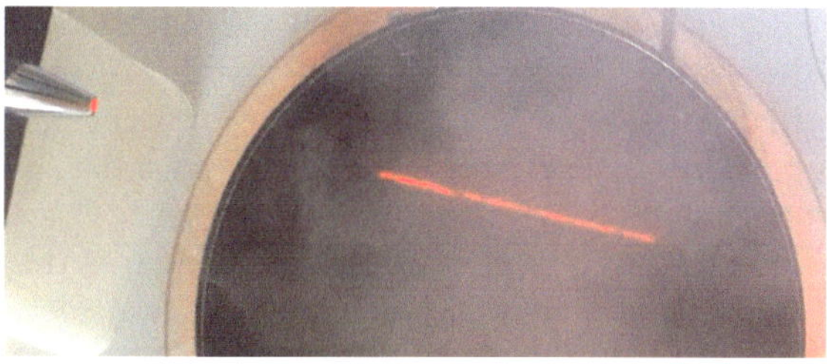

Figure 8.
A light ray made visible in a steam cloud. Over the distance of 0.25 m there are approximately 500,000 photons (= 0.25 m/500 × 10^{-9} m per wavelength) that illuminate the field in about 0.9 ns. Because the speed of the orbiting earth is small compared to light, each photon travels one after another forming a light ray where the paths of each successive photon essentially overlap. If the light source moved at near light speeds, the linear ray would be composed of photons having the same bearing but from shifted locations in space, where the photon arriving at the target on the right traveled the longest distance to arrive there, having left the source when at an earlier position. The photon on the left was produced last from the position where the source is now located. This is essential to understand why lateral moving light clocks do not prove that time dilates, but rather that travel distances for light depend on the relative motion of the source and target.

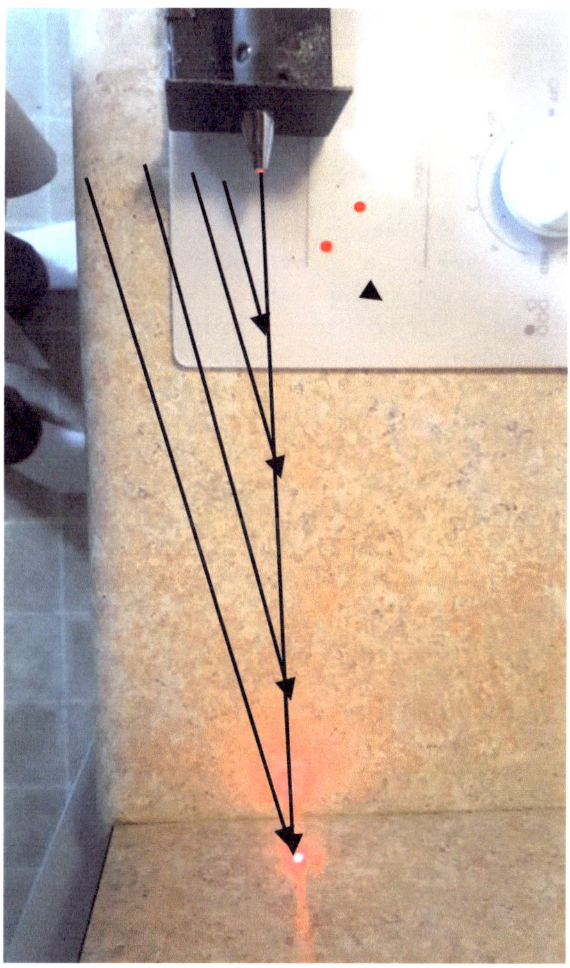

Figure 9.
Light produced by a lateral moving source actually forms a ray consisting of photons having slightly different travel path histories. Each photon departs from the moving source at different spatial coordinates. The photon shown arriving at the target actually left the source when it was at the leftward position. Photons produced from the source when the original photon arrives at the target are produced from the source in its pictured location and will arrive at the target when at its future shifted location further rightward.

shifted target because each photon departs from a different location during the lateral motion of the source [3]. A ray in a steam cloud (**Figure 8**) made continuously visible by reflection to an observer appears to have traveled in a direct, follow-the-leader path by an observer moving along with the source and target. A theoretic observer at some fixed coordinate could notice the actual travel path the photons all followed along linear but shifted diagonals if the source were moving laterally (if light could be made visible). Like an airplane that points at an angle skewed from a runway when a lateral wind is present, light photons would point toward a bearing other than the direction the ray follows. But since earth and planetary speeds are miniscule compared to that for light, this effect would not be observable but could be computed. Although Physics texts commonly claim that the illusion sensed by the moving observer means that time dilates for him, it is simply that a longer time is required for light to reach a shifting target, because the actual path traveled is determined by the photons, not the observer. Here each photon travels slightly further than 0.25 m because each leaves the source at propagation direction speed c from different coordinates, while forming a ray having a component velocity less than c [3]. The fact that photons in a linear ray would have distinct travel path histories if the source were moving laterally at near light speed (which is of course not actually possible for sources with mass) is diagrammed in **Figure 9**.

10. Intrinsic and relative velocity

From the photographs of the moon it is clear that the earth rotates on its axis 10° every 20 min. Since the earth latitude radius is 6372 km (3960 mi) in Southern California, then the tangential velocity of the observer due to earth rotation is $\mathbf{v} = \mathbf{r}\omega = 1036$ mi/h. However the relative motion between the moon and earth do not detect the additional velocity of the earth and moon system that co-orbits with the sun around their common barycenter near the edge of the sun, at 30 km/s. Further, the rotation of the Milky Way galaxy must add to the total velocity of the observer, and it is very possible that the entire universe of matter exhibits a translational velocity while drifting through space although this is not known for certain. Therefore the actual velocity of the observer with respect to some stationary point from which it travels is far different than the particular velocity due to earth's rotation alone. In most cases the total velocity of objects with mass are not actually known with certainty. However, as is evident from the above discussion, for light which has an intrinsic speed in its propagation direction of constant c in a given medium, since physical motions of its source cannot alter light speed c, this means that light velocity with respect to its spatial point of origin is fixed and known. This is the intrinsic speed of light c. The intrinsic velocity of light however is relative to the direction in which it is desired to be used, where component intrinsic velocities of light have magnitudes that are less than c.

Moreover, relative speed and velocity are different from c for light when detectors (not sources) move toward or away from the light front. Otis first reported that detectors moving toward light in its propagation direction detect a higher frequency of light, while the source does not change the wavelength of the light produced and the intrinsic speed of light remains c [6]. Thus from velocity (in the propagation direction) $c' = f\lambda$, a higher (or lower) frequency causes a higher ($c' = c + \mathbf{v}$) (or lower, $c' = c - \mathbf{v}$) relative velocity c' between the detector and light front due to the velocity $\mathbf{v}$ of the source. The simplistic notion that light speed cannot be exceeded also needs to be clarified. Two light beams traveling in opposite directions illuminate space at speed 2c, while each beam propagates at fried intrinsic speed c,

Figure 10.
Palomar Community College Library Learning Resource Center, San Marcos, CA.

as shown by experiment earlier [2]. Further, evidence has been presented that changing gravity magnitude may be sensed between two distant masses at a speed greater than c [7, 9].

The new Palomar Community College library pictured in **Figure 10** is about 300 feet long situated East–West. The time required for light to travel from one end to the other if the earth were stationary would be about 0.3 ms. Since the earth travels this Eastward direction at 65,000 miles/h at midnight, the time required to reach the other end is longer by 0.03 ns because the library retreats from the light 9.2 mm Eastward while light traverses the building. Because the earth also rotates on its own axis, the time required would be 0.03 ns less at noon when the earth orbits Westward, like a twirling figure skater who also orbits a rink. Moreover, rotation of the galaxy plus any translational motion of the universe of matter would also alter the actual time. These effects seem small but nevertheless emphasize that all matter in the universe is in constant motion with variable velocity components, while massless light is fixed at propagation speed c from a stationary coordinate from which it departs. A ray travels speed c across the library but has a vertical component of velocity $\mathbf{v}_y = 0$. A ray shined upward would travel up at speed c with a horizontal component of velocity $\mathbf{v}_x = 0$, where light velocity, but not speed in its travel direction, varies from −c to 0 to +c depending on the direction of interest used in a problem.

11. Conclusion

To avoid misunderstanding or false conclusions, relativity must be considered for most questions asked in Physics. Christians are to be grateful for Creation, and gratitude is expressed here for gravity which keeps us from drifting into deep space, and for light that allows us to see.

Acknowledgements

The author thanks the students at Palomar for discussions and all scientists who have done so much work in aiding our understanding of the physical world. The author would like to thank his Fontana High School, Fontana, CA, Physics teacher Gary Williams for taking them to Harvey Mudd College, Claremont, CA, where they learned relativity. Here, it was noticed that certain extrapolations needed to be corrected. The author thanks also his alma mater the University of California, San Diego, for valuable Physics instruction.

Author details

Richard Sauerheber
Palomar Community College, San Marcos, CA, United States

*Address all correspondence to: rsauerheber@palomar.edu

References

[1] Sauerheber R, Espinoza E. Characteristics of light: Velocity, massless energy, and special relativity. Optik. 2018;**168**:974-986. Available from: https://www.sciencedirect.com/science/article/pii/S0030402618305631

[2] Sauerheber R. The Truth Behind Relativity. Washington, DC: Library of Congress; 2009. Available from: www.lulu.com; http://physicsessays.org/my-orders/product/147-12-pdf-richard-d-sauerheber-on-the-nature-of-light-and-relativity.html

[3] Sauerheber R. On the nature of light and relativity. Physics Essays. 2014;**27**:116-125. Available from: https://www.researchgate.net/publication/274265997_Erratum_On_the_nature_of_light_and_relativity_Phys_Essays_27_116_2014; http://physicsessays.org/browse-journal-2/product/147-12-richard-d-sauerheber-on-the-nature-of-light-and-relativity.html

[4] The scientific website. Available from: www.Physicsmyths.org.uk

[5] Mueller G, Kneckebrodt K. 95 Years of Criticism of the Theory of Relativity. 1908-2003. Available from: http://www.ekkehard-friebe.de/95yearsrelativity.pdf

[6] Otis AS. Light velocity and relativity, the problem of light velocity, Einstein Theory found invalid, a classical theory of relativity, a challenge to Young Scintiss. Yonkers on Hudson, NY: Christian Burckel and Associates; 1963. Available from: https://www.goodreads.com/author/show/2886410.Arthur_S_Otils

[7] Sauerheber R. Gravity contrasted with light and other fields; energetics and solar system dynamics. 2016. Available from: http://jamesrobertdeal.org/wp-content/uploads/Sauerheber-Gravity-energy-solar-system-dynamics-8-26-16.pdf

[8] Sauerheber R. Thermodynamics and entropy in natural and artificial systems. American Research Journals of Chemistry. 2018;**2**(1):1-26. Available from: https://www.arjonline.org/papers/arjc/v2-i1/1.pdf

[9] Sauerheber R, Espinoza E. Perspectives on solar system dynamics. International Journal of Physics and Astronomy. 2018;**2**(3):224-229. Available from: https://medcraveonline.com/PAIJ/PAIJ-02-00090.pdf

Chapter 4

Radiation and Energy Flux of Electromagnetic Fields by a Segment of Relativistic Electron Beam Moving Uniformly in Vacuum

Sergey Prijmenko and Konstantin Lukin

Abstract

A finite-length segment of filamentous relativistic electron beam (REB), moving uniformly in vacuum, radiates hybrid electromagnetic waves, compound of potential and vortex electric fields, as well as a vortex magnetic field. The strengths of electric and magnetic fields radiated by the segment edges have the opposite signs. The electromagnetic fields in the wave zone are considered as superposition of the electromagnetic waves radiated by the beginning and the end of the REB segment, which, in particular, leads to formation of the field's interference components. In both the near and the intermediate zones, there is a flow of electrical energy due to the electric potential field and the field of displacement current.

Keywords: relativistic electron beam or REB segment, potential field, vortex field, radiation of EM waves, near field zone, intermediate zone and far field (wave) zone, EM energy flux

1. Introduction

The physics of charged particle beam is an area where relativistic effects manifest themselves substantially. Here, one has to deal with a moving object, so both a fixed (laboratory) coordinate system and a moving coordinate system are to be used. A charged particle moves relative to the laboratory coordinate system, while in the moving coordinate system, it is at rest. Hence, in a laboratory coordinate system, the problem is to be considered as an electrodynamical one, and in a moving coordinate system, the problem belongs to the area of electrostatics. Thus, electrostatic phenomena in a charged particle set at rest are transformed into electrodynamic ones when it moves. Electromagnetic fields in these two inertial reference systems are tied via the Lorentz transform ([1], p. 79).

In the wave zone, the dynamic component of the electric field strength and the axially symmetric magnetic field form both a constant flux into a given solid angle, i.e., electromagnetic radiation, and a flux per time unit directed along the normal to the conical surface of the solid angle. The potential component of the electric field, directed along the radius, and the axially symmetric magnetic field form a flux oriented along the polar direction, i.e., along the normal to the above conical surface. The fluxes crossing the conical surface do not depend on the distance between the source point and the observation point. In the wave zone, the radiations from the beginning and end of the REB segment are added up, while the fluxes through the above conical surface caused by dynamic and potential components of electric field, are subtracted.

To date, the issue of influence of the finite length of a charged particle beam, moving uniformly in vacuum on the radiation of electromagnetic fields remains poorly studied, with an exception of publication [2], where its experimental part deserves special attention.

This chapter presents the results of our theoretical analysis of the electromagnetic field radiated by a finite-length segment of filamentous relativistic electron beam (REB). The REB segment moves uniformly in vacuum along its own axis which we will address as the *longitudinal direction*. The stepped varying of the charge density at the edges of the REB segment creates point-like sources of the potential electric field; the strength of which is inversely proportional to the distance between the source point and the observation point. In addition, the time variation of the REB current density forms at the REB edges the point-like sources of both potential and vortex electric fields, as well as the vortex magnetic field, with their strengths being also inversely proportional to the distance between the source point and the observation point [3].

The filamentary REB edges are considered as relativistic point-like radiators of the electromagnetic energy propagating to the wave zone. The presence of a potential electric field in the wave zone is due to the fact that the electric scalar potential in the wave zone is proportional to the electric *monopole* moment ([4], p. 51), which equals to the total charge in the selected volume ([5], p. 280). As follows from the Jefimenko's generalization of the Coulomb law ([3], p. 246), the potential electric field strength in the wave zone is proportional to the time derivative of the electric monopole moment.

In the intermediate zone, there is a flow of electrical field energy, due to the electric potential field and the field of the displacement current. The electrical energy flux in the intermediate zone is due to the electric potential field and field of the displacement current. The REB part with a constant charge density between its edges forms a quasi-static electromagnetic field in the near zone.

Note that a similar problem has been considered in [6], but it was devoted to similarity of the solutions obtained with the help of two different methods: retarded field integral and transformation equations of the special theory of relativity. Unlike our work, it does not contain expressions for scalar and vector potentials, as well as the electromagnetic energy flux.

2. Formulation of the problem

Consider a filamentary REB segment of length L and electric charge density Q moving uniformly along its axis direction with velocity v_e. Charge density of the REB segment may be written as follows:

$$\rho(t, r(x,y,z))_L = \frac{Q}{L}\delta(x)\cdot\delta(y)\cdot[h(z - v_e t) - h(z - (v_e t + L))] \tag{1}$$

where $h(x)$ is Heaviside step function; $\delta(x)$ and $\delta(y)$ are Dirac delta functions of coordinates. The electric scalar potential $\psi(t,r)$ and vector potential $\vec{A}(t,r)$, taking into account Eq. (1), satisfy the wave equations [3, 7]:

$$\left[divgrad - \frac{1}{c^2}\frac{\partial^2}{\partial t^2}\right]\psi(t,r) = -\frac{\rho(t,r)}{\varepsilon_0}, \tag{2}$$

$$\left[graddiv - rotrot - \frac{1}{c^2}\frac{\partial^2}{\partial t^2}\right]\vec{A}(t,r) = -\mu_0\rho(t,r)v_e\vec{k}_0, \tag{3}$$

where ε_0 and μ_0 are the dielectric and magnetic permeability of vacuum, respectively; and $\vec{k}_0$ is the unit vector along the REB axis, the Oz axis.

3. Potentials

A potential part of the vector potential $\vec{A}^p(t,r)$ is related to the scalar potential by the Lorentz calibration [3, 7]:

$$div\,\vec{A}^p(t,r) = -\frac{1}{c^2}\frac{\partial}{\partial t}\psi(t,r), \tag{4}$$

Using the Green's function for the wave equation ([3], p. 243), we obtain:

$$\psi(t',x'=0,y'=0,v_et'<z'<v_et'+L;t,r(x,y,z)) =$$

$$= -\frac{Q}{L4\pi\varepsilon_0}\int\limits_{v_et'}^{v_et'+L}\frac{dz'}{\sqrt{x^2+y^2+(z-z')^2}}\Bigg|_{t'=t-\frac{\left|\vec{r}-\vec{r}'\right|}{c}}, \tag{5}$$

$$\vec{A}(t',x'=0,y'=0,v_et'<z'<v_et'+L;t,r(x,y,z)) =$$

$$= -\frac{Q\mu_0}{L4\pi}\int\limits_{v_et'}^{v_et'+L}\frac{dz'}{\sqrt{x^2+y^2+(z-z')^2}}\Bigg|_{t'=t-\frac{\left|\vec{r}-\vec{r}'\right|}{c}}, \tag{6}$$

where the hatched coordinates refer to the source point at the time instant t' of the field radiation, and the non-hatched coordinates refer to the observation point at the time instant t.

The formula for the scalar potential can be obtained in the closed form using the table integral ([8], p. 34):

$$\psi(t',x'=0,y'=0,v_et'<z'<v_et'+L;t,r(x,y,z))$$

$$= \frac{Q}{L4\pi\varepsilon_0}\,ln\left|(z-(v_et'+L))+\sqrt{x^2+y^2+(z-(v_et'+L))^2}\right|\Bigg|_{t'=t-\frac{\left|\vec{r}-\vec{r}'(t',z'=v_et'+L)\right|}{c}}$$

$$-\frac{Q}{L4\pi\varepsilon_0}\,ln\left|(z-v_et')+\sqrt{x^2+y^2+(z-v_et')^2}\right|\Bigg|_{t'=t-\frac{\left|\vec{r}-\vec{r}'(t',z'=v_et')\right|}{c}} \tag{7}$$

where the expressions in the first and second summands refer to the REB segment end and its beginning, respectively.

4. The electromagnetic field strengths

For estimation of the electric and magnetic fields, we use standard formulas ([7], p. 432):

$$\vec{E}(t', x' = 0, y' = 0, v_e t' < z' < v_e t' + L; t, r(x, y, z)) =$$
$$= -\frac{\partial \vec{A}(t', x' = 0, y' = 0, v_e t' < z' < v_e t' + L; t, r(x, y, z))}{\partial t} -$$
$$-grad_r \psi(t', x' = 0, y' = 0, v_e t' < z' < v_e t' + L; t, r(x, y, z)), \tag{8}$$
$$\vec{H}(t', x' = 0, y' = 0, v_e t' < z' < v_e t' + L; t, r(x, y, z)) =$$
$$= \frac{1}{\mu_0} rot_r \vec{A}(t', x' = 0, y' = 0, v_e t' < z' < v_e t' + L; t, r(x, y, z)), \tag{9}$$

where it is necessary to perform the differentiation over the coordinates of the observation point, taking into account the retardation effect ([7], p. 432) and ([4], p. 43) as well as the differentiation of integrals by the integration limits and by the parameter ([9], p. 58). Using Eqs. (5), (6), and (8), we get:

$$E_x^p(t', x' = 0, y' = 0, v_e t' < z' < v_e t' + L; t, r(x, y, z)) =$$
$$= \frac{Q v_e}{L 4\pi\varepsilon_0 c} \frac{\cos\left[\alpha_x(z' = v_e t')\right]}{\kappa(z' = v_e t')\left|\vec{r} - \vec{r}'(t', z' = v_e t')\right|}\Bigg|_{t' = t - \frac{\left|\vec{r} - \vec{r}'(t', z' = v_e t')\right|}{c}} -$$
$$- \frac{Q v_e}{L 4\pi\varepsilon_0 c} \frac{\cos\left[\alpha_x(z' = v_e t' + L)\right]}{\kappa(z' = v_e t' + L)\left|\vec{r} - \vec{r}'(t', z' = v_e t' + L)\right|}\Bigg|_{t' = t - \frac{\left|\vec{r} - \vec{r}'(t', z' = v_e t' + L)\right|}{c}} +$$
$$+ \frac{Q}{L 4\pi\varepsilon_0} \int_{v_e t'}^{v_e t' + L} \frac{\cos\left[\alpha_x(z')\right]}{\left|\vec{r} - \vec{r}'(t', z')\right|^2}\Bigg|_{t' = t - \frac{\left|\vec{r} - \vec{r}'(t', z')\right|}{c}}, dz' \tag{10}$$

$$E_y^p(t', x' = 0, y' = 0, v_e t' < z' < v_e t' + L; t, r(x, y, z)) =$$
$$= \frac{Q v_e}{L 4\pi\varepsilon_0 c} \frac{\cos\left[\alpha_y(z' = v_e t')\right]}{\kappa(z' = v_e t')\left|\vec{r} - \vec{r}'(t', z' = v_e t')\right|}\Bigg|_{t' = t - \frac{\left|\vec{r} - \vec{r}'(t', z' = v_e t')\right|}{c}} -$$
$$- \frac{Q v_e}{L 4\pi\varepsilon_0 c} \frac{\cos\left[\alpha_y(z' = v_e t' + L)\right]}{\kappa(z' = v_e t' + L)\left|\vec{r} - \vec{r}'(t', z' = v_e t' + L)\right|}\Bigg|_{t' = t - \frac{\left|\vec{r} - \vec{r}'(t', z' = v_e t' + L)\right|}{c}} +$$
$$+ \frac{Q}{L 4\pi\varepsilon_0} \int_{v_e t'}^{v_e t' + L} \frac{\cos\left[\alpha_y(z')\right]}{\left|\vec{r} - \vec{r}'(t', z')\right|^2}\Bigg|_{t' = t - \frac{\left|\vec{r} - \vec{r}'(t', z')\right|}{c}}, dz' \tag{11}$$

$$E_z(t', x' = 0, y' = 0, v_e t' < z' < v_e t' + L; t, r(x, y, z)) =$$
$$= \frac{Q v_e}{L 4\pi\varepsilon_0 c} \frac{\cos\left[\alpha_z(z' = v_e t')\right]}{\kappa(z' = v_e t')\left|\vec{r} - \vec{r}'(t', z' = v_e t')\right|}\Bigg|_{t' = t - \frac{\left|\vec{r} - \vec{r}'(t', z' = v_e t')\right|}{c}} -$$
$$- \frac{Q v_e}{L 4\pi\varepsilon_0 c} \frac{\cos\left[\alpha_z(z' = v_e t' + L)\right]}{\kappa(z' = v_e t' + L)\left|\vec{r} - \vec{r}'(t', z' = v_e t' + L)\right|}\Bigg|_{t' = t - \frac{\left|\vec{r} - \vec{r}'(t', z' = v_e t' + L)\right|}{c}} -$$

$$-\frac{Q\mu_0 {v_e}^2}{L4\pi}\frac{1}{\kappa(z'=v_et')\left|\vec{r}-\vec{r}'(t',z'=v_et')\right|}\Bigg|_{t'=t-\frac{\left|\vec{r}-\vec{r}'(t',z'=v_et')\right|}{c}}+$$

$$+\frac{Q\mu_0 {v_e}^2}{L4\pi}\frac{1}{\kappa(z'=v_et'+L)\left|\vec{r}-\vec{r}'(t',z'=v_et'+L)\right|}\Bigg|_{t'=t-\frac{\left|\vec{r}-\vec{r}'(t',z'=v_et'+L)\right|}{c}}+$$

$$+\frac{Q}{L4\pi\varepsilon_0}\int_{v_et'}^{v_et'+L}\frac{\cos\left[\alpha_z(z')\right]}{\left|\vec{r}-\vec{r}'(t',z')\right|^2}\Bigg|_{t'=t-\frac{\left|\vec{r}-\vec{r}'(t',z')\right|}{c}},dz' \tag{12}$$

where

$$\cos\left[\alpha_x(z'=v_et')\right]=\frac{x}{\left|\vec{r}-\vec{r}'(t',z'=v_et')\right|}\Bigg|_{t'=t-\frac{\left|\vec{r}-\vec{r}'(t',z'=v_et')\right|}{c}}, \tag{13}$$

$$\cos\left[\alpha_x(z'=v_et'+L)\right]=\frac{x}{\left|\vec{r}-\vec{r}'(t',z'=v_et'+L)\right|}\Bigg|_{t'=t-\frac{\left|\vec{r}-\vec{r}'(t',z'=v_et'+L)\right|}{c}}, \tag{14}$$

$$\cos\left[\alpha_y(z'=v_et')\right]=\frac{y}{\left|\vec{r}-\vec{r}'(t',z'=v_et')\right|}\Bigg|_{t'=t-\frac{\left|\vec{r}-\vec{r}'(t',z'=v_et')\right|}{c}}, \tag{15}$$

$$\cos\left[\alpha_y(z'=v_et'+L)\right]=\frac{y}{\left|\vec{r}-\vec{r}'(t',z'=v_et'+L)\right|}\Bigg|_{t'=t-\frac{\left|\vec{r}-\vec{r}'(t',z'=v_et'+L)\right|}{c}}, \tag{16}$$

$$\cos\left[\alpha_z(z'=v_et')\right]=\frac{(z-v_et')}{\left|\vec{r}-\vec{r}'(t',z'=v_et')\right|}\Bigg|_{t'=t-\frac{\left|\vec{r}-\vec{r}'(t',z'=v_et')\right|}{c}}, \tag{17}$$

$$\cos\left[\alpha_z(z'=v_et'+L)\right]=\frac{(z-(v_et'+L))}{\left|\vec{r}\quad\vec{r}'(t',z'=v_et'+L)\right|}\Bigg|_{t'=t-\frac{\left|\vec{r}-\vec{r}'(t',z'=v_et'+L)\right|}{c}}, \tag{18}$$

and

$$\kappa(z'=v_et')=[1-\frac{v_e}{c}\cos\left[\alpha_z(z'=v_et')\right], \tag{19}$$

$$\kappa(z'=v_et'+L)=[1-\frac{v_e}{c}\cos\left[\alpha_z(z'=v_et'+L)\right] \tag{20}$$

are the retardation factors ([3], p. 246).

The transverse components of the electric field strength $E_x^p(t',r'(x',y',z'(t'));t,r(x,y,z))$ and $E_y^p(t',r'(x',y',z'(t'));t,r(x,y,z))$ are potential relative to the space coordinates, and the longitudinal component $E_z(t',r'(x',y',z'(t'));t,r(x,y,z))$ consists of both a potential component relative to the space coordinates and a dynamic component.

The transverse components of the magnetic field strength $H_x(t',r'(x',y',z'(t'));t,r(x,y,z))$ and $H_y(t',r'(x',y',z'(t'));t,r(x,y,z))$, according to the Eq. (6) and (9), are:

$$H_x(t', x' = 0, y' = 0, v_e t' < z' < v_e t' + L; t, r(x, y, z)) =$$

$$= -\frac{Q{v_e}^2}{L4\pi c}\frac{\cos\left[\alpha_y(z' = v_e t')\right]}{\kappa(z' = v_e t')\left|\vec{r} - \vec{r}'(t', z' = v_e t')\right|}\Bigg|_{t'=t-\frac{\left|\vec{r}-\vec{r}'(t',z'=v_e t')\right|}{c}}$$

$$+\frac{Q{v_e}^2}{L4\pi c}\frac{\cos\left[\alpha_y(z' = v_e t' + L)\right]}{\kappa(z' = v_e t' + L)\left|\vec{r} - \vec{r}'(t', z' = v_e t' + L)\right|}\Bigg|_{t'=t-\frac{\left|\vec{r}-\vec{r}'(t',z'=v_e t'+L)\right|}{c}}$$

$$-\frac{Q v_e}{L4\pi}\int_{v_e t'}^{v_e t'+L} dz' \frac{\cos\left[\alpha_y(z')\right]}{\left|\vec{r} - \vec{r}'(t', z')\right|^2}\Bigg|_{t'=t-\frac{\left|\vec{r}-\vec{r}'(t',z')\right|}{c}} \tag{21}$$

$$H_y(t', x' = 0, y' = 0, v_e t' < z' < v_e t' + L; t, r(x, y, z)) =$$

$$= \frac{Q{v_e}^2}{L4\pi c}\frac{\cos\left[\alpha_x(z' = v_e t')\right]}{\kappa(z' = v_e t')\left|\vec{r} - \vec{r}'(t', z' = v_e t')\right|}\Bigg|_{t'=t-\frac{\left|\vec{r}-\vec{r}'(t',z'=v_e t')\right|}{c}} -$$

$$-\frac{Q{v_e}^2}{L4\pi c}\frac{\cos\left[\alpha_x(z' = v_e t' + L)\right]}{\kappa(z' = v_e t' + L)\left|\vec{r} - \vec{r}'(t', z' = v_e t' + L)\right|}\Bigg|_{t'=t-\frac{\left|\vec{r}-\vec{r}'(t',z'=v_e t'+L)\right|}{c}} +$$

$$+\frac{Q v_e}{L4\pi}\int_{v_e t'}^{v_e t'+L} dz' \frac{\cos\left[\alpha_x(z')\right]}{\left|\vec{r} - \vec{r}'(t', z')\right|^2}\Bigg|_{t'=t-\frac{\left|\vec{r}-\vec{r}'(t',z')\right|}{c}} \tag{22}$$

The strengths of the electric fields in Eqs. (10)–(12) and magnetic fields with Eqs. (21) and (22), formed by the ends and the main part of the beam, decrease inversely proportional to the first and second powers of the distance from the source point to the observation point.

5. Displacement current

We take into account that the displacement current density $\vec{j}_d(t, r)$ ([7], p. 87):

$$\vec{j}_d(t, r) = \frac{\partial}{\partial t}\vec{D}_d(t, r) = \frac{\partial}{\partial t}\varepsilon_0 \vec{E}(t, r), \tag{23}$$

where the $\vec{D}_d(t, r) = \varepsilon_0 \vec{E}(t, r)$ is the electric displacement vector. Taking into account the Eqs. (10)–(12) and (23), we get

$$j^p_{dx}(t', x' = 0, y' = 0, v_e t' < z' < v_e t' + L; t, r(x, y, z)) =$$

$$= \frac{Q{v_e}^2}{L4\pi c}\cos\left[\alpha_x(z' = v_e t')\right]\cdot\cos\left[\alpha_z(z' = v_e t')\right]\cdot\frac{1}{\kappa^2(z' = v_e t')\left|\vec{r} - \vec{r}'(t', z' = v_e t')\right|^2}$$

$$+\frac{Q{v_e}^3}{L4\pi c^2}\frac{\cos\left[\alpha_x(z' = v_e t')\right]}{\kappa^3(z' = v_e t')\left|\vec{r} - \vec{r}'(t', z' = v_e t')\right|^3}\left[\left|\vec{r} - \vec{r}'(t', z' = v_e t')\right|\right.$$

$$- \cos\left[\alpha_z(z' = v_e t')\right](z - v_e t')\Big] +$$

$$+ \frac{Q{v_e}^2}{L4\pi c} \frac{\cos\left[\alpha_x(z' = v_e t')\right] \cos\left[\alpha_z(z' = v_e t')\right]}{\kappa^2(z' = v_e t')\left|\vec{r} - \vec{r}'(t', z' = v_e t')\right|^2} -$$

$$- \frac{2Q{v_e}^2}{L4\pi c} \frac{\cos\left[\alpha_x(z' = v_e t' + L)\right] \cdot \cos\left[\alpha_z(z' = v_e t' + L)\right]}{\kappa^2(z' = v_e t' + L)\left|\vec{r} - \vec{r}'(t', z' = v_e t' + L)\right|^2} -$$

$$- \frac{Q{v_e}^3}{L4\pi c^2} \frac{\cos\left[\alpha_x(z' = v_e t' + L)\right]}{\kappa^3(z' = v_e t' + L)\left|\vec{r} - \vec{r}'(t', z' = v_e t' + L)\right|^3}$$

$$\cdot \Big[\left|\vec{r} - \vec{r}'(t', z' = v_e t' + L)\right| - \cos\left[\alpha_z(z' = v_e t' + L)\right] \cdot (z - (v_e t' + L))\Big] +$$

$$+ \frac{Q v_e}{L4\pi} \frac{\cos\left[\alpha_x(z' = v_e t' + L)\right]}{\kappa^2(z' = v_e t' + L)\left|\vec{r} - \vec{r}'(t', z' = v_e t' + L)\right|^2} -$$

$$- \frac{Q v_e}{L4\pi} \cdot \frac{\cos\left[\alpha_x(z' = v_e t')\right]}{\kappa^2(z' = v_e t')\left|\vec{r} - \vec{r}'(t', z' = v_e t')\right|^2} \tag{24}$$

$$j^p_{dy}(t', x' = 0, y' = 0, v_e t' < z' < v_e t' + L; t, r(x, y, z)) =$$

$$= \frac{Q{v_e}^2}{L4\pi c} \cos\left[\alpha_y(z' = v_e t')\right] \cdot \cos\left[\alpha_z(z' = v_e t')\right] \cdot$$

$$\cdot \frac{1}{\kappa^2(z' = v_e t')\left|\vec{r} - \vec{r}'(t', z' = v_e t')\right|^2} + \frac{Q{v_e}^3}{L4\pi c^2} \frac{\cos\left[\alpha_y(z' = v_e t')\right]}{\kappa^3(z' = v_e t')\left|\vec{r} - \vec{r}'(t', z' = v_e t')\right|^3}$$

$$\Big[\left|\vec{r} - \vec{r}'(t', z' = v_e t')\right| - \cos\left[\alpha_z(z' = v_e t')\right](z - v_e t')\Big] +$$

$$+ \frac{Q{v_e}^2}{L4\pi c} \frac{\cos\left[\alpha_y(z' = v_e t')\right] \cos\left[\alpha_z(z' = v_e t')\right]}{\kappa^2(z' = v_e t')\left|\vec{r} - = v_e t')\right](z - v_e t')](t', z' = v_e t')\right|^2}$$

$$- \frac{2Q{v_e}^2}{L4\pi c} \frac{\cos\left[\alpha_y(z' = v_e t' + L)\right] \cdot \cos\left[\alpha_z(z' = v_e t' + L)\right]}{\kappa^2(z' - v_e t' + L)\left|\vec{r} - \vec{r}'(t', z' = v_e t' + L)\right|^2} - \frac{Q{v_e}^3}{L4\pi c^2}$$

$$\frac{\cos\left[\alpha_y(z' = v_e t' + L)\right]}{\kappa^3(z' = v_e t' + L)\left|\vec{r} - \vec{r}'(t', z' = v_e t' + L)\right|^3} \Big[\left|\vec{r} - \vec{r}'(t', z' = v_e t' + L)\right|$$

$$- \cos\left[\alpha_z(z' = v_e t' + L)\right] \cdot (z - (v_e t' + L))\Big]$$

$$+ \frac{Q v_e}{L4\pi} \frac{\cos\left[\alpha_y(z' = v_e t' + L)\right]}{\kappa^2(z' = v_e t' + L)\left|\vec{r} - \vec{r}'(t', z' = v_e t' + L)\right|^2}$$

$$- \frac{Q v_e}{L4\pi} \frac{\cos\left[\alpha_y(z' = v_e t')\right]}{\kappa^2(z' = v_e t')\left|\vec{r} - \vec{r}'(t', z' = v_e t')\right|^2} \tag{25}$$

$$j_{dz}(t', x' = 0, y' = 0, v_e t' < z' < v_e t' + L; t, r(x, y, z)) =$$

$$= \frac{-Q{v_e}^2}{L4\pi c} \sin^2\left[\alpha_z(z' = v_e t')\right] \cdot$$

$$\cdot\frac{1}{\kappa^2(z'=v_et')\left|\vec{r}-\vec{r}'(t',z'=v_et')\right|^2}+\frac{Qv_e^3}{L4\pi c^2}\frac{\cos\left[\alpha_z(z'=v_et')\right]}{\kappa^3(z'=v_et')\left|\vec{r}-\vec{r}'(t',z'=v_et')\right|^3}\cdot$$

$$\cdot\left[\left|\vec{r}-\vec{r}'(t',z'=v_et')\right|-\cos\left[\alpha_z(z'=v_et')\right](z-v_et')\right]$$
$$+\frac{Qv_e^2}{L4\pi c}\frac{\cos^2\left[\alpha_z(z'=v_et')\right]}{\kappa^2(z'=v_et')\left|\vec{r}-\vec{r}'(t',z'=v_et')\right|^2}$$

$$-\frac{Qv_e^2}{L4\pi c}\frac{\sin^2\left[\alpha_z(z'=v_et'+L)\right]}{\kappa^2(z'=v_et'+L)\left|\vec{r}-\vec{r}'(t',z'=v_et'+L)\right|^2}+\frac{Qv_e^3}{L4\pi c^2}\cdot$$

$$\cdot\frac{\cos\left[\alpha_z(z'=v_et'+L)\right]}{\kappa^2(z'=v_et'+L)\left|\vec{r}-\vec{r}'(t',z'=v_et'+L)\right|^3}$$
$$\cdot\left[\left|\vec{r}-\vec{r}'(t',z'=v_et'+L)\right|-\cos\left[\alpha_z(z'=v_et'+L)\right]\cdot\right.$$

$$\left.\cdot(z-(v_et'+L))\right]+\frac{Qv_e^2}{L4\pi c}\frac{\cos^2\left[\alpha_z(z'=v_et'+L)\right]}{\kappa^2(z'=v_et'+L)\left|\vec{r}-\vec{r}'(t',z'=v_et'+L)\right|^2}-$$

$$\frac{Qv_e^4}{L4\pi c^3}\frac{1}{\kappa^3(z'=v_et')\left|\vec{r}-\vec{r}'(t',z'=v_et')\right|^3}\left[\left|\vec{r}-\vec{r}'(t',z'=v_et')\right|-\cos\left[\alpha_z(z'=v_et')\right](z-v_et')\right]-$$

$$-\frac{Qv_e^3}{L4\pi c^2}\frac{\cos\left[\alpha_z(z'=v_et')\right]}{\kappa^2(z'=v_et')\left|\vec{r}-\vec{r}'(t',z'=v_et')\right|^2}$$

$$+\frac{Qv_e^4}{L4\pi c^3}\frac{1}{\kappa^3(z'=v_et'+L)\left|\vec{r}-\vec{r}'(t',z'=v_et'+L)\right|^3}\cdot$$

$$\cdot\left[\left|\vec{r}-\vec{r}'(t',z'=v_et'+L)\right|-\cos\left[\alpha_z(z'=v_et'+L)\right](z-(v_et'+L))\right]+$$

$$+\frac{Qv_e^3}{L4\pi c^2}\frac{\cos\left[\alpha_z(z'=v_et'+L)\right]}{\kappa^2(z'=v_et'+L)\left|\vec{r}-\vec{r}'(t',z'=v_et'+L)\right|^2}$$
$$-\frac{Qv_e}{L4\pi}\frac{\cos\left[\alpha_z(z'=v_et')\right]}{\kappa(z'=v_et')\left|\vec{r}-\vec{r}'(t',z'=v_et')\right|^2} \tag{26}$$

The transverse components of the displacement current density $j^p_{dx}(t',r'(x',y',z'(t'));t,r(x,y,z))$ and $j^p_{dy}(t',r'(x',y',z'(t'));t,r(x,y,z))$ are potential with respect to space coordinates, and the longitudinal component $j_{dz}(t',r'(x',y',z'(t'));t,r(x,y,z))$ consists of potential and dynamic components. Displacement current densities are decreasing inversely proportional to the second power of the distance from the source point to the observation point.

6. Flux of electrical energy

The electrical energy flux density per unit time $\overrightarrow{S}^{\psi}(t, r)$, according to ([10], p. 125) Eq. (15) and [11] Eqs. (7) and (8), has the form

$$\overrightarrow{S}^{\psi}(t, r) = \psi(t, r) \cdot \overrightarrow{j}_d(t, r) \tag{27}$$

Taking into account the Eq. (5) or the Eq. (7) and the Eqs. (24)–(26), we can write

$$\begin{aligned} &S_x^{\psi}(t', x' = 0, y' = 0, v_e t' < z' < v_e t' + L; t, r(x, y, z)) = \\ &= \psi(t', x' = 0, y' = 0, v_e t' < z' < v_e t' + L; t, r(x, y, z)) \cdot \\ &j_{dx}^{p}(t', x' = 0, y' = 0, v_e t' < z' < v_e t' + L; t, r(x, y, z)) \end{aligned} \tag{28}$$

$$\begin{aligned} &S_y^{\psi}(t', x' = 0, y' = 0, v_e t' < z' < v_e t' + L; t, r(x, y, z)) = \\ &= \psi(t', x' = 0, y' = 0, v_e t' < z' < v_e t' + L; t, r(x, y, z)) \cdot \\ &\cdot j_{dy}^{p}(t', x' = 0, y' = 0, v_e t' < z' < v_e t' + L; t, r(x, y, z)) \end{aligned} \tag{29}$$

$$\begin{aligned} &S_z^{\psi}(t', x' = 0, y' = 0, v_e t' < z' < v_e t' + L; t, r(x, y, z)) = \\ &= \psi(t', x' = 0, y' = 0, v_e t' < z' < v_e t' + L; t, r(x, y, z)) \cdot \\ &j_{dz}^{p}(t', x' = 0, y' = 0, v_e t' < z' < v_e t' + L; t, r(x, y, z)) \end{aligned} \tag{30}$$

The electrical energy flux density $\overrightarrow{S}^{\psi}(t, r)$ decreases inversely proportional to the third power of the distance from the source point to the observation point. The electrical energy flux per unit time into a given solid angle decreases inversely proportional to the first power of the distance from the source point to the observation point. The flux takes place both in the near and the intermediate zones.

7. Pointing vector

The Poynting vector or the flux density of electromagnetic energy per unit time is determined by the formula ([3], p. 259)

$$\overrightarrow{S}(t, r) = \overrightarrow{E}(t, r) \times \overrightarrow{H}(t, r) \tag{31}$$

The Poynting vector along the Ox axis estimated according to Eq. (31) with the help of Eqs. (12) and (22) may be written as follows:

$$\begin{aligned} &S_x(t', x' = 0, y' = 0, v_e t' < z' < v_e t' + L; t, r(x, y, z)) = \\ &= -E_z(t', x' = 0, y' = 0, v_e t' < z' < v_e t' + L; t, r(x, y, z)) \cdot \\ &\cdot H_y(t', x' = 0, y' = 0, v_e t' < z' < v_e t' + L; t, r(x, y, z)) \\ &= -\{E_z^p(z' = v_e t') + E_z^p(z' = v_e t' + L) \\ &+ E_z^r(z' = v_e t') + E_z^r(v_e t' < z' < v_e t' + L)\} \cdot \\ &\left\{H_y(z' = v_e t') + H_y(z' = v_e t' + L) + + H_y^c(v_e t' < z' < v_e t' + L)\right\} \end{aligned} \tag{32}$$

where the summands in curly brackets are defined by Eq. (12) and Eq. (22), respectively. Rewriting the Eq. (32) in the following form:

$$\begin{aligned}&S_x(t', x' = 0, y' = 0, v_e t' < z' < v_e t' + L; t, r(x, y, z))\\&\quad = {}^iS_x(z' = v_e t', z' = v_e t' + L) + {}^{pi}S_x(z' = v_e t', z' = v_e t' + L, v_e t' < z' < v_e t' + L)\\&\qquad + {}^fS_x^c(v_e t' < z' < v_e t' + L),\end{aligned} \tag{33}$$

where the ${}^i\boldsymbol{S_x}(\boldsymbol{z}' = \boldsymbol{v_e t}', \boldsymbol{z}' = \boldsymbol{v_e t}' + \boldsymbol{L})$ there is a flux of electromagnetic energy in a unit time that goes into the wave zone, the ${}^{pi}\boldsymbol{S_x}(\boldsymbol{z}' = \boldsymbol{v_e t}', \boldsymbol{z}' = \boldsymbol{v_e t}' + \boldsymbol{L}, \boldsymbol{v_e t}' < \boldsymbol{z}' < \boldsymbol{v_e t}' + \boldsymbol{L})$ there is a flux of electromagnetic energy in the intermediate zone, the ${}^f\boldsymbol{S_x^c}(\boldsymbol{v_e t}' < \boldsymbol{z}' < \boldsymbol{v_e t}' + \boldsymbol{L})$ there is a flux of electromagnetic energy in the near zone. As this takes place

$$\begin{aligned}{}^iS_x(z' = v_e t', z' = v_e t' + L) = {}^iS_x(z' = v_e t') +\\ + {}^iS_x(z' = v_e t' + L) + {}^iS_x^{\psi A}(z' = v_e t', z' = v_e t' + L),\end{aligned} \tag{34}$$

$$\begin{aligned}{}^iS_x(z' = v_e t') = {}^iS_x^{\psi A}(z' = v_e t') + {}^iS_x^A(z' = v_e t') =\\ = -E_z^p(z' = v_e t') \cdot H_y(z' = v_e t') - E_z^r(z' = v_e t') \cdot H_y(z' = v_e t'),\end{aligned} \tag{35}$$

$$\begin{aligned}{}^iS_x(z' = v_e t' + L) = {}^iS_x^{\psi A}(z' = v_e t' + L) + {}^iS_x^A(z' = v_e t' + L) =\\ -E_z^p(z' = v_e t' + L) \cdot H_y(z' = v_e t' + L) - E_z^r(z' = v_e t' + L) \cdot H_y(z' = v_e t' + L),\end{aligned} \tag{36}$$

$$\begin{aligned}{}^iS_x^{\psi A}(z' = v_e t', z' = v_e t' + L) = {}&-E_z^p(z' = v_e t') \cdot H_y(z' = v_e t' + L) - E_z^p(z' = v_e t' + L)\\&\cdot H_y(z' = v_e t') - E_z^r(z' = v_e t') \cdot H_y(z' = v_e t' + L)\\&- E_z^r(z' = v_e t' + L) \cdot H_y(z' = v_e t').\end{aligned} \tag{37}$$

The energy fluxes, ${}^iS_x(z' = v_e t')$, ${}^iS_x(z' = v_e t' + L)$, ${}^iS_x^{\psi A}(z' = v_e t', z' = v_e t' + L)$, are determined by point sources of radiation at the REB segment beginning, the REB segment end, and the REB segment interference, respectively.

$$\begin{aligned}{}^{pi}S_x(z' = v_e t', z' = v_e t' + L, v_e t' < z' < v_e t' + L) =\\ -E_z^p(z' = v_e t') \cdot H_y^c(v_e t' < z' < v_e t' + L) - E_z^p(z' = v_e t' + L) \cdot H_y^c(v_e t' < z' < v_e t' + L)\\ -E_z^r(z' = v_e t') \cdot H_y^c(v_e t' < z' < v_e t' + L) - E_z^r(z' = v_e t') \cdot H_y^c(v_e t' < z' < v_e t' + L)\\ E_z^c(v_e t' < z' < v_e t' + L) \cdot H_y(z' = v_e t') - E_z^c(v_e t' < z' < v_e t' + L) \cdot H_y(z' = v_e t' + L).\end{aligned} \tag{38}$$

$${}^fS_x^c(v_e t' < z' < v_e t' + L) = -E_z^c(v_e t' < z' < v_e t' + L) \cdot H_y^c(v_e t' < z' < v_e t' + L). \tag{39}$$

The Poynting vector along the Oy axis, taking into account Eqs. (12), (21), (31), similarly to Eqs. (33)–(39), is represented by:

$$\begin{aligned}S_y(t', x' = 0, y' = 0, v_e t' < z' < v_e t' + L; t, r(x, y, z)) = {}^iS_y(z' = v_e t', z' = v_e t' + L) +\\ + {}^{pi}S_y(z' = v_e t', z' = v_e t' + L, v_e t' < z' < v_e t' + L) + {}^fS_y^c(v_e t' < z' < v_e t' + L),\end{aligned} \tag{40}$$

$$\begin{aligned}{}^iS_y(z' = v_e t', z' = v_e t' + L) = {}^iS_y(z' = v_e t') + {}^iS_y(z' = v_e t' + L) +\\ + {}^iS_y^{\psi A}(z' = v_e t', z' = v_e t' + L)\end{aligned} \tag{41}$$

$$
\begin{gathered}
{}^{i}S_y(z' = v_e t') = {}^{i}S_y^{\psi A}(z' = v_e t') + {}^{i}S_y^{A}(z' = v_e t') = \\
= E_z^p(z' = v_e t') \cdot H_x(z' = v_e t') + E_z^r(z' = v_e t') \cdot H_x(z' = v_e t')
\end{gathered} \tag{42}
$$

$$
\begin{gathered}
{}^{i}S_y(z' = v_e t' + L) = {}^{i}S_y^{\psi A}(z' = v_e t' + L) + {}^{i}S_y^{A}(z' = v_e t' + L) = \\
= E_z^p(z' = v_e t' + L) \cdot H_x(z' = v_e t' + L) + E_z^r(z' = v_e t' + L) \cdot H_x(z' = v_e t' + L),
\end{gathered} \tag{43}
$$

$$
\begin{gathered}
{}^{i}S_y^{\psi A}(z' = v_e t', z' = v_e t' + L) = E_z^p(z' = v_e t') \cdot H_x(z' = v_e t' + L) + E_z^p(z' = v_e t' + L) \cdot \\
\cdot H_x(z' = v_e t') + E_z^r(z' = v_e t') \cdot H_x(z' = v_e t' + L) + E_z^r(z' = v_e t' + L) \cdot H_x(z' = v_e t'),
\end{gathered} \tag{44}
$$

$$
\begin{gathered}
{}^{pi}S_y(z' = v_e t', z' = v_e t' + L, v_e t' < z' < v_e t' + L) = \\
E_z^p(z' = v_e t') \cdot H_x^c(v_e t' < z' < v_e t' + L) + E_z^p(z' = v_e t' + L) \cdot H_x^c(v_e t' < z' < v_e t' + L) + \\
+ E_z^r(z' = v_e t') \cdot H_x^c(v_e t' < z' < v_e t' + L) + E_z^r(z' = v_e t') \cdot H_x^c(v_e t' < z' < v_e t' + L) + \\
+ E_z^c(v_e t' < z' < v_e t' + L) \cdot H_x(z' = v_e t') + E_z^c(v_e t' < z' < v_e t' + L) \cdot H_x(z' = v_e t' + L),
\end{gathered} \tag{45}
$$

$$
{}^{f}S_y^c(v_e t' < z' < v_e t' + L) = -E_z^c(v_e t' < z' < v_e t' + L) \cdot H_x^c(v_e t' < z' < v_e t' + L). \tag{46}
$$

The Poynting vector along the Oz axis, taking into account Eqs. (10), (11), (21), (22), and (31), may be written as follows:

$$
\begin{gathered}
S_z(t', x' = 0, y' = 0, v_e t' < z' < v_e t' + L; t, r(x, y, z)) = {}^{i}S_z(z' = v_e t', z' = v_e t' + L) + \\
+ {}^{pi}S_z(z' = v_e t', z' = v_e t' + L, v_e t' < z' < v_e t' + L) + {}^{f}S_z^c(v_e t' < z' < v_e t' + L),
\end{gathered} \tag{47}
$$

$$
\begin{gathered}
{}^{i}S_z(z' = v_e t', z' = v_e t' + L) = {}^{i}S_z(z' = v_e t') + \\
+ {}^{i}S_z(z' = v_e t' + L) + {}^{i}S_z^{\psi A}(z' = v_e t', z' = v_e t' + L)
\end{gathered} \tag{48}
$$

$$
\begin{gathered}
{}^{i}S_z(z' = v_e t') = {}^{i}S_z^{\psi A}(z' = v_e t') = \\
= E_x^p(z' = v_e t') \cdot H_y(z' = v_e t') - E_y^p(z' = v_e t') \cdot H_x(z' = v_e t')
\end{gathered} \tag{49}
$$

$$
\begin{gathered}
{}^{i}S_z(z' = v_e t' + L) = {}^{i}S_z^{\psi A}(z' = v_e t' + L) = \\
= E_x^p(z' = v_e t' + L) \cdot H_y(z' = v_e t' + L) - E_y^p(z' = v_e t' + L) \cdot H_x(z' = v_e t' + L)
\end{gathered} \tag{50}
$$

$$
\begin{gathered}
{}^{i}S_z^{\psi A}(z' = v_e t', z' = v_e t' + L) = E_x^p(z' = v_e t') \cdot H_y(z' = v_e t' + L) + \\
+ E_x^p(z' = v_e t' + L) \cdot H_y(z' = v_e t') - E_y^p(z' = v_e t') \cdot H_x(z' = v_e t' + L) - \\
- E_y^p(z' = v_e t' + L) \cdot H_x(z' = v_e t')
\end{gathered} \tag{51}
$$

$$
\begin{gathered}
{}^{pi}S_z(z' = v_e t', z' = v_e t' + L, v_e t' < z' < v_e t' + L) = \\
E_x^p(z' = v_e t') \cdot H_y^c(v_e t' < z' < v_e t' + L) + E_x^p(z' = v_e t' + L) \cdot H_y^c(v_e t' < z' < v_e t' + L) + \\
+ E_x^c(v_e t' < z' < v_e t' + L) \cdot H_y(z' = v_e t') + E_x^c(v_e t' < z' < v_e t' + L) \cdot H_y(z' = v_e t' + L) - \\
- E_y^p(z' = v_e t') \cdot H_x^c(v_e t' < z' < v_e t' + L) - E_y^p(z' = v_e t' + L) \cdot H_x^c(v_e t' < z' < v_e t' + L) - \\
- E_y^c(v_e t' < z' < v_e t' + L) \cdot H_x(z' = v_e t') - E_y^c(v_e t' < z' < v_e t' + L) \cdot H_x(z' = v_e t' + L),
\end{gathered} \tag{52}
$$

$$
\begin{gathered}
{}^{f}S_z^c(v_e t' < z' < v_e t' + L) = E_x^c(v_e t' < z' < v_e t' + L) \cdot H_y^c(v_e t' < z' < v_e t' + L) - \\
- E_e^c(v_e t' < z' < v_e t' + L) \cdot H_x^c(v_e t' < z' < v_e t' + L).
\end{gathered} \tag{53}
$$

8. Numerical results

We have considered the filamentary REB of the length $L = 3\,m$, moving along the Oz axis with velocity $v_e = 0.94\,c$ (c is the speed of light) and having overall charge $Q = (-1) \cdot 10^{-10}C$.

In the laboratory coordinate system, the dependence of the electric field strength $E_x^p(t' = 0, x' = 0, y' = 0, z' = 0; t, r(x, y = 0, z = 0))$, radiated by the beginning of the REB segment $r'(x' = 0, y' = 0, z' = 0)$, on the transverse coordinate x was calculated using Eq. (10), (**Figure 1**). The signal radiation time t' was selected equal to zero $t' = 0$. The observation point $r(x, y = 0, z = 0)$ was selected in the cross section $z = 0$ at $y = 0$. The observation time t was determined by the formula $t = \frac{|x|}{c}$.

The dependence of the potential electric field strength $E_x^p(t', x' = 0, y' = 0,\ z' = v_e t'; t, r(x = 0.3m, y = 0, z = 0))$, radiated by the beginning of the REB segment $r'(x' = 0, y' = 0, z' = v_e t')$, on the signal generation time t' calculated with the help of Eq. (10), is represented in **Figure 2** where $r(x = 0.3m,\ y = 0, z = 0)$ is the observation point coordinates.

The dependence of the magnetic field strength $H_y(t' = 0, x' = 0, y' = 0,\ z' = L; t, r(x, y = 0, z = 0))$ radiated by the REB segment end $r'(x' = 0, y' = 0, z' = L)$ on the transverse coordinate x was calculated using Eq. (22) (**Figure 3**). The signal

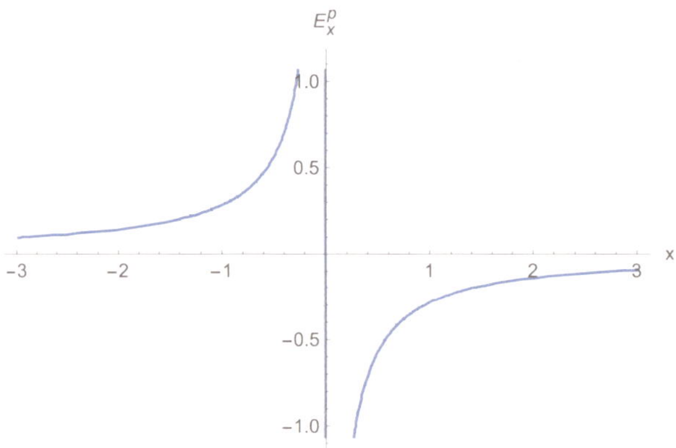

Figure 1.
The potential electric field strength $\mathbf{E_x^p(t' = 0, x' = 0, y' = 0, z' = 0; t, r(x, y = 0, z = 0))}$ *radiated by the REB segment beginning.*

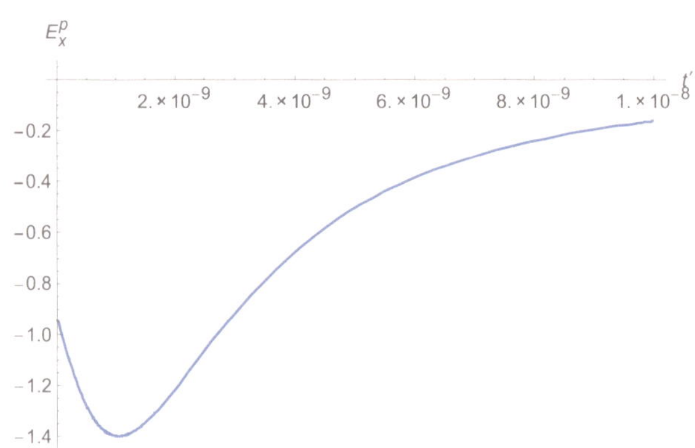

Figure 2.
The potential electric field strength $\mathbf{E_x^p(t', x' = 0, y' = 0, z' = v_e t'; t, r(x = 0.3m, y = 0, z = 0))}$ *radiated by the REB segment beginning.*

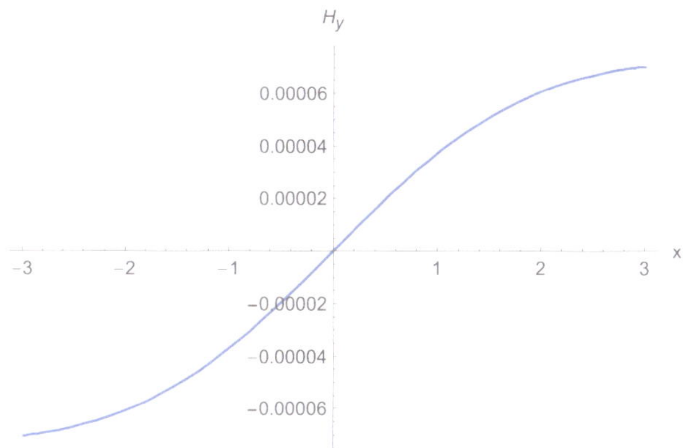

Figure 3.
Magnetic field strength $H_y(t' = 0, x' = 0, y' = 0, z' = L; t, r(x, y = 0, z = 0))$ *radiated by the REB segment end.*

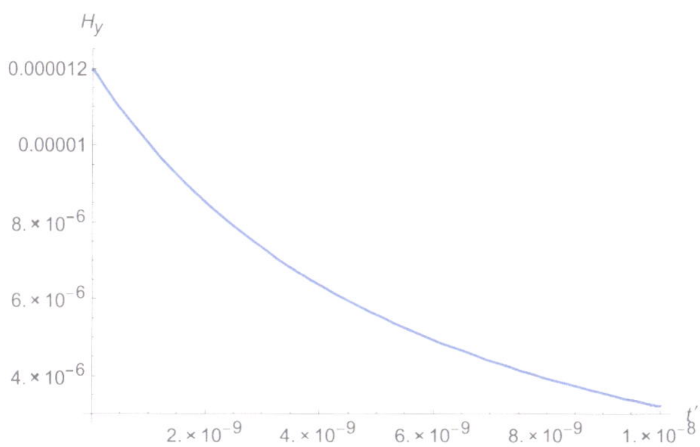

Figure 4.
Magnetic field strength $H_y(t', x' = 0, y' = 0, z' = v_e t' + L; t, r(x = 0.3m, y = 0, z = 0))$ *radiated by the REB segment end.*

generation time t' was selected equal to the zero, $t' = 0$ where $r(x = 0.3m, y = 0, z = 0)$ is the observation point coordinates. The observation time t was determined by the formula $t = \frac{\sqrt{x^2+L^2}}{c}$.

The dependence of the magnetic field strength $H_y(t', x' = 0, y' = 0, z' = v_e t' + L; t, r(x = 0.3m, y = 0, z = 0))$, radiated by the REB segment end $r'(x' = 0, y' = 0, z' = v_e t' + L)$, on the signal radiation time t' calculated using Eq.(22), is represented in **Figure 4** where $r(x = 0.3m,\ y = 0, z = 0)$ is the observation point coordinates.

The dependence of the electromagnetic energy flux ${}^{i}S_z(t' = 0, x' = 0, y' = 0, z' = 0; t, r(x, y = 0, z = 0))$, radiated by the REB segment beginning $r'(x' = 0, y' = 0, z' = 0)$, on the transverse coordinate x was calculated with the help of Eqs. (49), (10), (11), (21), and (22) (**Figure 5**). The signal generation time t' was selected equal to the zero, $t' = 0$. The $r(x, y = 0, z = 0)$ is the observation point coordinates. The observation time t was determined by the formula $t = \frac{|x|}{c}$.

The dependence of the electromagnetic energy flux ${}^{i}S_z(t', x' = 0, y' = 0, z' = v_e t'; t, r(x = 0.3m, y = 0, z = 0))$, radiated by the REB segment $r'(x' = 0, y' = 0, z' = v_e t')$, on the signal generation time t', calculated by Eqs. (49),

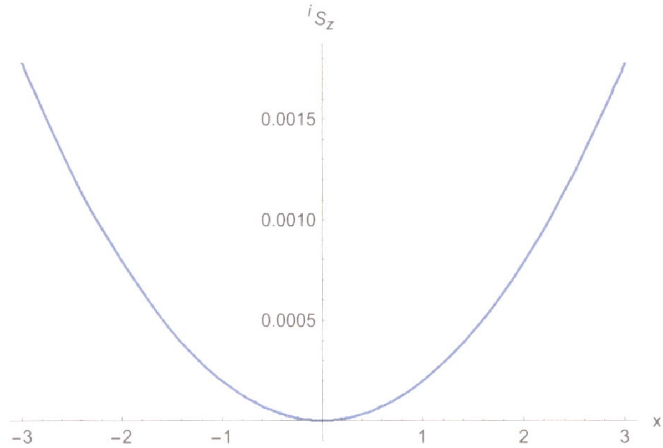

Figure 5.
The electromagnetic energy flux $^{i}S_z(t' = 0, x' = 0, y' = 0, z' = 0; t, r(x, y = 0, z = 0))$ *radiated by the REB segment beginning.*

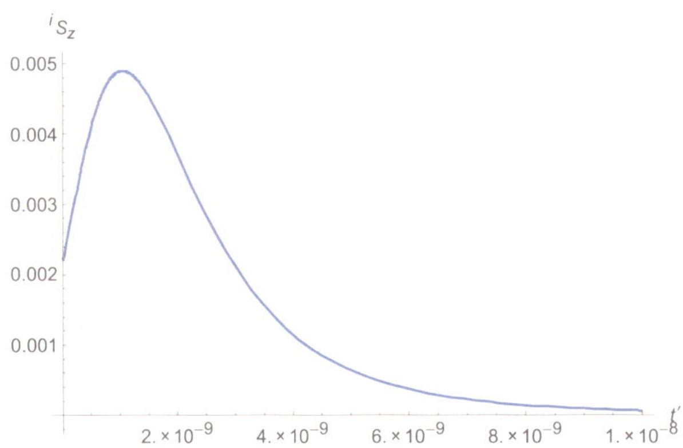

Figure 6.
The electromagnetic energy flux $^{i}S_z(t', x' = 0, y' = 0, z' = v_e t'; t, r(x = 0.3m, y = 0, z = 0))$ *radiated by the REB segment beginning.*

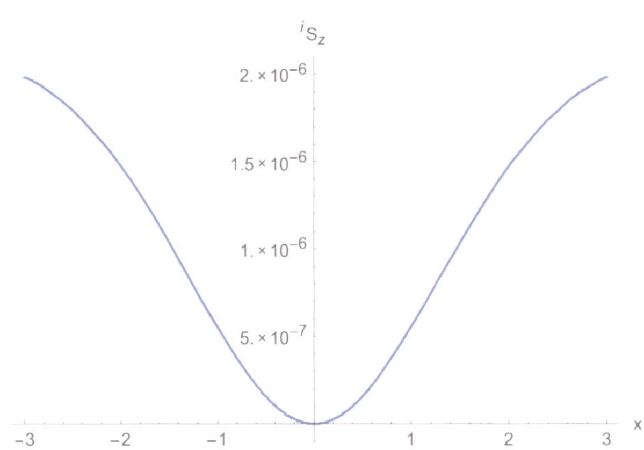

Figure 7.
The electromagnetic energy flux $^{i}S_z(t' = 0, x' = 0, y' = 0, z' = L; t, r(x, y = 0, z = 0))$ *radiated by the REB segment end.*

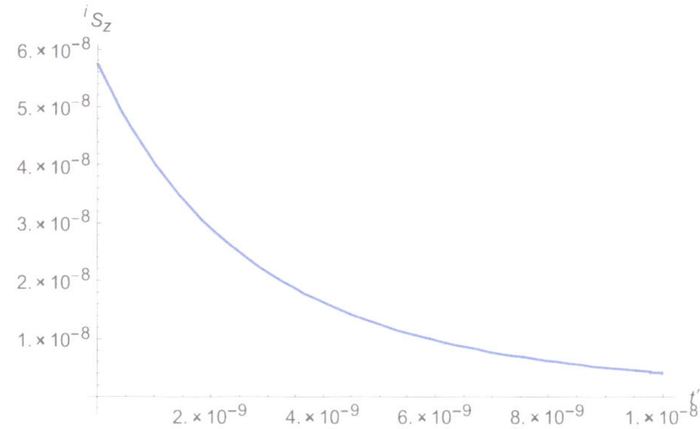

Figure 8.
The electromagnetic energy flux ${}^{i}S_z(t', x' = 0, y' = 0, z' = v_e t' + L; t, r(x = 0.3m, y = 0, z = 0))$ *radiated by the REB segment end.*

(10), (11), (21), (22), is shown in **Figure 6**. The observation point coordinate is $r(x = 0.3m, y = 0, z = 0)$.

The dependence of the electromagnetic energy flux ${}^{i}S_z(t' = 0, x' = 0, y' = 0, z' = L; t, r(x, y = 0, z = 0))$, radiated by the REB segment end $r'(x' = 0, y' = 0, z' = L)$, on the transverse coordinate x was calculated with the help of Eqs. (50), (10), (11), (21), (22) (**Figure 7**). The signal radiation time t' was selected equal to the zero $t' = 0$. The $r(x, y = 0, z = 0)$ is the observation point coordinates. The observation time t was determined by the formula $t = \frac{\sqrt{x^2+L^2}}{c}$.

The dependence of the electromagnetic energy flux ${}^{i}S_z(t', x' = 0, y' = 0, z' = v_e t' + L; t, r(x = 0.3m, y = 0, z = 0))$, radiated by the REB segment end $r'(x' = 0, y' = 0, z' = v_e t' + L)$, on the signal radiation time t', calculated according to Eqs. (50), (10), (11), (21), (22), is shown in **Figure 8** where $r(x = 0.3m, y = 0, z = 0)$ is the observation point coordinates.

9. Conclusions

The applicability of relativity in the physics of charged particle beams has been shown from the example of radiation by a filamentary REB segment uniformly moving in vacuum along a linear direction.

In electrodynamics, in a moving coordinate system, the relative distance between a charged object and an observer does not change. The phenomenon of relativity associated with the field dynamics degenerates to electrostatic processes. In rest, or laboratory, coordinate system, the relative distance is changing with time, the charge density also varies with the time, and as a result, the retardation phenomena came to the scene and the Poisson equation is to be substituted by the wave equation.

The expressions have been obtained to describe the strengths of the electric and magnetic fields and the electric and electromagnetic energy fluxes in all three zones: near field zone, intermediate, and wave zones. The filamentary REB edges are relativistic point-like sources of electromagnetic energy propagating in the wave zone. The REB edges form a potential component of the electric field strength, which is inversely proportional to the distance from the source point to the observation point. In the wave zone, strength of this field is comparable with that of the dynamic component of the electric field.

The dynamic component of the electric field strength and the axially symmetric magnetic field form both a constant flux into the given solid angle, i.e. electromagnetic radiation, and a flux per time unit directed along the normal to the conical surface of the above solid angle. The potential component of the electric field, directed along the radius, and the axially symmetric magnetic field form a flux oriented along the polar direction, i.e., along the normal to the conical surface. The fluxes crossing the above conical surface are independent of the distance between the source point and the observation point. In the wave zone, the radiations from the beginning and end of the REB segment are added up, while the fluxes through the above conical surface caused by dynamic and potential components of electric field, are subtracted.

Relativistic point-like sources create in the wave zone the vortex components of the magnetic field. The REB edges radiate hybrid electromagnetic waves, comprising of potential and vortex electric fields, as well as a vortex magnetic field. The electric and magnetic field strengths radiated by the REB segment edges have opposite signs. In the wave zone, the radiated electromagnetic field fluxes are compound of the electromagnetic energy fluxes, produced by both the REB segment beginning and its end, as well as of their interference components. In the intermediate zone, the electrical energy flux takes place due to the electric potential field and the displacement current. The REB segment, between the beam edges, having a constant charge density, produces a quasi-static electromagnetic field in the near zone.

Acknowledgements

This work was funded in part by NATO research project G5465 within frames of the Science for Peace and Security (SPS) program.

Author details

Sergey Prijmenko[1*] and Konstantin Lukin[2,3*]

1 Institute for Plasma Electronics and New Methods of Acceleration, National Science Center Kharkov Institute of Physics and Technology, National Academy of Science of Ukraine, Kharkov, Ukraine

2 Usikov Institute for Radiophysics and Electronics, National Academy of Science of Ukraine, Kharkov, Ukraine

3 Faculty of Electrical Engineering and Informatics, University of Pardubice, Pardubice, Czech Republic

*Address all correspondence to: sprijmenko@kipt.kharkov.ua and lukin.konstantin@gmail.com

References

[1] Landau LD, Lifshitz EM. The Classical Theory of Fields (Volume 2: A Course of Theoretical Physics). Pergamon Press; 1971

[2] de Sangro R, Finocchinro G, Patteri P, Piccolo M, Pizzella G. Measuring propagation speed of Coulomb fields. The European Physical Journal C. 2015;**75**:137. DOI: 10.1140/epic/s10052-015-3355-3

[3] Jackson JD. Classical Electrodynamics. New York: John Wiley; 1998. 537p

[4] Meshkov IN, Chirikov BV. Relativistic Electrodynamics. Novosibirsk: Vysshaya Shkola, NSU; 1982. 80p (in Russian)

[5] Purcell E. Electricity and Magnetism (Berkeley Physics Course, Vol. 2). 2nd ed. McGraw-Hill Science/Engineering/Math; 1984. 506p

[6] Jefimenko OD. Retardation and relativity: The case of a moving line charge. American Journal of Physics. 1995;**63**(5):454-459

[7] Akhiezer AI, Akhiezer IA. Electromagnetism and Electromagnetic Waves. Moscow: Vysshaya Shkola; 1985. 594p (in Russian)

[8] Dwight HB. Tables of Integrals and Other Mathematical Data. New York: The Macmillan Company; 1957. p. 172

[9] Madelung E. Mathematical Apparatus of Physics. Moscow: Nauka; 1968. 618p (in Russian)

[10] Stratton JA. Electromagnetic Theory. Moscow-Leningrad: Ogiz–Gostekhizdat; 1948. 539p (in Russian)

[11] Prijmenko SD, Lukin KA. The flow of electromagnetic energy in the presence of potential electric and magnetic fields. Applied Radio Electronics. 2018;**17**(1, 2):28-34 (in Russian)

Chapter 5

On the Nonuniqueness of the Hamiltonian for Systems with One Degree of Freedom

Sikarin Yoo-Kong

Abstract

The alternative Hamiltonians for systems with one degree of freedom are solved directly from the Hamilton's equations. These new Hamiltonians produce the same equation of motion with the standard one (called the Newtonian Hamiltonian). Furthermore, new Hamiltonians come with an extra-parameter, which can be used to recover the standard Hamiltonian.

Keywords: Hamiltonian, Lagrangian, nonuniqueness, variational principle, inverse problem of calculus of variations

1. Introduction

It was well known that the Lagrangian possesses the nonuniqueness property. It means that the constant can be added or multiplied into the Lagrangian: $L_N(\dot{x},x) \to \alpha L_N(\dot{x},x) + \beta$

Furthermore, the total derivative term can also be added to the Lagrangian without alternating the equation of motion: $L_N(\dot{x},x) \to \alpha L_N(\dot{x},x) + \beta + df/dt$, where $= f(x,t)$. This fact can be seen immediately from the variational principle with the action functional:

$$S[x] = \int_0^T dt\left(\alpha L_N(\dot{x},x) + \beta + \frac{df}{dt}\right) = \int_0^T dt(\alpha L_N(\dot{x},x) + \beta) + f(T) - f(0) \qquad (1)$$

Obviously, the last two terms contribute only at the boundary. Then the variation $x \to x + \delta x$ on the action and $\delta S[x] = 0$, with conditions $\delta x(0) = \delta x(T) = 0$, give us the same Euler-Lagrange equation:

$$\frac{\partial}{\partial x} L_N(\dot{x},x) - \frac{d}{dt}\frac{\partial}{\partial \dot{x}} L_N(\dot{x},x) = 0 \qquad (2)$$

The standard Lagrangian takes the form

$$L_N(\dot{x},x) = T(\dot{x}) - V(x) \qquad (3)$$

where $T(\dot{x})$ is the kinetic energy and $V(x)$ is the potential energy of the system. For a system with one degree of freedom, the kinetic energy is $T(\dot{x}) = m\dot{x}^2/2$. The equation of motion associated with the Lagrangian Eq. (1) is

IntechOpen

$$\ddot{x} = -(1/m)dV(x)/dx \stackrel{\text{def}}{=} Q \tag{4}$$

Recently, it has been found that actually there is an alternative form of the Lagrangian called the multiplicative form [1–3]: $L(\dot{x}, x) = F(\dot{x})G(x)$, where F and G are to be determined. Putting this new Lagrangian into the Euler–Lagrange Eq. (2), we obtained

$$L_\lambda(\dot{x}, x) = m\lambda^2 \left[e^{-\frac{E(x,\dot{x})}{m\lambda^2}} + \frac{\dot{x}}{\lambda^2} \int_0^{\dot{x}} e^{-\frac{E(x,\dot{q})}{m\lambda^2}} d\dot{q} \right] \tag{5}$$

where $E(x, \dot{x}) = m\dot{x}^2/2 + V(x)$ is the energy function and $m\lambda^2$ is in the energy unit. We find that under the limit λ which is very large $\lim_{\lambda\to\infty}(L_\lambda(\dot{x}, x) - m\lambda^2) = L_N(\dot{x}, x)$, we recover the standard Lagrangian. The derivation of Eq. (5) can be found in the Appendix. Interestingly, this new Lagrangian can be treated as a generating function producing an infinite hierarchy of the Lagrangian:

$$L_\lambda(\dot{x}, x) = \sum_{j=0}^{\infty} \frac{1}{j!} \left(\frac{-1}{m\lambda^2} \right)^{j-1} L_j(\dot{x}, x) \tag{6}$$

where

$$L_j(\dot{x}, x) = \sum_{k=0}^{j} \left(\frac{j! T^{j-k} V^k}{(j-k)! k! (2j - (2k+1))} \right) \tag{7}$$

These new Lagrangians $L_j(\dot{x}, x)$, however, produce the same equation of motion. Equations (6) and (7) provide an alternative way to modify the Lagrangian Eq. (3).

The problem studied in [1–3] that is actually related to the inverse problem of calculus of variations in the one-dimensional case. The well-known result can be dated back to the work of Sonin [4] and Douglas [5].

Theorem (Sonin): For every function Q, there exists a solution (g, L) of the equation:

$$g(Q - \ddot{x}) = \frac{\partial}{\partial x} L(\dot{x}, x) - \frac{d}{dt}\frac{\partial}{\partial \dot{x}} L(\dot{x}, x). \text{where} g = \frac{\partial^2}{\partial \dot{x}^2} L(\dot{x}, x) \neq 0 \tag{8}$$

What we did in [1–3] is that we went further to show that actually Eq. (8) admits infinite solutions.

In the present chapter, we will construct the Hamiltonian hierarchy for the system with one degree of freedom. In Section 2, the multiplicative Hamiltonian will be solved directly from Hamilton's equations. In Section 3, the physical meaning of the parameter λ will be discussed. In Section 4, the redundancy of the Hamiltonians and Lagrangians will be explained. In the last section, a summary will be delivered.

2. The multiplicative Hamiltonian

To obtain the Hamiltonian, we may use the Legendre transformation:

$$H_N(p, x) = p\dot{x} - L_N(\dot{x}, x) \tag{9}$$

where $p = \partial L/\partial \dot{x} = m\dot{x}$ is the momentum variable. The standard form of the Hamiltonian is

$$H_N(p,x) = \frac{p^2}{2m} + V(x) \tag{10}$$

which is nothing but the total energy of the system. The action is then

$$S[p,x] = \int_0^T dt(p\dot{x} - H_N(p,x)) \tag{11}$$

With the variations $x \to x + \delta x$ and $p \to p + \delta p$, with conditions $\delta x(0) = \delta x(T) = 0$, the least action principle $\delta S[x] = 0$ gives us

$$\dot{x} = \frac{\partial}{\partial p} H_N(p,x), \dot{p} = -\frac{\partial}{\partial x} H_N(p,x) \tag{12}$$

which are known as a set of Hamilton's equations.

We now introduce a new Hamiltonian, called the multiplicative Hamiltonian, in a form

$$\mathrm{H}(p,x) = K(p)W(x), \tag{13}$$

where $K(p)$ and $W(x)$ are to be determined. Equations (12) and (3) give us a new equation:

$$0 = \frac{1}{m}\frac{\partial}{\partial x} H_N(p,x) + \dot{p}\frac{\partial^2}{\partial p^2} H_N(p,x) + \frac{p}{m}\frac{\partial^2}{\partial x \partial p} H_N(p,x) \tag{14}$$

Replacing H_N by H and inserting Eqs. (13) into (14), we obtain

$$0 = \frac{d^2K}{dp^2} + \frac{1}{m\dot{p}W}\frac{dW}{dx}\left(p\frac{dK}{dp} + K\right) \tag{15}$$

Now we define

$$A \stackrel{\text{def}}{=} \frac{1}{m\dot{p}W}\frac{dW}{dx} \tag{16}$$

where A is a constant to be determined. Equation (16) can be immediately solved and result in

$$W(x) = ae^{-mAV(x)} \tag{17}$$

where a is a constant of integration. Substituting Eq. (17) into Eq. (15), we find that the function $K(p)$ is in the form

$$K(p) = be^{-\frac{Ap^2}{2}} \tag{18}$$

where b is another constant. Then the multiplicative Hamiltonian Eq. (13) becomes

$$H(p,x) = ce^{-\frac{Ap^2}{2} - mAV(x)} \tag{19}$$

where $c = ab$. If we now choose $c = -m\lambda^2$ and $A = 1/m^2\lambda^2$, the Hamiltonian Eq. (19) becomes

$$H_\lambda(p,x) = -m\lambda^2 e^{-\frac{H_N(p,x)}{m\lambda^2}} \tag{20}$$

Inserting Eq. (20) into Eq. (14), we find that

$$\begin{gathered} -\frac{dV(x)}{dx}\left(-\frac{p^2}{m^2\lambda^2}+1\right) = \dot{p}\left(-\frac{p^2}{m^2\lambda^2}+1\right) \\ -\frac{dV(x)}{dx} = \dot{p} \end{gathered} \tag{21}$$

which is the equation of motion of the system. Then this new Hamiltonian Eq. (20) gives us the same equation of motion as Eq. (10).

For the case $m\lambda^2 \gg H_N(p,x)$, we find that the multiplicative Hamiltonian

$$H_\lambda(p,x) \approx -m\lambda^2 + H_N(p,x) \tag{22}$$

gives back the standard Hamiltonian. The constant $-m\lambda^2$ does not alter the equation of the motion of the system.

We find that the multiplicative Hamiltonian Eq. (20) can also be directly obtained from the Legendre transformation:

$$H_\lambda(p,x) = p_\lambda \dot{x} - L_\lambda(\dot{x},x) \tag{23}$$

where

$$p_\lambda = \frac{\partial}{\partial \dot{x}} L_\lambda(\dot{x},x) = \frac{1}{\lambda^2}\int_0^p e^{-\frac{\zeta^2}{2m^2\lambda^2}} \frac{d\zeta}{m} \tag{24}$$

Inserting Eqs. (24) and (5) into Eq. (23), we obtain

$$\begin{aligned} H_\lambda(p,x) &= m\lambda^2 \left[\left(\frac{1}{\lambda^2}\int_0^p e^{-\frac{\zeta^2}{2m^2\lambda^2}} \frac{d\zeta}{m}\right)\frac{p}{m} \right. \\ &\quad \left. - m\lambda^2\left(e^{-\frac{p^2}{2m\lambda^2}} + \frac{p}{m^2\lambda^2}\int_0^p e^{-\frac{\zeta^2}{2m^2\lambda^2}} \frac{d\zeta}{m}\right)\right] e^{-\frac{V(x)}{m\lambda^2}} \\ &= -m\lambda^2 e^{-\frac{H_N(p,x)}{m\lambda^2}} \end{aligned} \tag{25}$$

which is identical to Eq. (20).

Furthermore, we can rewrite the multiplicative Hamiltonian Eq. (20) in terms of the series:

$$H_\lambda(p,x) = \sum_{j=0}^{\infty} \frac{1}{j!}\left(\frac{-1}{m\lambda^2}\right)^{j-1} H_j(p,x) \tag{26}$$

where $H_j(p,x) \equiv H_N^j = (p^2/2m + V(x))^j$. It is not difficult to see that $H_j(p,x)$ produces exactly the equation of motion Eq. (21).

From the structure of Eqs. (26) and (6), it must be a hierarchy of the Legendre transformation. To establish such hierarchy, we start to rewrite the momentum Eq. (24) in the form

$$p_\lambda = \frac{1}{\lambda^2}\int_0^p e^{-\frac{\zeta^2}{2m^2\lambda^2}}\frac{d\zeta}{m} = \sum_{j=0}^{\infty}\frac{1}{j!}\left(\frac{-1}{m\lambda^2}\right)^{j-1} p_j \tag{27}$$

where

$$p_j(p,x) = j!\left[p_{j-1}V(x) + \frac{p^{2j-1}}{(j-1)(2^{j-1})(2j-1)m^{j-1}}\right], j \geq 1 \text{ and } p = m\dot{x} \tag{28}$$

Then the Legendre transformation Eq. (23) becomes

$$0 = \sum_{j=0}^{\infty}\frac{1}{j!}\left(\frac{-1}{m\lambda^2}\right)^{j-1}\left[L_j(\dot{x},x) - p_j\dot{x} + H_j(p,x)\right] \tag{29}$$

Eq. (29) holds if

$$L_j(\dot{x},x) = p_j\dot{x} + H_j(p,x) \tag{30}$$

which are the Legendre transformations for each pair of the Hamiltonian $H_j(p,x)$ and Lagrangian $L_j(\dot{x},x)$ in the hierarchy.

Next, we consider the total derivative $dH_j(p,x) = d\left(p_j\dot{x}\right) - dL_j(\dot{x},x)$ resulting in

$$dx\left(\frac{\partial H_j}{\partial x} + \dot{p}\frac{\partial p_j}{\partial p}\right) + dp\left(\frac{\partial H_j}{\partial p} - \dot{x}\frac{\partial p_j}{\partial p}\right) = 0 \tag{31}$$

Eq. (31) holds if

$$\frac{\partial H_j}{\partial x} = -\dot{p}\frac{\partial p_j}{\partial p}, \frac{\partial H_j}{\partial p} = \dot{x}\frac{\partial p_j}{\partial p} \tag{32}$$

Eq. (32) can be considered as the modified Hamilton's equations for each $H_j(p,x)$ in the hierarchy. Obviously, for $j = 1$, we retrieve the standard Hamilton's Eq. (12), since $p_1 = p = m\dot{x}$.

From the structure of the multiplicative Hamiltonian Eq. (20), it seems to suggest that the exponential of the function, defined on phase space, is always a solution of the Eq. (14). Then we now introduce an ansatz form of the Hamiltonian as

$$H_{a,b}(p,x) = be^{aZ(p,x)} \tag{33}$$

where a and b are constants to be determined. Substituting Eq. (33) into Eq. (14), we obtain

$$0 = \frac{1}{m}\frac{\partial Z}{\partial x} + \dot{p}\frac{\partial^2 Z}{\partial p^2} + \frac{p}{m}\frac{\partial^2 Z}{\partial x \partial p} + a\left[\dot{p}\left(\frac{\partial Z}{\partial p}\right)^2 + \frac{p}{m}\frac{\partial Z}{\partial p}\frac{\partial Z}{\partial x}\right] \tag{34}$$

We find that if we take $H_N(p,x) = Z(p,x)$ to be the standard Hamiltonian, the first three terms in Eq. (34) give us back Eq. (14). Then the last bracket must vanish and gives us an extra-relation:

$$0 = \dot{p}\left(\frac{\partial H_N}{\partial p}\right)^2 + \frac{p}{m}\frac{\partial H_N}{\partial p}\frac{\partial H_N}{\partial x} \tag{35}$$

or

$$0 = \dot{p}\frac{\partial H_N}{\partial p} + \frac{p}{m}\frac{\partial H_N}{\partial x} \tag{36}$$

We immediately see that actually Eq. (36) is a consequence of the conservation of the energy of the system:

$$0 = \frac{dH_N}{dt} = \frac{\partial H_N}{\partial p}\frac{\partial p}{\partial t} + \frac{\partial H_N}{\partial x}\frac{\partial x}{\partial t} = \dot{p}\frac{\partial H_N}{\partial p} + \frac{p}{m}\frac{\partial H_N}{\partial x} \tag{37}$$

Then what we have here is another equation that can be used to determine for the Hamiltonian subject to the equation of motion Eq. (21). To see this, we may start with the standard form of the Hamiltonian $H_N(p,x) = T(p) + V(x)$, where $T(p)$ is a function of the momentum and to be determined. Inserting the Hamiltonian into Eq. (36), we obtain

$$0 = \dot{p}\frac{dT}{dp} + \frac{p}{m}\frac{dV}{dx} \tag{38}$$

Using Eqs. (21) and (38), it can be rewritten in the form

$$0 = \dot{p}\left(\frac{dT}{dp} - \frac{p}{m}\right) \tag{39}$$

Since $\dot{p} \neq 0$, it means that the term inside the bracket must be zero and

$$\int dT = \int \frac{p}{m} dp \rightarrow T(p) = \frac{p^2}{2m} + C \tag{40}$$

where C is a constant which can be chosen to be zero. So we successfully solved the standard Hamiltonian.

Next, we put $Z(p,x) = K(p)W(x)$ which is in the multiplicative form Eq. (13) into Eq. (36), and we obtain

$$0 = W\left[\dot{p}\left(W\frac{dK}{dp}\right) + \frac{p}{m}\left(K\frac{dW}{dx}\right)\right] \tag{41}$$

or

$$\frac{m}{Kp}\frac{dK}{dp} = \frac{1}{W}\frac{dW}{dV} \tag{42}$$

We see that both sides of Eq. (42) are independent to each other. Then Eq. (42) holds if both sides equal to a constant β. We have now for the left-hand side

$$\frac{m}{Kp}\frac{\partial K}{\partial p} = \beta$$

$$\int \frac{dK}{K} = \beta \int \frac{p}{m} dp \rightarrow K(p) = C_1 e^{\beta T(p)} \tag{43}$$

where C_1 is a constant to be determined. Next, we consider the right-hand side

$$\frac{1}{W}\frac{dW}{dV} = \beta$$

$$\int \frac{dW}{W} = \beta \int dV \rightarrow W(x) = C_2 e^{\beta V(x)} \tag{44}$$

where C_2 is a constant to be determined. Then finally the function $Z(p,x)$ becomes

$$Z(p,x) = C_1 C_2 e^{\beta H_N(p,x)} \tag{45}$$

where $H_N(p,x)$ is the standard Hamiltonian. If we now choose $C_1C_2 = -m\lambda^2$ and $\beta = -1/m\lambda^2$, the function $Z(p,x)$ is exactly the same with Eq. (20).

We see that with Eq. (36) the Hamiltonian can be easily determined. Here we come with the conclusion that in every function $Q' = -\frac{dV}{dx}$, there exist infinite Hamiltonians of equation

$$Q' - \dot{p} = \dot{p}\frac{\partial H}{\partial p} + \frac{p}{m}\frac{\partial H}{\partial x} \tag{46}$$

The existence of solutions of Eq. (46) implies that actually we can do an inverse problem of the Hamiltonian for the systems with one degree of freedom.

Remark: The perspective on nonuniqueness of Hamiltonian, as well as Lagrangian, here in the present work is quite different from those in Aubry-Mather theory [6, 7] (see also [8]). What they had been investigating is the modification of the Tonelli Lagrangian $L_\eta := L - \hat{\eta}$, where mechanical Lagrangian $L_N(\dot{x},x) = T(\dot{x}) - V(x)$ is one of Tonelli Lagrangians. Here $\hat{\eta} = <\eta(x),\dot{x}>: TM \rightarrow \mathbb{R}$ and $\eta(x)$ is a closed 1-from on the manifold M. This means that $\int dtL_\eta$ and $\int dtL$ will have the same extremals and therefore the same Euler-Lagrange evolution, since $\delta \int dt\hat{\eta} = 0$. Thus for a fixed L, the extreamise of the action will depend only on the de Rham cohomology class $c = [\eta] \in H^1(M,\mathbb{R})$. Then we have a family of modified Lagrangians, parameterized over $H^1(M,\mathbb{R})$. With the modified Tonelli Lagrangian L_η, one can easily find the associated Hamiltonian $H_\eta(x,p) = H(x,\eta(x)+p)$, where the momentum is altered: $p \rightarrow p + \eta(x)$. Then we also have a family of modified Hamiltonians, parameterized over $H^1(M,\mathbb{R})$. To make all this more transparent, we better go with a simplest example. Consider the modified Lagrangian $L_\epsilon := L_N + \epsilon\dot{x}$, where ϵ is a constant. We find that a new action differs from an old action by a constant depending on the endpoints, $\int_a^b dtL_\epsilon = \int_a^b dtL_N + \epsilon(x(b) - x(a))$, and they give exactly the same Euler-Lagrange equation (see also Eq. (1)). With this new Lagrangian L_ϵ, we can directly obtain the Hamiltonian $H_\epsilon(x,p) = H_N(x,p+\epsilon)$.

3. Harmonic oscillator

In this section, we give an explicit example, e.g., the harmonic example, and also give the physical interpretation of the parameter λ. The standard Hamiltonian for the harmonic oscillator reads

$$H(p,x) = \frac{p^2}{2m} + \frac{kx^2}{2} \tag{47}$$

Then the multiplicative Hamiltonian for the harmonic oscillator is

$$H_\lambda(p,x) = -m\lambda^2 e^{-\frac{1}{m\lambda^2}\left(\frac{p^2}{2m}+\frac{kx^2}{2}\right)} \tag{48}$$

Now we introduce $\eta = (x,p)$, and then we consider

$$\frac{d\eta}{dt_\lambda} = J\frac{\partial H_\lambda}{\partial \eta} \text{ where } \frac{\partial}{\partial \eta} = \begin{pmatrix} \partial/\partial x \\ \partial/\partial p \end{pmatrix} \tag{49}$$

where t_λ is a time variable associated with the multiplicative Hamiltonian and J is the symplectic matrix given by

$$J = \begin{pmatrix} 0 & 1 \\ -1 & 0 \end{pmatrix} \tag{50}$$

Inserting Eq. (48) into Eq. (49), we obtain

$$\frac{d\eta}{dt_\lambda} = \sum_{k=0}^{\infty} \frac{1}{k!}\left(\frac{-1}{m\lambda^2}\right)^{k-1} J\frac{\partial H^k}{\partial \eta} \overset{\text{def}}{=} \sum_{k=0}^{\infty} \frac{d\eta}{dt_k} \tag{51}$$

where

$$\frac{d}{dt_k} = \frac{E^{k-1}}{(k-1)!\left(m\lambda^2\right)^{k-1}} \frac{d}{dt} \tag{52}$$

where $E = T + V$ is the energy function and t is the standard time variable associated with the Hamiltonian Eq. (47). Equation (51) suggests that the λ-flow is comprised of infinite different flows on the same trajectory on the phase space (see **Figure 1**).

This means that we can choose any Hamiltonian in the hierarchy to work with. The physics of the system remains the same but with a different time scale. Then we may say that the parameter λ plays a role of scaling in the Hamiltonian flow on the phase space. From Eq. (52), we see that as $m\lambda^2 \to \infty$, only the standard flow survives, and of course we retrieve the standard evolution $t_1 = t$ of the system on phase space.

Next we consider the standard Lagrangian of the harmonic oscillator

$$L_N(\dot{x},x) = \frac{m\dot{x}^2}{2} - \frac{kx^2}{2} \tag{53}$$

and the multiplicative Lagrangian is

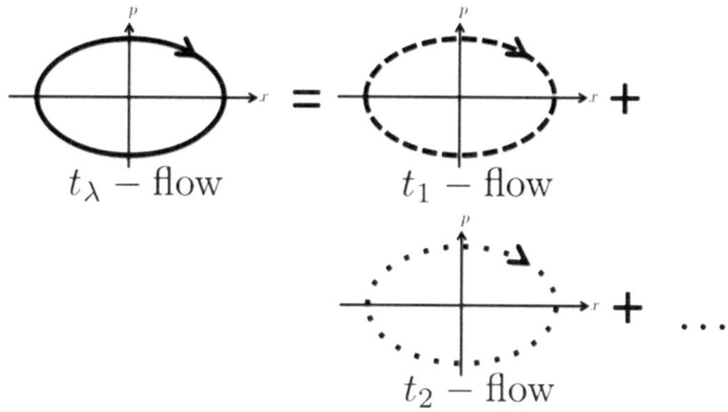

Figure 1.
Differential flows on the same trajectory on the phase space.

$$L_\lambda(\dot{x}, x) = m\lambda^2 \left[e^{-\frac{E(x,\dot{x})}{m\lambda^2}} + \frac{\dot{x}}{\lambda^2} \int_0^{\dot{x}} e^{-\frac{E(x,\dot{q})}{m\lambda^2}} d\dot{q} \right] \tag{54}$$

where $E(x, \dot{x}) = m\dot{x}^2/2 + kx^2/2$ is the energy function. We know that Lagrangian Eq. (54) can be rewritten in the form

$$L_\lambda(\dot{x}, x) = \sum_{j=0}^{\infty} \frac{1}{j!} \left(\frac{-1}{m\lambda^2} \right) L_j(\dot{x}, x) \tag{55}$$

where

$$L_j(\dot{x}, x) = \sum_{k=0}^{j} \left(\frac{j!(m\dot{x}^2/2)^{j-k}(kx^2/2)^k}{(j-k)!k!(2j-(2k+1))} \right) \tag{56}$$

The action of the system is given by

$$S[x] = \int_0^T dt L_\lambda(\dot{x}, x) = \sum_{j=1}^{\infty} \int_0^T dt_j L_j(\dot{x}, x) \tag{57}$$

where

$$dt_j = \frac{1}{j!} \left(\frac{1}{m\lambda^2} \right)^{k-1} dt \tag{58}$$

The variation $x \to x + \delta x$ with conditions $\delta x(0) = \delta x(T) = 0$ results in

$$\delta S[x] = \sum_{j=1}^{\infty} \int_0^T dt_j \left(\frac{\partial L_j}{\partial x} - \frac{d}{dt_j} \frac{\partial L_j}{\partial x_j} \right) \tag{59}$$

where $x_j = dx/dt_j$. Least action principle $\delta S[x] = 0$ gives infinite Euler-Lagrange equations

$$0 = \frac{\partial L_j}{\partial x} - \frac{d}{dt_j} \frac{\partial L_j}{\partial x_j}, j = 1, 2, 3 \tag{60}$$

which produce the equation of motions

$$\frac{d^2x}{dt_j^2} = -\frac{kx}{m} \tag{61}$$

associated with different time variables. Again in this case, we have the same structure of equation of motion for each Lagrangian in hierarchy but with a different time scale. From Eq. (58), we see that as $m\lambda^2 \to \infty$, only the standard flow survives, and of course we retrieve the standard evolution $t_1 = t$ of the system. Then the parameter λ also plays the role of scaling in the Lagrangian structure.

4. Redundancy

From previous sections, we see that there are many forms of the Hamiltonian that you can work with. One may start with the assumption that any new Hamiltonian is written as a function of the standard Hamiltonian H_N: $H = f(H_N)$. Inserting this new Hamiltonian into Hamilton's equations, we obtain

$$f'(E)\frac{\partial H}{\partial x} = -\frac{\partial p}{\partial \tau},\ f'(E)\frac{\partial H}{\partial p} = \frac{\partial x}{\partial \tau} \tag{62}$$

where $f'(E) = df(H_N)/H_N$ with fixing $H_N = E$ and $t = f'(H_N)\tau$ is the rescaling of time parameter. This result agrees with what we have in Section 3, rescaling the time evolution of the system. However, there are some major different features as follows. The first thing is that our new Hamiltonians contain a parameter λ, since the explicit forms of the Hamiltonian are obtained. With this parameter, it makes our rescaling much more interesting with the fact that the rescaling time variables depend on also the parameter (see Eq. (52)). Then it means that we know how to move from one scale to another scale and of course we know how to obtain the standard time evolution by playing with the limit of the parameter λ. Without explicit form of the new Hamiltonian, which contains a parameter, we cannot see this fine detail of family of rescaling time variables, since there is only a fixed parameter E. The second thing is that actually the new Hamiltonian Eq. (20), which is a function of the standard Hamiltonian, can be obtained from the Lagrangian Eq. (5) by means of Legendre transformation. What we have seen is that Lagrangian Eq. (5) is nontrivial and is not a function of the standard Lagrangian. Again this new Lagrangian contains a parameter λ, the same with the one in the new Hamiltonian. With this parameter, the Lagrangian hierarchy Eq. (7) is obtained. What we have here is a family of nontrivial Lagrangians to work with, producing the same equation of motion, as a consequence of nonuniqueness property. An importance thing is that there is no way you can guess the form of this family of Lagrangian without our mechanism in the appendix. This means that the Hamiltonian in the form $H = f(H_N)$ cannot deliver all these fine details. The explicit form of the Hamiltonian Eq. (20) allows us to study in more detail and is definitely richer than the standard one.

5. Summary

We show that actually there exist infinite Hamiltonian functions for the systems with one degree of freedom. We may conclude that there exists the reverse engineering of the calculus of variation on phase space (see Eq. (44)). Furthermore, the solution of Eq. (44) exists not only as one but infinite. Interesting fact here is that these new Hamiltonians come with the extra-parameter called λ. We give the interpretation that the term $m\lambda^2$ involves the time scaling of the system. This means that we can pick any Hamiltonian or Lagrangian to study the system, but the evolution will be in different scales.

In the case of many degrees of freedom, the problem turns out to be very difficult. Even in the case of two degrees of freedom, the problem is already hard to solve from scratch. We may start with an anzast form of the Lagrangian: $L(\dot{x},\dot{y},x,y) = F(\dot{x},\dot{y})G(x,y)$. This difficulty can be seen from the fact that we have to solve a non-separable coupled equation. A mathematical trig or further assumptions might be needed for solving $F(\dot{x},\dot{y})$ and $G(x,y)$. The investigation is now monitored.

Furthermore, promoting the Hamiltonian Eq. (20) to be a quantum operator in the context of Schrodinger's equation is also an interesting problem. This seems to suggest that an alternative form of the wave function for a considering system is possibly obtained. This can be seen as a result from that fact that with new Hamiltonian operator, we need to solve a different eigenvalue equation, and of course a new appropriate eigenstate is needed. From the Lagrangian point of view, extension to the quantum realm in the context of Feynman path integrals is quite natural to address. However, this problem is not easy to deal with since the Lagrangian multiplication is not in the quadratic form. Then a common procedure for deriving the propagator is no longer applicable. Further study is on our program of investigation.

Acknowledgements

The author would like to thank Kittikun Sarawuttinack, Saksilpa Srisukson, and Kittapat Ratanaphupha for their interests and involving themselves in the investigation on this topic.

Notes

The content in this chapter is collected from a series of papers [1–3].

Appendix

In this section, we will demonstrate how to solve the multiplicative Lagrangian Eq. (5). We introduce here again the Lagrangian $L(\dot{x},x) = F(\dot{x})G(x)$, where F and G are to be determined. Inserting the Lagrangian into the action and performing the variation $x \to x + \delta x$, with conditions $\delta x(0) = \delta x(T) = 0$, we obtain

$$\delta S[x] = \int_0^T dt\left(-\frac{d}{dt}\left(G\frac{dF}{d\dot{x}}\right) + F\frac{dG}{dx}\right)\delta x \tag{63}$$

The least action principle states that the system will follow the path which $\delta S = 0$ resulting in

$$-\frac{d}{dt}\left(G\frac{dF}{d\dot{x}}\right)+F\frac{dG}{dx}=0 \tag{64}$$

Eq. (64) can be rewritten in the form

$$\frac{d^2F}{d\dot{x}^2}-\frac{1}{\ddot{x}G}\frac{dG}{dx}\left(F-\dot{x}\frac{dF}{d\dot{x}}\right)=0 \tag{65}$$

Using equation of motion, we observe that the coefficient of the second term depends only x variable. Then we may set

$$\frac{1}{\ddot{x}G}\frac{dG}{dx}\overset{\text{def}}{=}A\rightarrow\frac{1}{G}\frac{dG}{dx}=-\frac{A}{m}\frac{dV}{dx} \tag{66}$$

We find that it is not difficult to see that the function G that satisfies Eq. (66) is

$$G(x)=\alpha_1 e^{-AV(x)/m} \tag{67}$$

where α_1 is a constant to be determined. Inserting Eq. (66) into Eq. (65), we obtain

$$\frac{d^2F}{d\dot{x}^2}-A\left(F-\dot{x}\frac{dF}{d\dot{x}}\right)=0 \tag{68}$$

and the solution F is given by

$$F(\dot{x})=\alpha_2\dot{x}-\alpha_3\left(e^{-A\dot{x}^2/2}+\dot{x}A\int_0^{\dot{x}}dve^{-Av^2/2}\right) \tag{69}$$

where α_2 and α_3 are constants. Then the multiplicative Lagrangian is

$$L(\dot{x},x)=\left[k_1\dot{x}-k_2\left(e^{-A\dot{x}^2/2}+\dot{x}A\int_0^{\dot{x}}dve^{-Av^2/2}\right)\right]e^{-AV(x)/m} \tag{70}$$

where $k_1=\alpha_1\alpha_2$ and $k_2=\alpha_1\alpha_3$ are new constants to be determined. We find that if we choose $k_1=0, A=1/\lambda^2$ which is a unit of inverse velocity squared and $k_2=-m\lambda^2$ which is in energy unit, Lagrangian Eq. (70) can be simplified to

$$\lim_{\lambda\to\infty}\left(L_\lambda(\dot{x},x)-m\lambda^2\right)=\frac{m\dot{x}^2}{2}-V(x)=L_N(\dot{x},x) \tag{71}$$

the standard Lagrangian at the limit λ approaching to infinity. Therefore, the Lagrangian Eq. (70) is now written in the form

$$L_\lambda(\dot{x},x)=m\lambda^2\left(e^{-\dot{x}^2/2\lambda^2}+\frac{\dot{x}}{\lambda^2}\int_0^{\dot{x}}dve^{-v^2/2\lambda^2}\right)e^{-V(x)/m\lambda^2} \tag{72}$$

which can be considered as the one-parameter extended class of the standard Lagrangian.

Author details

Sikarin Yoo-Kong
The Institute for Fundamental Study (IF), Naresuan University, Phitsanulok, Thailand

*Address all correspondence to: sikariny@nu.ac.th

References

[1] Sarawuttinack K, Yoo-Kong S, Tanasittikosil M. Multiplicative form of the Lagrangian. Theoretical and Mathematical Physics. 2016;**189**(3): 1693-1711. DOI: 10.1134/s0040577916120023

[2] Srisukson S, Sarawuttinack K, Yoo-Kong S. The multiplicative Hamiltonian and its hierarchy. Journal of Physics: Confernce Series-Siam Physics Congress 2017(SPC2017). 2018;**901**:012167. DOI: 10.1008/1742-6596/901/1/012167

[3] Srisukson S, Ratanaphupha K, Yoo-Kong S. Hamiltonian Zoo for the System with One Degree of Freedom. arXiv: 1802.06370. 2018

[4] Sonin NJ. About determining maximal and minimal properties of plane curves (in Russian) [English translation]. Lepage Research Institute Archive. 2012;**1**(1–2):1-68

[5] Douglas J. Solution of the inverse problem of the calculus of variations. Transactions of the American Mathematical Society. 1941;**50**(1): 71-128. DOI: 10.2307/1989912

[6] Aubry A, Le Daeron YP. The discrete Frenkel-Kontorova model and its extensions. Physica D: Nonlinear Phenomena. 1983;**8**(3):381-422

[7] Mather NJ. Existence of quasi-periodic orbits for twist homeomorphisms of the annulus. Topology. 1982;**21**(4):457-467

[8] Fathi A. Weak Kam theorem and Lagrangian. Dynamics (Pembroke, Ont.). 2007. Available from: https://www.math.u-bordeaux.fr/~pthieull/Recherche/KamFaible/Publications/Fathi2008_01.pdf

Chapter 6

From Relativity to Creation of Temporal (t > 0) Universe

Francis T.S. Yu

Abstract

One of the important aspects of science must be the substantiated physical realities, which were built by the fundamental laws of physics that cannot be simply substituted by unsubstantiated virtual reality. In writing this chapter we have mostly based on the constraints of the current laws of physics to illustrate the enigmatic time as the origin for creating our physical space (i.e., temporal universe). The differences between physical reality and virtual reality are that physical reality is existing within the rule of time and supported by the laws of science, while virtual reality is created without the constraints of time and mostly not substantiated by the laws of physics. One of the important aspects of temporal (i.e., $t > 0$) space is that any emerging science has to be proven to exist within our temporal universe; otherwise it is fictitious and virtual as mathematics is.

Keywords: relativity theory, Einstein energy equation, temporal space, creation of universe, time and space

1. Introduction

One of the most intriguing variables in science must be time. Without time, there would be no physical substances, no space, and no life. In other words, time and substance have to coexist. In the chapter, I will start with Einstein's relativity theory to show his famous energy equation, derived from in which we will show that energy and mass can be traded. Since mass is equivalent to energy and energy is equivalent to mass, we see that mass can be treated as an energy reservoir. We will show any physical space cannot be embedded in an absolute empty space and it cannot have any absolute empty subspace in it and empty space is a timeless (i.e., $t = 0$) space. We will show that every physical space has to be fully packed with substances (i.e., energy and mass), and we will show that our universe is a subspace within a more complex space. We see that our universe could have been one of the many universes outside our universal boundary. We will also show that it takes time to create a subspace, and it cannot bring back the time that has been used for the creation. Since all physical substances exist with time, all subspaces are created by time and substances (i.e., energy and mass). This means that our cosmos was created by time with a gigantic energy explosion, for which every subspace coexists with time. This means that without time the creation of substances would not have happened. We see that our universe is in a temporal (i.e., $t > 0$) space, and it is still expanding based on current observation. This shows that our universe has not reached its half-life yet, as we have accepted the big bang creation. We are not alone with almost absolute certainty. Someday, we may find a planet that once upon a time had harbored a civilization for a period of light-years. In short, the burden of a scientific postulation is to prove a solution exists within our temporal universe; otherwise it is not real or virtual as mathematics is.

Professor Hawking was a world renowned astrophysicist, a respected cosmic scientist, and a genius who passed away last year on March 14, 2018. As you will see, our creation of universe was started with the same root of the big bang explosion, but it is not a sub-universe of Hawking's. You may see from this chapter that the creation of temporal universe is somewhat different from Hawking's creation.

2. Relativity to Einstein energy equation

The essence of Einstein's special theory of relativity [1] is that time is a relative quantity with respect to velocity as given by

$$\Delta t' = \frac{\Delta t}{\sqrt{1 - v^2/c^2}} \tag{1}$$

where $\Delta t'$ is the relativistic time window as compared with a standstill subspace, Δt is the time window of the standstill subspace, v is the velocity of a moving subspace, and c is the velocity of light.

We see that the time window $\Delta t'$ of a moving subspace, with respect to the time window Δt of a standstill subspace, appears to be wider as velocity of the moving subspace increases. In other words, velocity of a moving subspace changes the relative time speed as with respect to a standstill subspace. For instance, the time speed goes slower for a moving subspace as with respect to a standstill subspace. We see that time speed within the subspaces is invariant or constant. In other words, the speed of time goes as it is within the subspaces but is relatively different between the subspaces at different velocities. As a matter of fact, the speed of time within a subspace is governed by the speed of light (such as 1 s, 2 s, etc.) as will be seen in how our temporal universe was created.

Equivalently, Einstein's relativity equation can be shown in terms of relative mass as given by

$$m = \frac{m_0}{\sqrt{(1 - v^2/c^2)}} = m_0\left(1 - v^2/c^2\right)^{-1/2} \tag{2}$$

where m is the effective mass (or mass in motion) of a particle, m_o is the rest mass of the particle, v is the velocity of the moving particle, and c is the speed of light. In other words, the effective mass (or mass in motion) of a particle increases at the same amount with respect to when the relative time window increases.

With reference to the binomial expansion, Eq. (2) can be written as

$$m = m_0\left(1 + \frac{1}{2} \bullet \frac{v^2}{c^2} \quad + terms\ of\ order\ \frac{v^4}{c^4}\right) \tag{3}$$

By multiplying the preceding equation with the velocity of light c^2 and noting that the terms with the orders of v^4/c^2 are negligibly small, the above equation can be approximated by

$$m \approx m_0 + \frac{1}{2} m_0 v^2 \frac{1}{c^2} \tag{4}$$

which can be written as

$$(m - m_0)c^2 \approx \frac{1}{2} m_0 \mathrm{v}^2 \tag{5}$$

The significance of the preceding equation is that $m - m_o$ represents an increase in mass due to motion, which is the kinetic energy of the rest mass m_o. And $(m - m_o)c^2$ is the extra energy gain due to motion.

What Einstein postulated, as I remembered, is that there must be energy associated with the mass even at rest. And this was exactly what he had proposed:

$$\varepsilon \approx mc^2 \tag{6}$$

where ε represents the total energy of the mass and

$$\varepsilon_o \approx m_o c^2 \tag{7}$$

the energy of the mass at rest, where $\mathrm{v} = 0$ and $m \approx m_o$.

We see that Eq. (6) or equivalently Eq. (7) is the well-known Einstein energy equation.

3. Time and energy

One of the most enigmatic variables in the laws of science must be "time." So what is time? Time is a variable and not a substance. It has no mass, no weight, no coordinate, and no origin, and it cannot be detected or even be seen. Yet time is an everlasting existing variable within our known universe. Without time there would be no physical matter, no physical space, and no life. The fact is that every physical matter is associated with time which includes our universe. Therefore, when one is dealing with science, time is one of the most enigmatic variables that are ever present and cannot be simply ignored. Strictly speaking, all the laws of science as well every physical substance cannot exist without the existence of time.

On the other hand, energy is a physical quantity that governs every existence of substance which includes the entire universe. In other words without the existence of energy, there would be no substance and no universe! Nonetheless based on our current laws of science, all the substances were created by energy, and every substance can also be converted back to energy. Thus energy and substance are exchangeable, but it requires some physical conditions (e.g., nuclei and chemical interactions and others) to make the conversion start. Since energy can be derived from mass, mass is equivalent to energy. Hence every mass can be treated as an energy reservoir. The fact is that our universe is compactly filled with mass and energy. Without the existence of time, the trading (or conversion) between mass and energy could not have happened.

4. Time-dependent energy equation

Let us now start with Einstein's energy equation which was derived by his special theory of relativity [1] as given by

$$\varepsilon \approx mc^2 \tag{8}$$

where m is the rest mass and c is the velocity of light.

Since all the laws in science are approximations, for which we have intentionally used an approximated sign. Strictly speaking the energy equation should be more appropriately presented with an inequality sign as described by

$$\varepsilon < mc^2 \tag{9}$$

This means that in practice, the total energy should be smaller or at most approaching to the rest mass m times square of light speed (i.e., c^2).

In view of Einstein's energy equation of Eq. (8), we see that it is a singularity-point approximation and timeless equation (i.e., t = 0). In other words, the equation needs to convert into a temporal (i.e., t > 0) representation or time-dependent equation for the conversion to take place from mass into energy. We see that, without the inclusion of time variable, the conversion would not have taken place. Nonetheless, Einstein's energy equation represents the total amount of energy that can be converted from a rest mass m. Every mass can be viewed as an energy reservoir. Thus by incorporating with the time variable, the Einstein's energy equation can be represented by a partial differential equation as given by [2]

$$\frac{\partial \varepsilon(t)}{\partial t} = c^2 \frac{\partial m(t)}{\partial t}, \ \mathrm{t} > 0 \tag{10}$$

where $\partial \varepsilon(t)/\partial t$ is the rate of increasing energy conversion, $\partial m(t)/\partial t$ is the corresponding rate of mass reduction, c is the speed of light, and t > 0 represents a forward time variable. We see that a time-dependent equation exists at time t > 0, representing a forward time variable that only occurs after time excitation at t = 0. Incidentally, this is a well-known causality constraint (i.e., t > 0) [3] as imposed by our universe.

5. Trading mass and energy

One of the important aspects in Eq. (10) must be that energy and mass can be traded, for which the rate of energy conversion from a mass can be written in terms of electromagnetic (EM) radiation or Radian Energy as given by [4]

$$\frac{\partial \varepsilon}{\partial t} = c^2 \frac{\partial m}{\partial t} = [\nabla \cdot S(\mathrm{v})] = -\frac{\partial}{\partial t}\left[\frac{1}{2} \in E^2(\mathrm{v}) + \frac{1}{2}\mu H^2(\mathrm{v})\right], \ \mathrm{t} > 0 \tag{11}$$

where ϵ and μ are the permittivity and the permeability of the physical space, respectively, v is the radian frequency variable, $E^2(v)$ and $\mathrm{H}^2(v)$ are the respective electric and magnetic field intensities, the negative sign represents the outflow energy per unit time from a unit volume, (∇) is the divergent operator, and S is known as the Poynting vector or energy vector of an electromagnetic radiator [4] as can be shown by $S(v) = E(v) \times H(v)$. Again we note that it is a time-dependent equation with t > 0 added to present the causality constraint. In view of the preceding equation, we see that radian energy (i.e., radiation) diverges from the mass, as mass reduces with time. In other words we see that Eq. (11) is not just a piece of mathematical formula; it is a symbolic representation, a description, a language, a picture, or even a video as can be seen that it has transformed from a point-singularity approximation to a three-dimensional representation and it is continually expanding as time moves on.

Similarly the conversion from energy to mass can also be presented as

$$\frac{\partial m}{\partial t} = \frac{1}{c^2}\frac{\partial \varepsilon}{\partial t} = -\frac{1}{c^2}[\nabla \bullet S(\text{v})] = \frac{1}{c^2}\frac{\partial}{\partial t}\left[\frac{1}{2}\epsilon E^2(\text{v}) + \frac{1}{2}\mu H^2(\text{v})\right], \text{t} > 0 \tag{12}$$

The major difference of this equation, as compared with Eq. (11), must be the energy convergent operator $-\nabla \cdot \mathrm{S}(v)$, where we see that the rate of energy as in the form of EM radiation converges into a small volume for the mass creation, instead of diverging from the mass. Since mass creation is inversely proportional to c^2, it requires a huge amount of energy to produce a small quantity of mass. Nevertheless in view of the cosmological environment, availability of huge amount of energy has never been a problem.

Incidentally, black hole [5, 6] can be considered as one of the energy convergent operators. Instead the convergent force is relied more on the black hole's intense gravitational field. The black hole still remains an intriguing physical substance to be known. Its gravitational field is so intense even light cannot be escaped.

By the constraints of the current laws of science, the observation is limited by the speed of light. If light is totally absorbed by the black hole, it is by no means that the black hole is an infinite energy sink [6]. Nonetheless, every black hole can actually be treated as an energy convergent operator, which is responsible for the eventuality in part of energy to mass conversion, where an answer remained to be found.

6. Physical substances and subspaces

In our physical world, every matter is a substance which includes all the elemental particles; electric, magnetic, and gravitation fields; and energy. The reason is that they were all created by means of energy or mass. Our physical space (e.g., our universe) is fully compacted with substances (i.e., mass and energy) and left no absolute empty subspace within it. As a matter of fact, all physical substances exist with time, and no physical substance can exist forever or without time, which includes our universe. Thus, without time there would be no substance and no universe. Since every physical substance described itself as a physical space and it is constantly changing with respect to time. The fact is that every physical substance is itself a temporal space (or a physical subspace), as will be discussed in the subsequent sections.

In view of physical reality, every physical substance cannot exist without time; thus if there is no time, all the substances which include all the building blocks in our universe and the universe itself cannot exist. On the other hand, time cannot exist without the existence of substance or substances. Therefore, time and substance must mutually coexist or inclusively exist. In other words, substance and time have to be simultaneously existing (i.e., one cannot exist without the other). Nonetheless, if our universe has to exist with time, then our universe will eventually get old and die. So the aspects of time would not be as simple as we have known. For example, for the species living in a far distant galaxy moving closer to the speed of light, their time goes somewhat slower relatively to ours [1]. Thus, we see that the relativistic aspects of time may not be the same at different subspaces in our universe (e.g., at the edge of our universe).

Since substances (i.e., mass) were created by energy, energy and time have to simultaneously exist. As we know every conversion, either from mass to energy or from energy to mass, cannot get started without the inclusion of time. Therefore, time and substance (i.e., energy and mass) have to simultaneously exist. Thus we see that all the physical substances, including our universe and us, are coexisting with time (or function of time).

7. Absolute empty and physical subspaces

Let us define various subspaces in the following, as they will be used in the subsequence sections:

An absolute empty space has no time, no substance, and no coordinate and is not event bounded or unbounded. It is a virtual space and timeless space (i.e., $t = 0$), and it does not exist in practice.

A physical space is a space described by dimensional coordinates, which existed in practice, compactly filled with substances, supported by the current laws of science and the rule of time (i.e., time can only move forward and cannot move backward; $t > 0$). Physical space and absolute empty space are mutually exclusive. In other words, a physical space cannot be embedded in an absolute empty space, and it cannot have absolute empty subspace in it. In other words, physical space is a temporal space in which time is a forward variable (i.e., $t > 0$), while absolute empty space is a timeless space (i.e., $t = 0$) in which nothing is in it.

A temporal space is a time-variable physical space supported by the laws of science and rule of time (i.e., $t > 0$). In fact, all physical spaces are temporal spaces (i.e., $t > 0$).

A spatial space is a space described by dimensional coordinates and may not be supported by the laws of science and the rule of time (e.g., a mathematical virtual space).

A virtual space is an imaginary space, and it is generally not supported by the laws of science and the rule of time. Only mathematicians can make it happen.

As we have noted, absolutely empty space cannot exist in physical reality. Since every physical space needs to be completely filled with substances and left no absolutely empty subspace within it, every physical space is created by substances. For example, our universe is a gigantic physical space created by mass and energy (i.e., substances) and has no empty subspaces in it. Yet, in physical reality all the masses (and energy) existed with time. Without the existence of time, then there would be no mass, no energy, and no universe. Thus, we see that every physical substance coexists with time. As a matter of fact, every physical subspace is a temporal subspace (i.e., $t > 0$), which includes us and our universe.

Since a physical space cannot be embedded within an absolute empty space and it cannot have any absolute empty subspace in it [7], our universe must be embedded in a more complex physical space. If we accepted our universe is embedded in a more complex space, then our universe must be a bounded subspace.

How about time? Since our universe is embedded in a more complex space, the complex space may share the same rule of time (i.e., $t > 0$). However, the complex space that embeds our universe may not have the same laws of science as ours but may have the same rule of time (i.e., $t > 0$); otherwise our universe would not be bounded. Nevertheless, whether our universe is bounded or not bounded is not the major issue of our current interest, since it takes a deeper understanding of our current universe before we can move on to the next level of complex space revelation. It is however our aim, abiding within our current laws of science, to investigate the essence of time as the enigma origin of our universe.

8. Time and physical space

One of the most intriguing questions in our life must be the existence of time. So far, we know that time comes from nowhere, and it can only move forward, not backward, not even stand still (i.e., $t = 0$). Although time may somewhat relatively slow down, based on Einstein's special theory of relativity [1], so far time cannot move backward and cannot even stand still. As a matter of fact, time is moving at a constant rate within our subspace, and it cannot move faster or slower. We stress that time moves at the same rate within any subspace within the universe even closer the boundary of our universe, but the difference is the relativistic time. Since time is ever existing, then how do we know there is a physical space? One answer is that there is a profound connection between time and physical space. In other words, if there is no time, then there would be no physical space. A physical space is in fact a temporal (i.e., $t > 0$) space, in contrast to a virtual space. Temporal space can be described by time, while virtual space is an imaginary space without the constraint of time. Temporal space is supported by the laws of science, while virtual space is not.

A television video image is a typical example of trading time for space. For instance, each TV displayed an image of (dx, dy) which takes an amount of time to be displayed. Since time is a forward-moving variable, it cannot be traded back at the expense of a displayed image (dx, dy). In other words, it is time that determines the physical space, and it is not the physical space that can bring back the time that has been expended. And it is the size (or dimension) of space that determines the amount of time required to create the space (dx, dy). Time is distance and distance is time within a temporal space. Based on our current constraints of science, the speed of light is the limit. Since every physical space is created by substances, a physical space must be described by the speed of light. In other words, the dimension of a physical space is determined by the velocity of light, where the space is filled with substances (i.e., mass and energy). And this is also the reason that speed of time (e.g., 1 s, 2 s, etc.) is determined by the speed of light.

Another issue is why the speed of light is limited. It is limited because our universe is a gigantic physical space that is filled with substances that cause a time delay on an EM wave's propagation. Nevertheless, if there were physical substances that travel beyond the speed of light (which remains to be found), their velocities would also be limited, since our physical space is fully compacted with physical substances and it is a temporal (i.e., $t > 0$) space. Let me further note that a substance can travel in space without a time delay if and only if the space is absolutely empty (i.e., timeless; $t = 0$), since distance is time (i.e., $d = ct$, $t = 0$). However, absolute empty space cannot exist in practice, since every physical space (including our universe) has to be fully filled with substances (i.e., energy and mass), with no empty subspace left within it. Since every physical subspace is temporal (i.e., $t > 0$), in which we see that timeless and temporal spaces are mutually exclusive.

9. Electromagnetic and laws of physics

Strictly speaking, all our laws of physics are evolved within the regime of EM science. Besides, all physical substances are part of EM-based science, and all the living species on Earth are primarily dependent on the source of energy provided by the sun. About 78% of the sunlight that reaches the surface of our planet is well concentrated within a narrow band of visible spectrum. In response to our species' existence, which includes all living species on Earth, a pair of visible eyes (i.e., antennas) evolved in us humans, which help us for our survival. And this narrow band of visible light led us to the discovery of an even wider band of EM spectral distribution in nature. It is also the major impetus allowing us to discover all the physical substances that are part of EM-based physics. In principle, all physical substances can be observed or detected with EM interaction, and the speed of light is the current limit.

Then there is question to be asked, why is the speed of light limited? A simple answer is that our universe is filled with substances that limit the speed of light. The energy velocity of an electromagnetic wave is given by [3]

$$\mathrm{v} = \frac{1}{\sqrt{\mu\varepsilon}} \tag{13}$$

where (μ, ε) are the permeability and the permittivity of the medium. We see that the velocity of light is shown by

$$c = \frac{1}{\sqrt{\mu_0 \varepsilon_0}} \tag{14}$$

where (μ_0, ε_0) are the permeability and the permittivity of the space.

In view of Eq. (13), it is apparent that the velocity of electromagnetic wave (i.e., speed of light) within an empty subspace (i.e., timeless space) is instant (or infinitely large) since distance is time (i.e., d = ct, t = 0).

A picture that is worth more than a thousand words [8] is a trivial example to show that EM observation is one of the most efficient aspects in information transmission. Yet, the ultimate physical limitation is also imposed by limitation of the EM regime, unless new laws of science emerge. The essence of Einstein's energy equation shows that mass and energy are exchangeable. It shows that energy and mass are equivalent and energy is a form of EM radiation in view of Einstein's equation. We further note that all physical substances within our universe were created from energy and mass, which include the dark energies [9] and dark matter [10]. Although the dark substances may not be observed directly using EM interaction, we may indirectly detect their existence, since they are basically energy-based substances (i.e., EM-based science). It may be interesting to note that our current universe is composed of 72% dark energy, 23% dark matter, and 5% other physical substances. Although dark matter contributes about 23% of our universe, it represents a total of 23% of gravitational fields. With reference to Einstein's energy equation (Eq. (8)), dark energy and dark matter dominate the entire universal energy reservation, well over 95%. Furthermore, if we accept the big bang theory for our universe creation [11], then creation could have been started with Einstein's time-dependent energy formula of Eq. (11) as given by

$$\frac{\partial\varepsilon}{\partial t} = c^2\frac{\partial M}{\partial t} = [\nabla \cdot S(\mathrm{v})] = -\frac{\partial}{\partial t}\left[\frac{1}{2}\in E^2(\mathrm{v}) + \frac{1}{2}\mu H^2(\mathrm{v})\right], \mathrm{t} > 0 \tag{15}$$

where $[\nabla \cdot S(\mathrm{v})]$ represents a divergent energy operation. In this equation, we see that a broad spectral band intense radian energy diverges (i.e., explodes) at the speed of light from a compacted matter M, where M represents a gigantic mass of energy reservoir. It is apparent that the creation is ignited by time and the exploded debris (i.e., matter and energy) starts to spread out in all directions, similar to an expanding air balloon. The boundary (i.e., radius of the sphere) of the universe expands at the speed of light, as the created debris is disbursed. It took about 15 billion chaotic light-years [12–14] to come up with the present state of constellation, in which the boundary is still expanding at the speed of light beyond the current observation. With reference to a recent report using the Hubble Space Telescope, we can see galaxies about 15 billion light-years away from us. This means that the creation process is not stopping yet and at the same time the universe might have started to de-create itself, since the big bang started, due to intense convergent gravitational forces from all the newly created debris of matter (e.g., galaxies and dark matter). To wrap up this section, we would stress that one of the viable aspects of Eq. (15) is the transformation from a spatially dimensionless equation to a space-time function (i.e., ∇·S); it describes how our universe was created with a huge explosion. Furthermore, the essence of Eq. (15) is not just a piece of mathematical formula; it is a symbolic representation, a description, a language, a picture, or even a video as may be seen from its presentation. We can visualize how our universe was created, from the theory of relativity to Einstein's energy equation and then to temporal space creation.

10. Trading time and subspace

Let us now take one of the simplest connections between physical subspace and time [15]:

$$d = \mathrm{v}t \tag{16}$$

where d is the distance, v is the velocity, and t is the time variable. Notice that this equation may be one of the most profound connections between time and physical space (or temporal space). Therefore, a three-dimensional (Euclidean) physical (or temporal) subspace can be described by

$$(dx, dy, dz) = (\mathrm{v}x, \mathrm{v}y, \mathrm{v}z)t \tag{17}$$

where ($\mathrm{v}x$, $\mathrm{v}y$, $\mathrm{v}z$) are the velocities' vectors and t is the time variable. Under the current laws of science, the speed of light is the limit. Then, by replacing the velocity vectors equal to the speed of light c, a temporal space can be written as

$$(dx, dy, dz) = (ct, ct, ct) \tag{18}$$

Thus, we see that time can be traded for space and space cannot be traded for time, since time is a forward variable (i.e., $t > 0$). In other words, once a section of time Δt is expended, we cannot get it back. Needless to say, a spherical temporal space can be described by

$$\mathrm{r} = ct \tag{19}$$

where radius r increases at the speed of light. Thus, we see that the boundary (i.e., edge) of our universe is determined by radius r, which is limited by the light speed, as illustrated in a composite temporal space diagram of **Figure 1**. In view of this figure, we see that our universe is expanding at the speed of light well beyond the current observable galaxies. **Figure 2** shows a discrete temporal space diagram, in which we see that the size of our universe is continuously expanding as time moves forward (i.e., $t > 0$). Assuming that we have already accepted the big bang creation, sometime in the future (i.e., billions of light-years later), our universe will eventually stop expanding and then start to shrink back, preparing for the next cycle of the big bang explosion. The forces for the collapsing universe are mainly due to the intense gravitational field, mostly from giant black holes and matter that were derived from merging (or swallowing) with smaller black holes and other debris (i.e., physical substances). Since a black hole's gravitational field is so intense, even light cannot escape; however, a black hole is by no means an infinite energy reservoir. Eventually, the storage capacity of a black hole will reach a limit for explosion, as started for the mass to energy and debris creation.

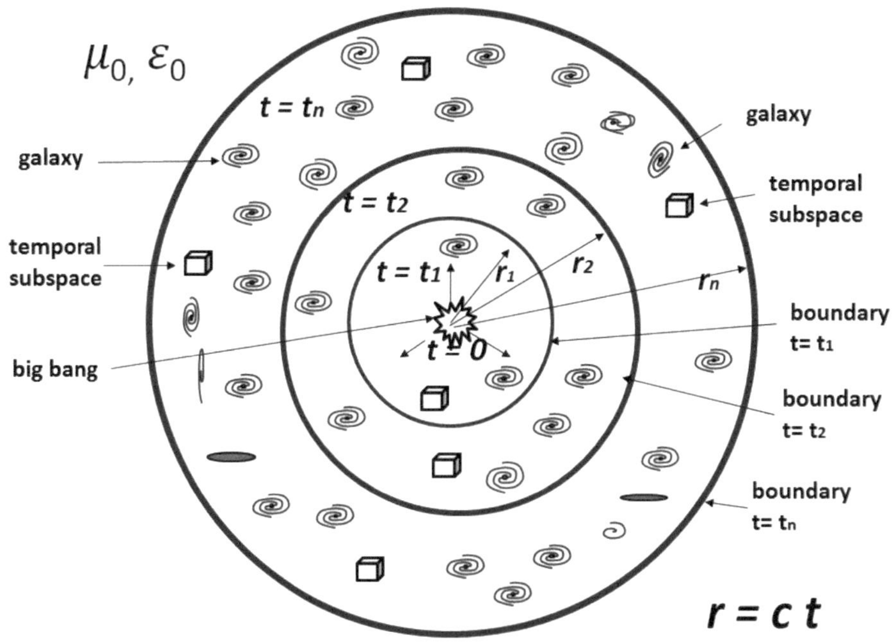

Figure 1.
Composite temporal space universe diagram. r = ct, r is the radius of our universe, t is time, c is the velocity of light, and ε_0 and μ_0 are the permittivity and permeability of the space.

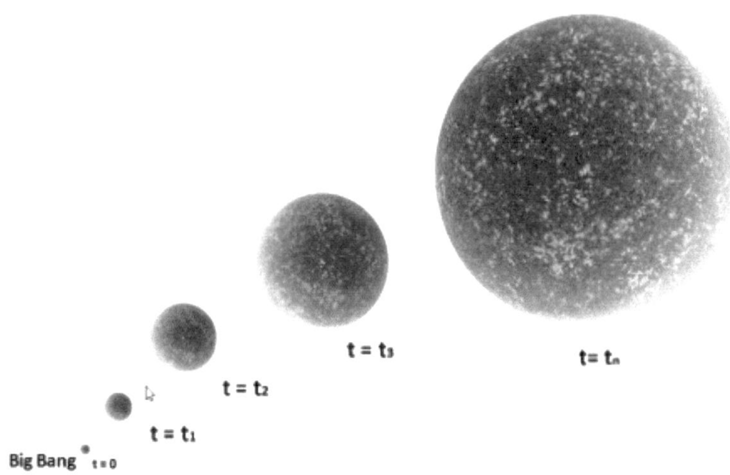

Figure 2.
Discrete temporal universe diagrams; t is time.

In other words, there will be one dominant giant black hole within the shrinking universe, to initiate the next cycle of universe creation. Therefore, every black hole can be treated as a convergent energy sink, which relies on its intense gravitation field to collect all the debris of matter and energies. Referring to the big bang creation, a gigantic energy explosion was the major reason for the universe's creation. In fact, it can be easily discerned that the creating process has never slowed down since the birth of our universe, as we see that our universe is still continuingly expanding even today. This is by no means an indication that all the debris created came from the big bang's energy (e.g., mc^2); there might have been some leftover debris from a preceding universe. Therefore, the overall energy within our universe cannot be restricted to just the amount that came from the big bang creation. In fact, the conversion processes between mass and energy have never been totally absent since the birth of our universe, but they are on a much smaller scale. In fact, right after birth, our universe started to slow down the divergent process due to the gravitational forces produced by the created matter. In other words, the universe will eventually reach a point when overall divergent forces will be weaker than the convergent forces, which are mostly due to gravitational fields coming from the newly created matter, including black holes. As we had mentioned earlier, our universe currently has about 23% dark matter, which represents about 23% of the gravitational fields within the current universe. The intense localized gravitational field could have been produced from a group or a giant black hole, derived from merging with (or swallowing up) some smaller black holes, nearby dark matter, and debris. Since a giant black hole is not an infinite energy sink, eventually it will explode for the next cycle of universal creation. And it is almost certain that the next big bang creation will not occur at the same center of our present universe. One can easily discern that our universe will never shrink to a few inches in size, as commonly speculated. It will, however, shrink to a smaller size until one of the giant black holes (e.g., swallowed-up sufficient physical debris) reaches the big bang explosive condition to release its gigantic energy for the next cycle of universal creation. The speculation of a possible collapsing universe remains to be observed. Nonetheless, we have found that our universe is still expanding, as observed by the Doppler shifts of the distant galaxies at the edge of our universe, about 15 billion light-years away [12–14]. This tells us that our universe has not reached its half-life yet. In fact, the expansion has never stopped since the birth of our universe, and our universe has also been started to de-create since the big bang started, which is primarily due to convergent gravitational forces from the newly created debris (e.g., galaxies, black holes, and dark matter).

11. Relativistic time and temporal (t > 0) space

Relativistic time at a different subspace within a vast universal space may not be the same as that based on Einstein's special theory of relativity [1]. Let us start with the relativistic time dilation as given by

$$\Delta t' = \frac{\Delta t}{\sqrt{1 - v^2/c^2}} \tag{20}$$

where $\Delta t'$ is the relativistic time window, compared with a standstill subspace, Δt is the time window of a standstill subspace, v is the velocity of a moving subspace, and c is the velocity of light. We see that time dilation $\Delta t'$ of the moving subspace, relative to the time window of the standstill subspace Δt, appears to be wider as velocity increases. For example, a 1-s time window Δt is equivalent to the 10-s relative time window $\Delta t'$. This means that a 1-s time expenditure within the moving subspace is relative to about a 10-s time expenditure within the standstill subspace. Therefore, for the species living in an environment that travels closer to the speed of light (e.g., at the edge of the universe), their time appears to be slower than ours, as illustrated in **Figure 3**. In this figure, we see an old man traveling at a speed closer to the velocity of light; his relative observation time window appears to be wider as he is looking at us, and the laws of science within his subspace may not be the same as ours.

Two of the most important pillars in modern physics must be Einstein's relativity theory and Schrödinger's quantum mechanics [15]. One is dealing with very large objects (e.g., universe), and the other is dealing with very small particles

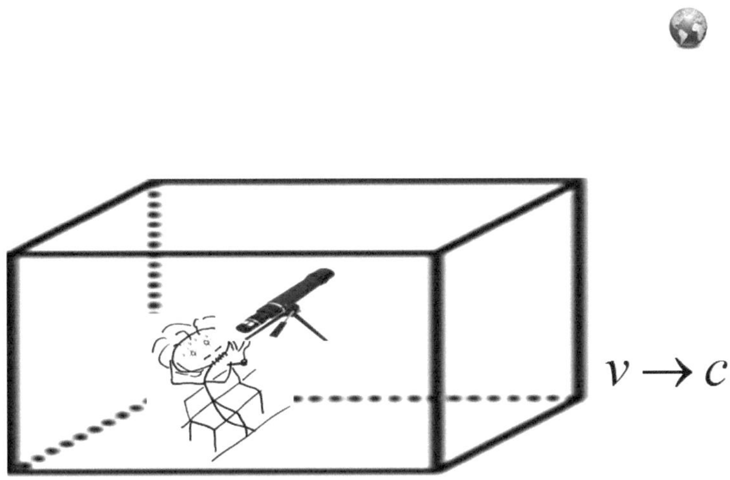

Figure 3.
Effects on relativistic time.

(e. g., atoms). Yet, there exists a profound connection between them, by means of the Heisenberg's uncertainty principle [16]. In view of the uncertainty relation, we see that every temporal subspace takes a section of time Δt and an amount of energy ΔE to create. Since we cannot create something from nothing, everything needs an amount of energy ΔE and a section of time Δt to make it happen. By referring to the Heisenberg uncertainty relation as given by

$$\Delta E \cdot \Delta t \geq h \tag{21}$$

where h is the Planck's constant. We see that every subspace is limited by ΔE and Δt. In other words, it is the h region, but not the shape, that determines the boundary condition. For example, the shape can be either elongated or compressed, as long as it is larger than the h region.

Incidentally, the uncertainty relationship of Eq. (21) is also the limit of reliable bit information transmission as pointed out by Gabor in [17]. Nonetheless, the connection with the special theory of relativity is that the creation of a subspace near the edge of our universe will take a short relative time with respect to our planet earth, since Δt' > Δt. The "relativistic" uncertainty relationship within the moving subspace, as with respect to a standstill subspace, can be shown as

$$\Delta E \cdot \Delta t'\left[1-(v/c)^2\right]^{½} \geq h \tag{22}$$

where we see ΔE energy is conserved. Thus a narrower time-width can be achieved as with respect to standstill subspace. It is precisely possible that one can exploit for time-domain digital communication, as from ground station to satellite information transmission.

On the other hand, as from satellite to ground station information transmission, we might want to use digital bandwidth (i.e., Δv) instead. This is a frequency-domain information transmission strategy, as in contrast with time domain, which has not been exploited yet. The "relativistic" uncertainty relationship within the standstill subspace as with respect to the moving subspace can be written as

$$\frac{\Delta E \Delta t}{\sqrt{1-\left(\frac{v}{c}\right)^2}} \geq h \tag{23}$$

DOI: http://dx.doi.org/10.5772/intechopen.86785

Or equivalently we have

$$\frac{\Delta v \bullet \Delta t}{\sqrt{1-\left(\frac{v}{c}\right)^2}} \geq 1 \tag{24}$$

in which we see that a narrower bandwidth Δv can be in principle use for frequency-domain communication.

12. Time and physical space

Every physical (or temporal) subspace is created by substances (i.e., energy and mass), and substances coexist with time. In this context, we see that our universe was essentially created by time and energy and the universe is continuously evolving (i.e., changing) with time. Although relativistic time may not be the same at the different subspaces within our universe, the rule of time may remain the same. As for the species living closer to the speed of light, relativistic time may not be noticeable to them, but their laws of science within their subspace may be different from ours. Nonetheless, our universe was simultaneously created by time with a gigantic energy explosion. Since our universe cannot be embedded in an empty space, it must be embedded in a more complex space that remains to be found. From an inclusive point of view, mass is energy or energy is mass, which was discovered by Einstein almost a century ago [1]. And it is this basic fundamental law of physics that we have used for investigating the origin of time. Together with a huge energy explosion (i.e., big bang theory [11]), time is the igniter for the creation of our universe. As we know, without the existence of time, the creation of our universe would not have happened. As we have shown, time can be traded for space, but space cannot be traded for time. Our universe is in fact a temporal physical subspace, and it is continuously evolving or changing with time (i.e., $t > 0$). Although every temporal subspace is created by time (and substances), it is not possible for us to trade any temporal subspace for time. Since every physical substance has a life, our universe (a gigantic substance) cannot be excluded. With reference to the report from a recent Hubble Space Telescope observation [12–14], we are capable of viewing galaxies about 15 billion light-years away and have also learned that our universe is still by no means slowing down in expansion. In other words, our universe has still not reached its half-life, based on our estimation. As we have shown, time ignited the creation of our universe, yet the created physical substances presented to us the existence of time.

13. Essence of our temporal (i.e., t > 0) universe

In view of the preceding discussion, we see that our universe is a time-invariant system (i.e., from system theory stand point); as in contrast with an empty space, it is a not a time-invariant system and it is a timeless or no-time space. We see that timeless solution cannot be directly implemented within our universe. Since science is a law of approximation and mathematics is an axiom of absolute certainty, using exact math to evaluate inexact science cannot guarantee its solution to exist within our temporal (i.e., $t > 0$) universe. One important aspect of temporal universe is that one cannot get something from nothing: There is always a price to pay; every piece of temporal subspace (or every bit of information [7]) takes an amount of energy (i.e., ΔE) and a section of time (i.e., Δt) to create. And the subspace [i.e., $f(x, y, z; t), t > 0$] is a forward time-variable function. In other words, time and subspace coexist or are mutually inclusive. This is the boundary condition and constrain of our temporal universe [i.e., $f(x, y, z; t), t > 0$], in which every existence within our universe has to comply with this condition. Otherwise it is not existing within our universe, unless new law emerges since laws are made to be broken. Thus we see that any emerging science has to be proven to exist within our temporal universe [i.e., $f(x, y, z; t), t > 0$]. Otherwise it is a fictitious science, unless it can be validated by repeated experiments.

In mathematics, we see that the burden of a postulation is first to prove if there exists a solution and then search for a solution. Although we hardly have had, there is an existent burden in science. Yet, we need to prove that a scientific postulation is existing within our temporal universe [i.e., $f(x, y, z; t), t > 0$]; otherwise it is not real or virtual as mathematics is. For example such as the superposition principle in quantum mechanics, in which we have proven [18] it is not existed within our temporal universe (i.e., $t > 0$), since Schrödinger's quantum mechanics is timeless as mathematics is.

There is however an additional constrain as imposed by our temporal universe which is the affordability. As we have shown that everything (e.g., any physical subspace) existed within our universe has a price tag, in terms of an amount of energy ΔE and a section of time Δt (i.e., ΔE, Δt). To be precise, the price tag also includes an amount of "intelligent" information ΔI or an equivalent amount of entropy ΔS (i.e., ΔE, Δt, ΔI) [7]. For example, creation of a piece of simple facial tissue will take a huge amount of energy ΔE, a section of time Δt, and an amount of information ΔI (i.e., equivalent amount of entropy ΔS). We note that on this planet Earth, only humans can make it happen. Thus we see that every physical subspace (or equivalently substance) within our universe has a price tag (i.e., ΔE, Δt, ΔS), and the question is that can we afford it?

14. Are we not alone?

Within our universe, we can easily estimate there were billions and billions of civilizations that had emerged and faded away in the past 15 billion light-years. Our civilization is one of the billions and billions of current consequences within our universe, and it will eventually disappear. We are here, and will be here, for just a very short moment. Hopefully, we will be able to discover substances that travel well beyond the limit of light before the end of our existence, so that a better observational instrument can be built. If we point the new instrument at the right place, we may see the edge of our universe beyond the limit of light. We are not alone with almost absolute certainty. By using the new observational equipment, we may find a planet that once upon a time had harbored a civilization for a period of twinkle thousands of (Earth) years.

15. Remarks

We have shown that time is one of the most intriguing variables in the universe. Without time, there would be no physical substances, no space, and no life. With reference to Einstein's energy equation, we have shown that energy and mass can be traded. In other words, mass is equivalent to energy, and energy is equivalent to mass, for which all mass can be treated as an energy reservoir. We have also shown that a physical space cannot be embedded in an absolute empty space or a timeless (i.e., $t = 0$) space, and it cannot even have any absolute empty subspace in it. In reality, every physical space has to be fully packed with physical substances (i.e., energy and mass). Since no physical space can be embedded in an absolute empty space, it is reasonable to assume that our universe is a subspace within a more complex space, which remains to be found. In other words, our universe could have been one of the many universes outside our universal boundary, which comes and goes like bubbles. We have also shown that it takes time to create a physical space and the time that has been used for the creation cannot be brought back. Since all physical substances exist with time, all physical spaces are created by time and substances (i.e., energy and mass). This means that our cosmos was created by time and a gigantic energy explosion, in which we see that every substance coexists with time. That is, without time, the creation of physical substances would not have happened. We have further noted that our universe is in a temporal space and it is still expanding based on current observation. This shows that our universe has not reached its half-life yet, as we have accepted the big bang creation. And it is noted that we are not alone with almost absolute certainty. Someday, we may find a planet that once upon a time had harbored a civilization for a period of light-years. We have further shown that the burden of a scientific postulation is to prove it exists within our temporal universe [i.e., $f(x, y, z; t)$, $t > 0$]; otherwise it is not real or virtual as mathematics is.

Finally, I would like to take this opportunity to say a few words on behalf of Professor Stephen Hawking, who passed away last year on March 14, 2018. Professor Hawking was a world-renowned astrophysicist, a respected cosmic scientist, and a genius. Although the creation of temporal universe started with the same root of the big bang explosion, it is not a subspace of Professor Hawking's universe. You may see from the preceding presentation that the creation of temporal universe is somewhat different from Hawking's creation. One of the major differences may be at the origin of big bang creation. My temporal universe was started with a big bang creation within a "non-empty" space, instead within of an empty space which was normally assumed.

Author details

Francis T.S. Yu
Emeritus Evan Pugh (University), Penn State University, University Park, Pennsylvania, USA

*Address all correspondence to: fty1@psu.edu

References

[1] Einstein A. Relativity, the Special and General Theory. New York: Crown Publishers; 1961

[2] Yu FTS. Gravitation and radiation. Asian Journal of Physics. 2016;**25**(6): 789-795

[3] Einstein A. Zur elektrodynamik bewegter koerper. Annals of Physics. 1905;**17**:891-921

[4] Kraus JD. Electro-Magnetics. New York: McGraw-Hill Book Company; 1953. p. 370

[5] Bartrusiok M. Black Hole. New Haven, CT: Yale University Press; 2015

[6] Abell GO, Morrison D, Wolff SC. Exploration of the Universe. 5th ed. New York: Saunders College Publishing; 1987. pp. 47-88

[7] Yu FTS. Science and the myth of information. Asian Journal of Physics. 2015;**24**(24):1823-1836

[8] Yu FTS. Optics and Information Theory. New York: Wiley-Interscience; 1976

[9] Amendola L, Tsujikawa S. Dark Energy: Theory and Observation. Cambridge: Cambridge University Press; 2010

[10] Bertone G, editor. Particle Dark Matter: Observation, Model and Search. Cambridge: Cambridge University Press; 2010

[11] Bartrusiok M, Rubakov VA. Introduction to the Theory of the Early Universe: Hot Big Bang Theory. Princeton, NJ: World Scientific Publishing; 2011

[12] Bennett JO, Donahue MO, Voit M, Schneider N. The Cosmic Perspective Fundamentals. Cambridge, MA: Addison Wesley Publishing; 2015

[13] Zimmerman R. The Universe in a Mirror: The Saga of the Hubble Space Telescope. Princeton, NJ: Princeton Press; 2016

[14] Yu FTS. Time, space, information and life. Asian Journal of Physics. 2015; **24**(2):217-223

[15] Schrödinger E. An undulatory theory of the mechanics of atoms and molecules. Physical Review. 1926;**28**(6): 1049

[16] Heisenberg W. Über den anschaulichen Inhalt der quantentheoretischen Kinematik und Mechanik. Zeitschrift für Physik. 1927; **43**:172

[17] Gabor D. Communication theory and physics. Philosophical Magazine. 1950;**41**(7):1161

[18] Yu FTS. The fate of Schrodinger's Cat. Asian Journal of Physics. 2019; **28**(1):63-70

Chapter 7

Hot Compression Tests Using Total Lagrangian SPH Formulation in Energy-Based Framework

Kadiata Ba

Abstract

Limitations of the finite element method (FEM) in some cases involving large deformations as in forging or high compression tests are overcome nowadays by meshless methods such as the smoothed particle hydrodynamic (SPH) method. This paper presents a corrected total Lagrangian SPH formulation for problems encountering large deformations in solid mechanics. The continuum is modeled as a Hamiltonian system of particles (energy-based framework). The total Lagrangian formulation proposed overcomes some problems faced by standard SPH in simulating solid mechanic problems such as tensile instability. Numerical applications compared with experimental results are presented to show the capabilities of this novel formulation.

Keywords: SPH, Hamiltonian system, total Lagrangian, thermomechanical, solid mechanics

1. Introduction

The use of the smoothed particle hydrodynamic (SPH) method [1–5] (**Figure 1**) in solid mechanics is quite recent (in the 1990s) compared to the SPH fluid formulation. Libersky and Petschek [6] and Libersky et al. [1] are cited as the first to use SPH in solid mechanics, for impacts modeling at high speeds and phenomena of rupture, perforation, and fragmentation. As SPH is a meshless method, there is no mesh to distort; it can efficiently handle large deformations. SPH is an efficient numerical method for applications in forging processes [7], machining [8–10], and welding [11]. Classical approach is widely used to describe SPH equations but faces drawbacks such as lack of completeness and tensile instability (numerical fragmentation). Total Lagrangian corrected SPH formulation is then used to fix the cited problems. In this paper, a Hamiltonian formulation is proposed for dynamic and steady-state problem simulation focusing on numerical efficiency such as accuracy and simulation time. The governing equations are derived following a Lagrangian variational principle leading to a Hamiltonian system of particles [12–14]. With the Hamiltonian SPH formulation, local conservation of momentum between particles is respected, and they remain locally ordered during the process as wanted in solid mechanic problems.

Total Lagrangian formulation reduces also the computational time by avoiding the search for neighboring particles for the construction of the kernel function at

IntechOpen

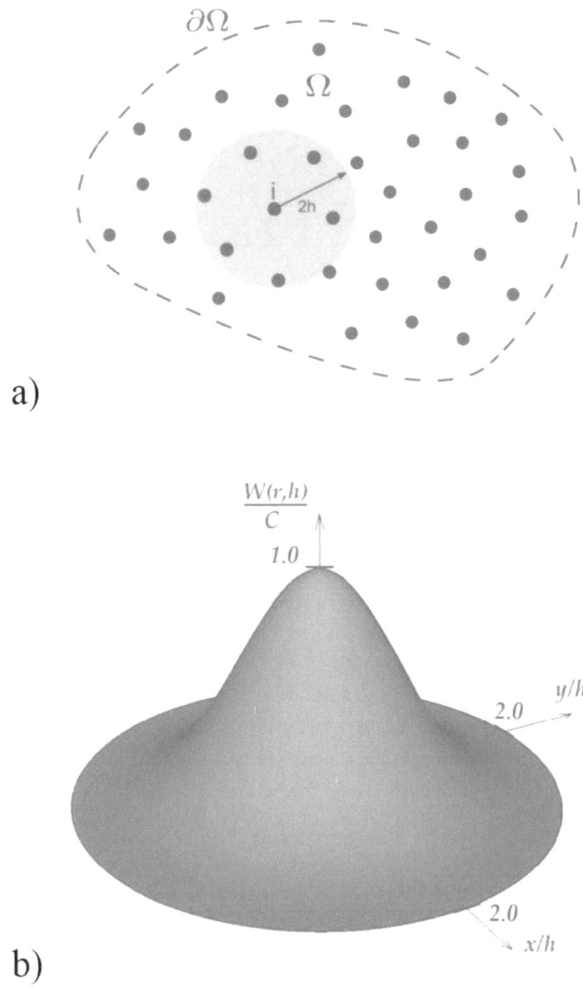

Figure 1.
(a) Schematic representation of the discretization of the domain Ω by a set of points i [15] and (b) seen in the space of a B-spline [16].

each time step. Through axial and lateral compression tests, the accuracy of the new formulation is shown. Results are compared to a classical formulation based on differential equations for solid mechanic applications.

2. Discrete equations of motion from energy-based formulation

The governing equations are derived following a Lagrangian variational principle leading to a Hamiltonian system of particles (energy-based) [12, 17–19] where the motion of each particle is given by the classical Lagrange equations. Therefore, as explained by Bonet et al. [18], constants of the motion such as linear and angular momentum are conserved.

For each particle, the physical quantities are calculated through interpolation over neighbor particles. Every particle is considered as a moving thermodynamic subsystem [12]. The volume of each particle is given by

$$V_i = m_i/\rho_i \tag{1}$$

where m_i is the mass of the particle and ρ_i its density.

To proceed with a variational formulation of the equations of motion of the continuum, the kinetic K, internal $\prod_{int}$, and external energy $\prod_{ext}$ of the system need to be defined.

With dissipative effects such as plasticity, the equations of motion of the system of particles representing the continuum can be evaluated following the classical Lagrangian formalism. For more details, readers can refer to Ba and Gakwaya [12]:

$$\frac{d}{dt}\frac{\partial L}{\partial \mathbf{v}_i} - \frac{\partial L}{\partial \mathbf{x}_i} = -\frac{\partial \Pi_{dis}}{\partial \mathbf{v}_i} \tag{2}$$

where $\mathbf{x}_i$ and $\mathbf{v}_i$ are the spatial position and velocity of the particle i, Π_{dis} is the dissipative energy, and L is the Lagrangian given by

$$L(\mathbf{x}_i, \mathbf{v}_i) = K(\mathbf{v}_i) - \Pi_{ext}(\mathbf{x}_i) - \Pi_{int}(\mathbf{x}_i) \tag{3}$$

By substituting Eq. (3) into Eq. (2), it leads to

$$\frac{d}{dt}\frac{\partial K}{\partial \mathbf{v}_i} = -\frac{\partial \Pi_{ext}}{\partial \mathbf{x}_i} - \frac{\partial \Pi_{int}}{\partial \mathbf{x}_i} - \frac{\partial \Pi_{dis}}{\partial \mathbf{v}_i} \tag{4}$$

The total kinetic energy of the system can be approximated as the sum of the kinetic energy of each particle:

$$K = \frac{1}{2}\sum_i m_i(\mathbf{v}_i.\mathbf{v}_i) \tag{5}$$

For a common case where the external forces result from a gravitational field **g**, the total external energy is

$$\Pi_{ext} = -\sum_i m_i(\mathbf{x}_i.\mathbf{g}) \tag{6}$$

The total internal energy can be expressed as the sum of the products of particle masses by the amount of energy accumulated per unit mass π that depends on the deformation, density, or other constitutive parameters:

$$\Pi_{int} = \sum_i m_i \pi(\rho_i, \ldots) \tag{7}$$

The dissipative energy can also be expressed as

$$\Pi_{disp} = \sum_{i=1} m_i \pi_{disp}(\mathbf{d}) \tag{8}$$

where π_{disp} is the dissipative energy per unit mass and **d** is the rate of deformation tensor.

$$\mathbf{d} = \frac{1}{2}\left(\nabla\mathbf{v} + \nabla\mathbf{v}^T\right) \tag{9}$$

with the velocity gradient given by

$$\nabla\mathbf{v}_i = \sum_j \frac{m_j}{\rho_j}(\mathbf{v}_j - \mathbf{v}_i)\nabla W(\mathbf{x}_i - \mathbf{x}_j, h) \tag{10}$$

where $W(\mathbf{x}_i - \mathbf{x}_j, h)$ is the SPH kernel function and h is the smoothing length.

3. Corrected total Lagrangian SPH formulation for solid mechanics

Total Lagrangian formulation [20, 21] is well suited for solid mechanic problems as the SPH particles change less often their neighbors than in fluid mechanics [12]. The SPH kernels and their gradients are then expressed in the initial configuration (material coordinates $\mathbf{X}$ are used). The proposed corrected kernel is to address the lack of completeness and interpolation consistency; the smoothing length h is considered as a functional variable in the calculation of the gradient of the kernel function ∇W [19].

Lagrangian and spatial coordinates are connected through the gradient of deformation tensor $\mathbf{F}$:

$$\mathbf{F} = \frac{\partial \mathbf{x}}{\partial \mathbf{X}} = \frac{\partial(\mathbf{X} + \mathbf{u})}{\partial \mathbf{X}} \tag{11}$$

where $\mathbf{u}$ is the displacement of a material point.

The expression of the corrected gradient of deformation tensor $\mathbf{F}$, in total Lagrangian formulation, is given by

$$\langle \mathbf{F}_i \rangle = \left(-\sum_j (\mathbf{u}_j - \mathbf{u}_i) \otimes \nabla_{\mathbf{X}_j} W(\mathbf{X}_i - \mathbf{X}_j, h_0) V_j^0 \right) \mathbf{B} + I \tag{12}$$

where $\nabla_{\mathbf{X}_j}$ is the gradient with respect to a material point $\mathbf{X}$, V_j^0 is the initial volume of particle j, h_0 is the initial smoothing length, and $\mathbf{I}$ is the identity matrix.

$\mathbf{B}$ is the expression of the correction of the gradient expressed as [20]

$$\mathbf{B} = \left(\sum_j \frac{m_j}{\rho_j} (\mathbf{X}_i - \mathbf{X}_j, h_0) \otimes \nabla_{\mathbf{X}_i} W(\mathbf{X}_i - \mathbf{X}_j, h_0) \right)^{-1} \tag{13}$$

The corrected mass conservation equation for particle i is

$$\rho_{0i} = \rho_i J = \rho_i \det \mathbf{F}_i \tag{14}$$

where J and ρ_0 are the Jacobian and the initial density.

The corrected momentum equation for a particle i is

$$\langle \mathbf{a}_i \rangle = \left(-\sum_j (\mathbf{P}_j - \mathbf{P}_i) \otimes \nabla_{\mathbf{X}_j} \tilde{W}(\mathbf{X}_i - \mathbf{X}_j, h_0) V_j^0 + \mathbf{f}_i \right) : \mathbf{B} \tag{15}$$

where $\mathbf{a}$, $\tilde{W}$, and $\mathbf{f}_i$ are the acceleration, the normalized smoothing function, and the body force.

$\mathbf{P}$ is the first Piola-Kirchhoff stress.

$$\mathbf{P} = J \boldsymbol{\sigma} \mathbf{F}^{-T} \tag{16}$$

where $\boldsymbol{\sigma}$ is the Cauchy stress tensor and $\mathbf{F}^{-T}$ is the inverse of the transpose of the gradient of deformation tensor.

The corrected energy conservation equation for particle i is

$$\langle \dot{\mathbf{e}}_i \rangle = \mathbf{P}_j : \left[\begin{pmatrix} -\sum_j \frac{m_j}{\rho_i \rho_j} (\mathbf{v}_j - \mathbf{v}_i, h_0) \otimes \nabla_{\mathbf{X}_i} W(\mathbf{X}_i - \mathbf{X}_j, h_0) V_j^0 \\ +k\nabla T_i + r_{pl} \end{pmatrix} \mathbf{B} \right] \tag{17}$$

where $\dot{\mathbf{e}}$ is the energy rate, r_{pl} is the mechanical contribution (heat generated by the plastic dissipation), k is the conductivity, and T is the temperature of the particle.

The equation of motion and the equation of the thermal energy of each particle can be put after discretization and evaluation of all the interactions in the following forms:

$$\mathbf{a}_i = \ddot{\mathbf{u}}_i = \frac{1}{\mathbf{m}_i} \left(\mathbf{f}_{ext(i)} - \mathbf{f}_{\text{int}(i)} \right) \tag{18}$$

$$\dot{\mathbf{T}}_i = \frac{1}{\mathbf{C}_i} \left(\mathbf{h}^k_{\text{ext}(i)} - \mathbf{h}^k_{\text{int}(i)} \right) \tag{19}$$

where $\mathbf{f}_{\text{int}(i)}$ and $\mathbf{f}_{ext(i)}$ are the internal and external forces, $\mathbf{h}^k_{\text{int}(i)}$ and $\mathbf{h}^k_{\text{ext}(i)}$ are the internal and the external heat flux, and $\mathbf{C}$ is the capacitance matrix.

The expression for the internal force for a given particle can be expressed by differentiating the internal energy per unit mass with respect to the nodal positions as

$$\mathbf{f}_{\text{int}(i)} = \sum_{j=1}^{n} V_j^0 \mathbf{P}_j \mathbf{G}_i(\mathbf{X}_j) \tag{20}$$

where $\mathbf{G}$ is the gradient function and contains the corrected kernel gradients $\nabla^{\sim W}$ at the initial reference configuration.

$$\mathbf{G}_i(\mathbf{X}_j) = V_i \overset{\sim 0}{\nabla} \mathbf{W}_i(\mathbf{X}_j) \tag{21}$$

Internal heat flux can be expressed as

$$\mathbf{h}_{\text{int}(i)} - \mathbf{k}_i \mathbf{T}_i \tag{22}$$

where $\mathbf{k}$ is the heat conductivity matrix and $\mathbf{T}$ the vector of nodal temperatures.

Explicit finite difference method is used to solve numerically the differential equation presented in this section through explicit dynamic algorithm to update the velocity, position, and temperature of each SPH particle.

4. Temporal integration scheme

A typical integration scheme used for integrating SPH equations is the leapfrog algorithm (**Figure 2**), an extension of the Verlet algorithm with low storage memory during computation.

The heat transfer equations are integrated using the explicit forward-difference time integration rule [22].

$$\mathbf{T}_{(t+\Delta t)} = \mathbf{T}_{(t)} + \Delta t_{(t+1)} \dot{\mathbf{T}}_{(t)} \tag{23}$$

$\dot{\mathbf{T}}_t$ is computed at the beginning of the increment by

$$\dot{\mathbf{T}}_{(t)} = \mathbf{C}^{-1}\left(\mathbf{h}^{t}_{ext} - \mathbf{h}^{t}_{\text{int}}\right) \tag{24}$$

The stability time is given by

$$\Delta t_T \approx \frac{\Delta r^2_{\text{min}}}{2\alpha} \tag{25}$$

where Δr_{min} is the smallest interparticle distance and α is the diffusivity of the material.

$$\alpha = \frac{k}{\rho s} \tag{26}$$

where k is the conductivity and s is the specific heat.

For the mechanical part, an explicit central-difference integration rule is used to integrate the equation of motion. The nodal accelerations $\ddot{u}$ at time t is given by

$$\ddot{\mathbf{u}}_{(t)} = \mathbf{M}^{-1}\left(\mathbf{P}_{(t)} - \mathbf{I}_{(t)}\right) \tag{27}$$

where $\mathbf{M}$, $\mathbf{P}_{(t)}$, and $\mathbf{I}_{(t)}$ represent the mass matrix and the external and internal forces.

The integration leads to the nodal velocity $\dot{\mathbf{u}}$ (Eq. 28) and the nodal displacement $\mathbf{u}$ (Eq. 29).

$$\dot{\mathbf{u}}_{\left(t+\frac{\Delta t}{2}\right)} = \dot{\mathbf{u}}_{\left(t-\frac{\Delta t}{2}\right)} + \frac{\left(\Delta t_{(t+\Delta t)} + \Delta t_{(t)}\right)}{2}\ddot{\mathbf{u}}_{(t)} \tag{28}$$

$$\mathbf{u}_{(t+\Delta t)} = \mathbf{u}_{(t)} + \Delta t_{(t+\Delta t)}\dot{\mathbf{u}}_{\left(t+\frac{\Delta t}{2}\right)} \tag{29}$$

The stable time is calculated as follows:

$$\Delta t = \min\left(\frac{L_e}{c_d}\right) \tag{30}$$

where L_e and c_d are, respectively, the characteristic length of the element and the dilatational wave speed of the material.

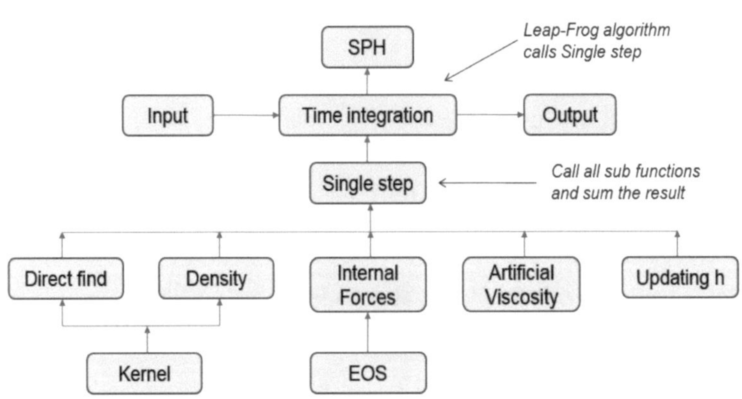

Figure 2.
SPH code structure [23].

The whole thermomechanical problem is solved by explicit coupling; both the forward-difference (for the thermal problem) and central-difference (for the mechanical problem) integrations are explicit.

The structure of the SPH code is described below (**Figure 2**). The time integration routine is the main subroutine. It calculates the new variables (density, acceleration, external force, internal force).

5. Material behavior

Johnson-Cook model [24–26] is used in this work, and the flow stress is expressed as follows:

$$\sigma_f = [A + B(\varepsilon_p)^n]\left[1 + C\ln\left(\frac{\dot{\varepsilon_p}}{\dot{\varepsilon_{p0}}}\right)\right]\left(1 - \left(\frac{T - T_r}{T_m - T_r}\right)^m\right) \tag{31}$$

where ε is the plastic strain, $\dot{\varepsilon}$ (s^{-1}) is the plastic strain rate, $\dot{\varepsilon}_0(s^{-1})$ is the reference plastic strain rate, T_m is the melting temperature, T_r is the reference temperature, T is the current temperature, A is the yield stress, B is the coefficient of strain hardening, C is the coefficient of strain rate hardening, n is the strain hardening exponent, and m is the thermal softening exponent.

The material used for the simulations (see Section 6) is an Al-Zn-Mg-Cu aluminum alloy. The Johnson-Cook material parameters are shown in **Table 1**.

A (MPa)	B (MPa)	c	n	m	$\dot{\varepsilon}_0$ (s^{-1})	Tm (°C)	Tr (°C)
420	465	0.862	0.5088	0.081	0.1	641	25

Table 1.
Johnson-Cook material parameters [12].

6. Applications

6.1 Axial compression test

A cylindrical sample (diameter, 25.4 mm; length, 25.4 mm) was subjected to the uniaxial compression test at constant velocity (2.54 mm s^{-1}) and high temperature (400°C). Both experimental and numerical tests were performed (**Figure 3**). The aim of this test is to demonstrate the efficiency of the proposed total Lagrangian SPH formulation. We compared the numerical stress-strain curve with the

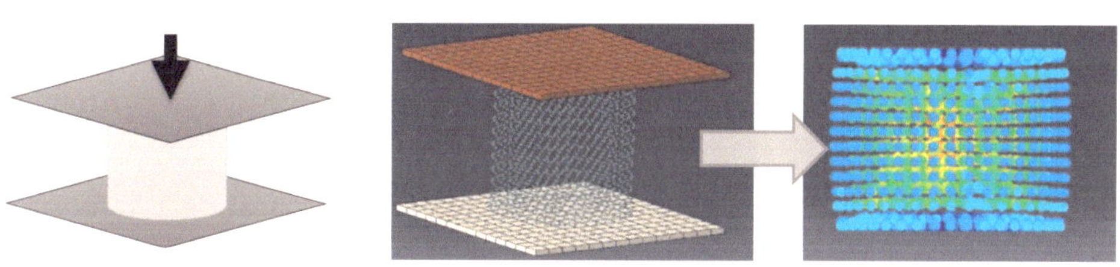

Figure 3.
Axial compression test setup in SPH.

experimental ones to verify the accuracy and the stability of the code (see **Figure 4**). To confirm the validity of the experimental result, the tests were repeated three times.

6.2 Lateral compression test

Lateral compression test is performed in this section (**Figure 5**). Eulerian and Lagrangian simulation time are compared, and tensile instability is verified. The test is carried out at 30 mm s^{-1}, and cylindrical sample (diameter 25.4 mm, length 25.4 mm) with 5313 particles was used. This is a case of a large deformation test; the

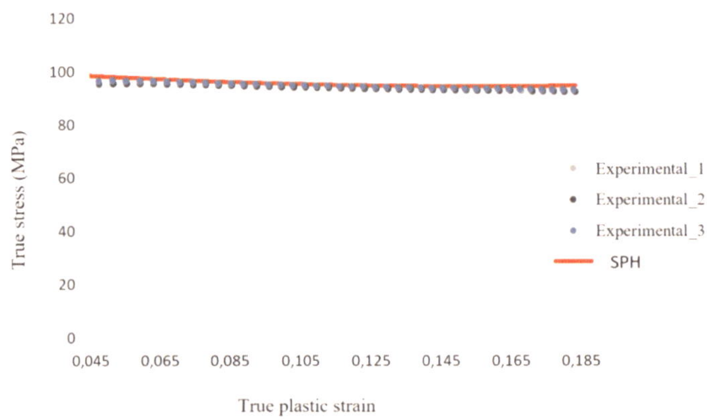

Figure 4.
Stress-strain curves: experimental vs. SPH.

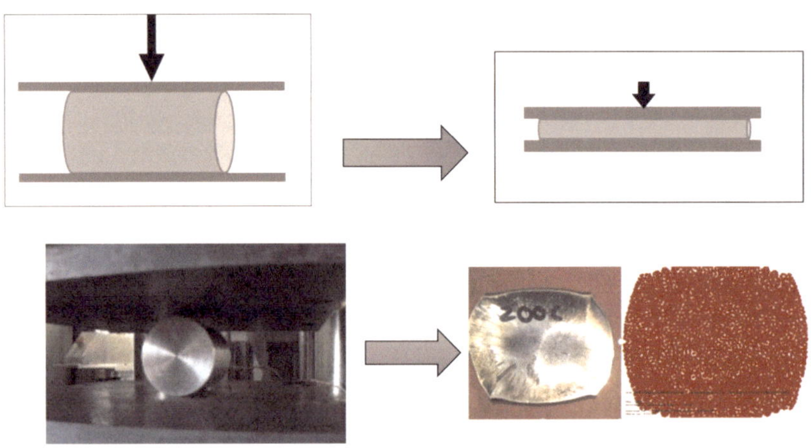

Figure 5.
Lateral compression test: Eulerian vs. Lagrangian.

	Simulation time	Number of particles
Eulerian SPH	4 h 04 min	5313
Lagrangian SPH	1 h 36 min	5313

Table 2.
Simulation time comparison.

initial diameter of the sample was reduced over 50% during the test. See comparison of results at **Table 2**.

6.3 Discussion

Figures 4 and 5 and **Table 2** gather the tests results. From **Figure 4** (axial compression test), we can see that the SPH result is very accurate compared to the experimental ones. Less than 5% of error is noted between the curves. The simulated sample shows no clustered particles, meaning there is no tensile instability.

Figure 5 and **Table 2** show the results of the lateral compression test and confirm the previous result. Even in very large deformation test, particles keep their initial neighbors and do not suffer from tensile instability. In addition, the simulation time is very interesting compared to classical SPH formulation; simulation time is reduced drastically (from 4 h 04 min to 1 h 36 min); a good numerical efficiency is reached.

7. Conclusion

A corrected SPH particle approximation in energy-based framework is presented. Stability (no tensile instability), accuracy, and fast result production are shown leading to the conclusion that the total Lagrangian SPH formulation is very well suited to simulate solid mechanic problems. This is particularly interesting in simulating large deformation problems with physical fragmentation where the numerical fragmentation (tensile instability) will not corrupt the results.

Acknowledgements

The author wishes to acknowledge Augustin Gakwaya for his appreciated help.

Author details

Kadiata Ba
Department of Applied Sciences, Université du Québec à Chicoutimi (UQAC), Québec, Canada

*Address all correspondence to: kadiata_ba@uqac.ca

References

[1] Fries TP, Matthies HG. Classification and Overview of Meshfree Methods. Brunswick, Germany: Technical University Braunschweig; 2004

[2] Li S, Liu WK. Meshfree and particle methods and their applications. Applied Mechanics Reviews. 2002;**55**(1):1-34

[3] Liu GR, Liu MB. Smoothed Particle Hydrodynamics, a Meshfree Particle Method. Livre: Springer; 2005

[4] Libersky L, Petschek A, Carney T, Hipp J, Allahdadi F. High strain Lagrangian hydrodynamics. Journal of Computational Physics. 1993;**109**:67-75

[5] Wolf S. Méthode Sans Maillage. Laboratoire de mathématiques appliquées Aux systèmes; 2007

[6] Libersky L, Petschek A. Smooth particle hydrodynamics with strength of materials. In: Advances in the Free-Lagrange Method Including Contributions on Adaptive Gridding and the Smooth Particle Hydrodynamics Method. Springer; 1991. pp. 248-257

[7] Cleary PW, Prakash M, Das R, Ha J. Modelling of metal forging using SPH. Applied Mathematical Modelling. 2012; **36**:3836-3855

[8] Limido J, Espinosa C. Modélisation numérique de la Coupe Orthogonale en Ugv. National Conference Proceedings; 2005

[9] Limido J, Espinosa C, Salaun M, Lacome JL. SPH method applied to high speed cutting modelling. International Journal of Mechanical Sciences. 2007; **49**(7):898-908

[10] Limido J. Étude de l'effet de l'usinage grande vitesse sur la tenue en fatigue de pièces aéronautiques [Thèse de doctorat]. Université de Toulouse; 2008

[11] Timesli A. Simulation du soudage par friction et malaxage à l'aide de méthodes sans maillage [PhD dissertation]. Université de Lorraine; 2013

[12] Ba K, Gakwaya A. Thermomechanical total Lagrangian SPH formulation for solid mechanics in large deformation problems. Computer Methods in Applied Mechanics and Engineering. 2018;**342**(1):458-473

[13] Grenier N, Antuono M, Colagrossi A, Le Touzé D, Alessandrini B. An Hamiltonian interface SPH formulation for multi-fluid and free surface flows. Journal of Computational Physics. 2009; **228**:8380-8393

[14] Hu XY, Adams NA. A multi-phase SPH method for macroscopic and mesoscopic flows. Journal of Computational Physics. 2006;**213**: 844-861

[15] Caleyron F, Combescure A, Faucher V, Potapov S. Une méthode sans maillage pour la modélisation des interactions fluide-structure: Application à la rupture d'un réservoir sous impact. 10e Colloque National en Calcul des Structures, 9-13 Mai 2011, Presqu'île de Giens (Var); 2011

[16] Maurel B. Modélisation par la méthode SPH de l'impact d'un réservoir rempli de fluide [thèse de doctorat]. Institut National des Sciences Appliquées de Lyon; 2008

[17] Bonet J, Rogríguez-Paz MX. Hamiltonian formulation of the variable-h SPH equations. Journal of Computational Physics. 2005;**209**: 541-558

[18] Bonet J, Kulasegaram S, Rodriguez-Paz MX, Profit M. Variational formulation for the smooth particle hydrodynamics (SPH) simulation of fluid and solid problems. Computer

Methods in Applied Mechanics and Engineering. 2004;**193**:1245-1256

[19] Lavoie MA, Gakwaya A, Nejad Ensan M. Variable-h and energy conserving SPH formulation with application in aerospace engineering. Journal Mathematics in Engineering, Science and Aerospace. 2010;**1**(1):27-70

[20] Reveles J. Development of a total lagrangian SPH code for the simulation of solids under dynamic loading [PhD dissertation]. Cranfield University; 2007

[21] Vignjevic R. Review of development of the smooth particle hydrodynamics (SPH) method. In: Predictive Modeling of Dynamic Processes. UK: Cranfield University; 2009. pp. 367-396

[22] Abaqus documentation v6.14

[23] Loosveldt C, Watrin D. Alternative Numerical Methods in Continuum Mechanics, Smoothed Particle Hydrodynamics. Université de Liège; 2010

[24] Johnson GR, Cook WH. A constitutive model and data for metals subjected to large strains, high strain rates and high temperatures. In: Seventh International Symposium on Ballistics; The Hague, The Netherlands. 1983. pp. 541-547

[25] Hor A. Simulation physique des conditions thermomécanique de forgeage et d'usinage-caractérisation et modélisation de la rhéologie et de l'endommagement [Thèse de doctorat]. École Nationale Supérieure d'Arts et Métiers; 2011

[26] Lin YC, Li L-T, Fu Y-X, Jiang Y-Q. Hot compressive deformation behavior of 7075 Al alloy under elevated temperature. Journal of Materials Science. 2012;**47**(3):1306-1318

Chapter 8

Dynamics of Biostructures on a Fractal/Multifractal Space-Time Manifold

Maricel Agop, Calin Buzea, Decebal Vasincu and Daniel Timofte

Abstract

A theory of space-time is built on a fractal/multifractal variety. Thus, considering that both the spatial coordinates and the time are fractal/multifractal, it is shown that both the energy and the non-differentiable mass of any biostructure depend on both the "state" of the biostructure and a speed limit of constant value. For the dynamics on Peano fractal/multifractal curves and Compton scale resolutions, it is shown that our results are reduced to those of Einstein relativity. In such a context, it has been shown that the "chameleon effect" of cholesterol corresponds to the HDL-LDL state transfer dictated by the spontaneous symmetry breaking through a fractal/multifractal tunnel effect. Then both HDL and LDL become distinct states of the same biostructure as in nuclear physics where proton and neutron are distinct states of the same nucleon.

Keywords: fractal/multifractal tunnel effect, biostructures, cholesterol, spontaneous symmetry breaking, chameleon effect

1. Mathematical model

1.1 Time as a fractal/multifractal

Analyzing the nonrelativistic dynamics of a particle in a fractal/multifractal space [1–4], we observe a big discrepancy between the space coordinates and the temporal one (considered as affine parameter of motion curve). If the space coordinates are fractal/multifractal, the temporal coordinate is not a fractal/multifractal. This discrepancy has an important consequence: the particle travels on an infinite length curve in a finite time span, and so, it has an infinite velocity. In order to eliminate this contradiction, in the following we will assume that not only the space coordinates are fractal/multifractal but also the temporal one is a fractal/multifractal. Practically, we shall build dynamics of biostructures on a non-differentiable space-time manifold. In this framework, the most important elements from the nonrelativistic approach of scale relativity theory with arbitrary constant fractal dimension, as described in [5–7], remain valid, but the time differential element dt is now replaced by the proper time differential element $d\tau$. In this way, not only the space but the entire space-time continuum is considered to be non-differentiable and, therefore, fractal/multifractal.

IntechOpen

1.2 Consequences of non-differentiability on a space-time manifold

Let us suppose that on a space-time manifold, the motions of biostructures take place on continuous but non-differentiable curves (in particular fractal/multifractal curves). The non-differentiability of motion curves implies the following [2]:

(i) Any continuous but non-differentiable curve is explicitly scale dependent (which will be referred as $\delta\tau$). In other words, its length tends to infinity when its proper time interval, $\Delta\tau$, tends to zero (an extension of the Lebesgue theorem on a space-time manifold). Consequently, in this limit, a curve in a space-time manifold is zigzagged as one can imagine. Thus, it exhibits the property of self-similarity in all its points of a space-time manifold, which can be translated into an extension property of holography (every part of a space-time manifold reflects the whole of the same space-time manifold).

Then a continuous but non-differentiable space-time is fractal/multifractal in Mandelbrot's sense:

(ii) The differential proper time reflection invariance of any variable is broken. For example, the proper time derivative of four-coordinate X^μ, where $\mu = 0$, 1, 2, 3, can be written two fold:

$$\left[\frac{dX^\mu}{d\tau}\right]_+ = \lim_{\Delta\tau\to 0_+}\frac{X^\mu(\tau+\Delta\tau)-X^\mu(\tau)}{\Delta\tau}$$
$$\left[\frac{dX^\mu}{d\tau}\right]_- = \lim_{\Delta\tau\to 0_-}\frac{X^\mu(\tau)-X^\mu(\tau-\Delta\tau)}{\Delta\tau} \tag{1}$$

These relations are equivalent in the differentiable case, $\Delta\tau \to -\Delta\tau$. In the non-differentiable case, the previous definitions fail since the limits $\Delta\tau \to 0\pm$ are no longer defined. In the approach of the non-differentiable model, the biophysical phenomena are related to the behavior of the function during the "zoom" operation on the proper time resolution $\delta\tau$: then, by means of the substitution principle, $\delta\tau$ will be identified with the differential element $d\tau$, i.e., $\delta\tau \equiv d\tau$, and will be considered as independent variable. Thus, every classical variable $Q(\tau)$ is replaced by the non-differentiable variable $Q(\tau,d\tau)$ explicitly dependent on the proper time resolution interval whose derivative is undefined only in the limit, $\Delta\tau \to 0$. As a consequence, two derivatives of every non-differentiable variable as explicit functions of τ and $d\tau$ will be defined. For example, the two derivatives of the four-coordinate $X^\mu(\tau,\Delta\tau)$ takes the following form:

$$\frac{d_+X^\mu}{d\tau} = \lim_{\Delta\tau\to 0_+}\frac{X^\mu(\tau+\Delta\tau,\Delta\tau)-X^\mu(\tau,\Delta\tau)}{\Delta\tau}$$
$$\frac{d_-X^\mu}{d\tau} = \lim_{\Delta\tau\to 0_-}\frac{X^\mu(\tau,\Delta\tau)-X^\mu(\tau-\Delta\tau,\Delta\tau)}{\Delta\tau} \tag{2}$$

The sign + corresponds to the forward biophysical process and the sign − to the backward one:

(iii) The differential of four-coordinate $dX^\mu(\tau,\Delta\tau)$ can be expressed as the sum of two differentials, one not scale dependent (differentiable part $d_\pm x^\mu(\tau)$) and other scale dependent (non-differentiable part $d_\pm\xi^\mu(\tau,d\tau)$), i.e.,

$$d_\pm X^\mu(\tau,\Delta\tau) = d_\pm x^\mu(\tau) + d_\pm\xi^\mu(\tau,\Delta\tau) \tag{3}$$

(iv) The non-differentiable part of the four-coordinate satisfies the non-differentiable equation

$$d_{\pm}\xi^{\mu}(\tau, \Delta\tau) = \lambda_{\pm}^{\mu}(d\tau)^{1/D_F} \tag{4}$$

where $\lambda_{\pm}{}^{\mu}$ are constant coefficients whose statistical significance will be given in what follows and D_F is the fractal dimension of the motion curves from the space-time manifold.

In our opinion, the complexity of the biophysical processes implies dynamics on geodesics with various fractal dimensions. Precisely, $D_F = 2$ is a characteristic to the biophysical processes of quantum type, $D_F < 2$ is a characteristic to the biophysical processes of correlative type, while $D_F > 2$ is a characteristic to the biophysical processes of non-correlative type. Since such dynamics simultaneously are operational on a given biophysical system, the space-time manifold will exhibit multifractal type properties [2].

(v) The differential proper time reflection invariance is recovered by combining the derivatives $d_+/d\tau$ and $d_-/d\tau$ in the non-differentiable operator:

$$\frac{\hat{d}}{d\tau} = \frac{1}{2}\left(\frac{d_+ + d_-}{d\tau}\right) - \frac{i}{2}\left(\frac{d_+ - d_-}{d\tau}\right) \tag{5}$$

This specific procedure is called, according to [8], "differentiability by extension in complex on a space-time manifold" (Cresson's theorem). Applying now the non-differentiable operator to the four-coordinate X^{μ} yields the complex velocity:

$$\hat{V}^{\mu} = \frac{\hat{d}X^{\mu}}{d\tau} = \frac{1}{2}\left(\frac{d_+X^{\mu} + d_-X^{\mu}}{d\tau}\right) - \frac{i}{2}\left(\frac{d_+X^{\mu} - d_-X^{\mu}}{d\tau}\right)$$
$$= \frac{1}{2}\left(\frac{d_+x^{\mu} + d_-x^{\mu}}{d\tau} + \frac{d_+\xi^{\mu} + d_-\xi^{\mu}}{d\tau}\right) - \frac{i}{2}\left(\frac{d_+x^{\mu} - d_-x^{\mu}}{d\tau} + \frac{d_+\xi^{\mu} - d_-\xi^{\mu}}{d\tau}\right) = V^{\mu} - iU^{\mu} \tag{6}$$

with

$$V^{\mu} = \frac{1}{2}\left(v_+^{\mu} + v_-^{\mu}\right), U^{\mu} = \frac{1}{2}\left(v_+^{\mu} - v_-^{\mu}\right), v_+^{\mu} = \frac{d_+x^{\mu} + d_+\xi^{\mu}}{d\tau}, v_-^{\mu} = \frac{d_-x^{\mu} + d_-\xi^{\mu}}{d\tau} \tag{7}$$

The real part V^{μ} is differentiable and scale resolution independent, while the imaginary one U^{μ} is non-differentiable and scale resolution dependent.

(vi) An infinite number of geodesics can be found relating any pair of points of a space-time manifold, and this is true on all scale resolutions of the dynamics of biostructures. Then, in the space-time manifold, all the entities of the biostructures are substituted with the geodesics themselves so that any external constraint can be interpreted as a selection of geodesics in the same space-time manifold. The infinity of geodesics in the bundle, their non-differentiability, the two values of the derivative, etc., imply a generalized statistical fluidlike description (fractal/multifractal fluid). In this way, one provides the fractalization/multifractalization type through stochastic processes. From such a perspective, averages, variances, covariances, etc. of the fractal/multifractal fluid variables (by means of which now we can describe the dynamics of the biostructures) must be considered in the sense of the stochastic process associated with fractalization/multifractalization. In such a context, the choice of the average of $d_{\pm}X^i$ in the form

$$\langle d_{\pm}X^{i}\rangle \equiv d_{\pm}x^{i} \tag{8}$$

implies through (3)

$$\langle d_{\pm}\xi^{i}\rangle = 0 \tag{9}$$

1.3 Motion non-differentiable operator on a space-time manifold

Let us now consider that the movement curves (continuous and non-differentiable) are immersed in the space-time and that X^{μ} are the four coordinates of a point on the curve. We also consider a variable $Q(X^{\mu},\tau)$ and the following Taylor expansion, up to the second order

$$d_{\pm}Q(X^{\mu},\tau,d\tau) = \partial_{\tau}Qd\tau + \partial_{\mu}Qd_{\pm}X^{\mu} + \frac{1}{2}\partial_{\mu}\partial_{\nu}Qd_{\pm}X^{\mu}d_{\pm}X^{\nu} \tag{10}$$

where

$$\partial_{\tau} = \frac{\partial}{\partial_{\tau}},\ \partial_{\mu} = \frac{\partial}{\partial X^{\mu}},\ \partial_{\mu}\partial_{\nu} = \frac{\partial^{2}}{\partial X^{\mu}\partial X^{\nu}}$$

Relations (10) are valid in any point of the space-time manifold and more for the points "X^{μ}" on the non-differentiable curve which we have selected in relation (10).

From here, forward and backward average values of (10) become

$$\langle d_{\pm}Q(X^{\mu},\tau,d\tau)\rangle = \langle\partial_{\tau}Qd\tau\rangle + \langle\partial_{\mu}Qd_{\pm}X^{\mu}\rangle + \frac{1}{2}\partial_{\mu}\partial_{\nu}Q\langle d_{\pm}X^{\mu}d_{\pm}X^{\nu}\rangle \tag{11}$$

We make the following stipulations: the average values of the variables $Q(X^{\mu},\tau,d\tau)$ and its derivatives coincide with themselves, and the differentials $d_{\pm}X^{\mu}$ and $d\tau$ are independent. Therefore, the average of their products coincides with the product of their averages. In these conditions, (11) takes the form

$$d_{\pm}Q(X^{\mu},\tau,d\tau) = \partial_{\tau}Qd\tau + \partial_{\mu}Q\langle d_{\pm}X^{\mu}\rangle + \frac{1}{2}\langle\partial_{\mu}\partial_{\nu}Qd_{\pm}X^{\mu}d_{\pm}X^{\nu}\rangle \tag{12}$$

or using (3), (8), and (9)

$$d_{\pm}Q(X^{\mu},\tau,d\tau) = \partial_{\tau}Qd\tau + \partial_{\mu}Qd_{\pm}x^{\mu} + \frac{1}{2}\partial_{\mu}\partial_{\nu}Q(d_{\pm}x^{\mu}d_{\pm}x^{\nu} + \langle d_{\pm}\xi^{\mu}d_{\pm}\xi^{\nu}\rangle) \tag{13}$$

Even the average values of the 4-non-differentiable coordinate $d_{\pm}\xi^{\mu}$ is null, for the higher order of the four-non-differentiable coordinate average, the situation can be different. Let us focus now on the mean $<d_{\pm}\xi^{\mu}d_{\pm}\xi^{\nu}>$. Using (4), we can write

$$\langle d_{\pm}\xi^{\mu}d_{\pm}\xi^{\nu}\rangle = \pm\lambda_{\pm}^{\mu}\lambda_{\pm}^{\nu}(d\tau)^{(2/D_{F}-1)}d\tau \tag{14}$$

using the convention that the sign + corresponds to $d\tau > 0$, while the sign − corresponds to $d\tau < 0$.

Then (13) takes the form:

$$d_{\pm}Q(X^{\mu},\tau,d\tau) = \partial_{\tau}Qd\tau + \partial_{\mu}Qd_{\pm}x^{\mu} + \frac{1}{2}\partial_{\mu}\partial_{\nu}Qd_{\pm}x^{\mu}d_{\pm}x^{\nu} \pm \frac{1}{2}\partial_{\mu}\partial_{\nu}Q\lambda_{\pm}^{\mu}\lambda_{\pm}^{\nu}(d\tau)^{(2/D_{F}-1)}d\tau \tag{15}$$

If we divide by $d\tau$ and neglect the terms that contain differential factors, using the method from [5–7], we obtain:

$$\frac{d_{\pm}Q(X^{\mu},\tau,d\tau)}{d\tau}=\partial_{\tau}Q+\nu_{\pm}^{\mu}\partial_{\mu}Q\pm\frac{1}{2}\lambda_{\pm}^{\mu}\lambda_{\pm}^{\nu}(d\tau)^{(2/D_F-1)}\partial_{\mu}\partial_{\nu}Q \tag{16}$$

These relations also allow us to define the operators:

$$\frac{d_{\pm}}{d\tau}=\partial_{\tau}+\nu_{\pm}^{\mu}\partial_{\mu}\pm\frac{1}{2}\lambda_{\pm}^{\mu}\lambda_{\pm}^{\nu}(d\tau)^{(2/D_F-1)}\partial_{\mu}\partial_{\nu} \tag{17}$$

Under these circumstances, let us calculate $\hat{d}/d\tau$. Taking into account (5), (6), and (17), we obtain:

$$\begin{aligned}\frac{\hat{d}Q}{d\tau}&=\frac{1}{2}\left[\left(\frac{d_{+}Q+d_{-}Q}{d\tau}\right)-i\left(\frac{d_{+}Q-d_{-}Q}{d\tau}\right)\right]\\&=\partial_{\tau}Q+\hat{V}^{\mu}\partial_{\mu}Q+\frac{1}{4}(d\tau)^{(2/D_F-1)}D^{\mu\nu}\partial_{\mu}\partial_{\nu}Q\end{aligned} \tag{18}$$

where

$$\begin{gathered}D^{\mu\nu}=d^{\mu\nu}-i\bar{d}^{\mu\nu}\\ d^{\mu\nu}=\lambda_{+}^{\mu}\lambda_{+}^{\nu}-\lambda_{-}^{\mu}\lambda_{-}^{\nu},\bar{d}^{\mu\nu}=\lambda_{+}^{\mu}\lambda_{+}^{\nu}+\lambda_{-}^{\mu}\lambda_{-}^{\nu},i=\sqrt{-1}\end{gathered} \tag{19}$$

The relation also allows us to define the motion non-differentiable operator:

$$\frac{\hat{d}}{d\tau}=\partial_{\tau}+\hat{V}^{\mu}\partial_{\mu}+\frac{1}{4}(d\tau)^{(2/D_F-1)}D^{\mu\nu}\partial_{\mu}\partial_{\nu} \tag{20}$$

If the non-differentiability of motion curves is realized through Markov type stochastic process [2, 4].

$$\lambda_{+}^{\mu}\lambda_{+}^{\nu}=\lambda_{-}^{\mu}\lambda_{-}^{\nu}=-\lambda\eta^{\mu\nu} \tag{21}$$

where $\eta^{\mu\nu}$ is the Minkowski metric and λ is the coefficient associated with the differentiable-non-differentiable transition, then the motion non-differentiable operator takes the form

$$\frac{\hat{d}}{d\tau}=\partial_{\tau}+\hat{V}^{\mu}\partial_{\mu}+i\frac{\lambda}{2}(d\tau)^{(2/D_F-1)}\partial_{\mu}\partial^{\mu} \tag{22}$$

If the non-differentiability of motion curves is realized through non-Markov type stochastic process [2, 4].

$$\begin{aligned}\lambda_{+}^{\mu}\lambda_{+}^{\nu}-\lambda_{-}^{\mu}\lambda_{-}^{\nu}&=\lambda_1\eta^{\mu\nu}\\ \lambda_{+}^{\mu}\lambda_{+}^{\nu}+\lambda_{-}^{\mu}\lambda_{-}^{\nu}&=-\lambda_2\eta^{\mu\nu}\end{aligned} \tag{23}$$

where λ_1 and λ_2 are two coefficients associated with the differentiable-non-differentiable transition, then the motion non-differentiable operator takes the form

$$\frac{\hat{d}}{d\tau}=\partial_{\tau}+\hat{V}^{\mu}\partial_{\mu}+\frac{1}{4}(\lambda_1+i\lambda_2)(d\tau)^{(2/D_F-1)}\partial_{\mu}\partial^{\mu} \tag{24}$$

1.4 Non-differentiable geodesics on a space-time manifold

In what follows, let us consider the functionality of the scale covariance principle [5–7]: the biophysics laws are simultaneously invariant both with respect to the four-coordinate transformation and with respect to scale transformations. Then the passage from differentiable biophysics in a space-time manifold to the non-differentiable biophysics in a same space-time, manifold which is considered here, can be implemented by replacing the standard derivative $d/d\tau$ by the non-differentiable operator $\hat{d}/d\tau$. This operator plays the role of a "covariant derivative operator," namely, it is used to write the fundamental equations of dynamics of biostructures under the same form as in the classical (differentiable) case. Thus, applying the operator (20) to the complex velocity (6), the geodesics equation becomes:

$$\frac{\hat{d}\hat{V}^{\mu}}{d\tau} = \partial_{\tau}\hat{V}^{\mu} + \hat{V}^{\nu}\partial_{\nu}\hat{V}^{\mu} + \frac{1}{4}(d\tau)^{(2/D_F-1)}D^{\alpha\beta}\partial_{\alpha}\partial_{\beta}\hat{V}^{\mu} \equiv 0 \tag{25}$$

or, also using (6), through separation of motions on scale resolutions (the real part from the imaginary one):

$$\begin{aligned}\frac{\hat{d}V^{\mu}}{d\tau} &= \partial_{\tau}V^{\mu} + V^{\nu}\partial_{\nu}V^{\mu} - U^{\nu}\partial_{\nu}U^{\mu} + \frac{1}{4}(d\tau)^{(2/D_F-1)}d^{\alpha\beta}\partial_{\alpha}\partial_{\beta}V^{\mu}\\ &\quad - \frac{1}{4}(d\tau)^{(2/D_F-1)}\overline{d}^{\alpha\beta}\partial_{\alpha}\partial_{\beta}U^{\mu} = 0\\ \frac{\hat{d}U^{\mu}}{d\tau} &= \partial_{\tau}U^{\mu} + V^{\nu}\partial_{\nu}U^{\mu} + U^{\nu}\partial_{\nu}V^{\mu} + \frac{1}{4}(d\tau)^{(2/D_F-1)}d^{\alpha\beta}\partial_{\alpha}\partial_{\beta}U^{\mu}\\ &\quad + \frac{1}{4}(d\tau)^{(2/D_F-1)}\overline{d}^{\alpha\beta}\partial_{\alpha}\partial_{\beta}V^{\mu} = 0\end{aligned} \tag{26}$$

For motions on non-differentiable curves realized through Markov type stochastic process [1, 2, 4], the geodesics equation takes the form

$$\frac{\hat{d}\hat{V}^{\mu}}{d\tau} = \partial_{\tau}\hat{V}^{\mu} + \hat{V}^{\nu}\partial_{\nu}\hat{V}^{\mu} - i\frac{\lambda}{2}(d\tau)^{(2/D_F-1)}\partial^{\nu}\partial_{\nu}\hat{V}^{\mu} = 0 \tag{27}$$

or through separation of motions on scale resolutions:

$$\begin{aligned}\frac{\hat{d}V^{\mu}}{d\tau} &= \partial_{\tau}V^{\mu} + V^{\nu}\partial_{\nu}V^{\mu} - \left(U^{\nu} - \frac{\lambda}{2}(d\tau)^{(2/D_F-1)}\partial^{\nu}\right)\partial_{\nu}U^{\mu} = 0\\ \frac{\hat{d}U^{\mu}}{d\tau} &= \partial_{\tau}U^{\mu} + V^{\nu}\partial_{\nu}U^{\mu} + \left(U^{\nu} - \frac{\lambda}{2}(d\tau)^{(2/D_F-1)}\partial^{\nu}\right)\partial_{\nu}V^{\mu} = 0\end{aligned} \tag{28}$$

For motions on non-differentiable curves realized through non-Markov type stochastic process [1, 2, 4], the geodesics equation becomes

$$\frac{\hat{d}\hat{V}^{\mu}}{d\tau} = \partial_{\tau}\hat{V}^{\mu} + \hat{V}^{\nu}\partial_{\nu}\hat{V}^{\mu} + \frac{1}{4}(\lambda_1 + i\lambda_2)^{(2/D_F-1)}\partial_{\nu}\partial^{\nu}\hat{V}^{\mu} = 0 \tag{29}$$

or through separation of motions on scale resolutions:

$$\frac{\hat{d}V^{\mu}}{d\tau} = \partial_{\tau}V^{\mu} + V^{\nu}\partial_{\nu}V^{\mu} - \left(U^{\nu} - \frac{\lambda_2}{4}(d\tau)^{(2/D_F-1)}\partial^{\nu}\right)\partial_{\nu}U^{\mu} + \frac{\lambda_1}{4}(d\tau)^{(2/D_F-1)}\partial_{\nu}\partial^{\nu}V^{\mu} = 0$$
$$\frac{\hat{d}U^{\mu}}{d\tau} = \partial_{\tau}U^{\mu} + V^{\nu}\partial_{\nu}U^{\mu} + \left(U^{\nu} - \frac{\lambda_2}{4}(d\tau)^{(2/D_F-1)}\partial^{\nu}\right)\partial_{\nu}V^{\mu} + \frac{\lambda_1}{4}(d\tau)^{(2/D_F-1)}\partial_{\nu}\partial^{\nu}U^{\mu} = 0 \tag{30}$$

1.5 Non-differentiable geodesics in terms of the scalar complex field on a space-time manifold

Let us choose $\hat{V}^{\mu}$ in terms of the scalar complex field Ψ:

$$\hat{V}^{\alpha} = i\lambda(d\tau)^{(2/D_F-1)}\partial_{\alpha}\ln\Psi \tag{31}$$

Then the geodesics equation (27) becomes

$$\frac{\hat{d}\hat{V}_{\alpha}}{d\tau} = \lambda(d\tau)^{(2/D_F-1)}\partial_{\tau}\partial_{\alpha}\ln\Psi$$
$$+\left[i\lambda(d\tau)^{(2/D_F-1)}\partial^{\mu}\ln\Psi + i\frac{\lambda}{2}(d\tau)^{(2/D_F-1)}\partial^{\mu}\right]\partial_{\mu}\partial_{\alpha}\left[i\lambda(d\tau)^{(2/D_F-1)}\ln\Psi\right] = 0 \tag{32}$$

Since

$$\partial_{\alpha}\left(\partial_{\mu}\ln\Psi\partial^{\mu}\ln\Psi\right) = 2\partial^{\mu}\ln\Psi\partial_{\alpha}\partial_{\mu}\ln\Psi$$
$$\partial_{\alpha}\partial_{\mu}\partial^{\mu}\ln\Psi = \partial^{\mu}\partial_{\mu}\partial_{\alpha}\ln\Psi$$
$$\partial_{\alpha}\left(\partial_{\mu}\ln\Psi\partial^{\mu}\ln\Psi + \partial_{\mu}\partial^{\mu}\ln\Psi\right) = \partial_{\alpha}\left(\frac{\partial_{\mu}\partial^{\mu}\Psi}{\Psi}\right) \tag{33}$$

Equation (32) takes the form:

$$i\lambda(d\tau)^{(2/D_F-1)}\partial_{\tau}\partial_{\alpha}\ln\Psi + \lambda^2(d\tau)^{(4/D_F-2)}\partial_{\alpha}\left(\frac{\partial_{\mu}\partial^{\mu}\Psi}{\Psi}\right) = 0 \tag{34}$$

By integrating the above relation, we obtain:

$$\lambda^2(d\tau)^{(4/D_F-2)}\partial_{\mu}\partial^{\mu}\Psi + i\lambda(d\tau)^{(2/D_F-1)}\partial_{\tau}\Psi + F^2(\tau)\Psi = 0 \tag{35}$$

where $F^2(\tau)$ is an arbitrary function depending on τ.

Consequently, the non-differentiable geodesics (35) in terms of Ψ are well defined up to an arbitrary function $F^2(\tau)$ depending on τ. A particular form of $F^2(\tau)$ can be obtained, for instance, based on a correspondence with the standard Klein-Gordon equation.

1.6 Non-differentiable geodesics in terms of Klein-Gordon equation of fractal/multifractal type

If Ψ is independent on τ, i.e., $\partial_{\tau}\Psi = 0$ and $F^2(\tau) = V_0{}^2$ = const., with V_0 a limit velocity with constant value, the geodesics (35) become the Klein-Gordon equation of fractal/multifractal type

$$\partial_\mu \partial^\mu \Psi + \frac{1}{\overline{\Lambda}^2} \Psi \equiv 0 \tag{36}$$

with

$$\overline{\Lambda} = \overline{\Lambda}_0 (d\tau)^{(2/D_F - 1)}, \Lambda_0 = \frac{\lambda}{V_0} \tag{37}$$

From (37) it results in a scale resolution dependence of the fundamental length $\overline{\Lambda}$, where $\overline{\Lambda}_0$ is the fundamental unscaled length. For relativistic motions on Peano curves, D_F = 2, at Compton scale $\overline{\Lambda}_0 = \lambda / V_0 \equiv \ /(m_0 c), \lambda = \ /m_0, V_0 \equiv c$ with $\hbar$ the reduced Planck constant, m_0 the rest mass of the biophysical system entity, and c the velocity light in vacuum, (37) takes the usual form of Klein-Gordon equation:

$$\partial_\mu \partial^\mu \Psi + \left(\frac{m_0 c}{\hbar} \right)^2 \Psi \equiv 0$$

1.7 Non-differentiable specific potential force and energy

Using the explicit form of the function, $\Psi = \sqrt{\rho} e^{iS}$, where $\sqrt{\rho}$ is an amplitude and S is a phase, the expression of U_α becomes

$$U_\alpha = -\lambda \partial_\alpha \ln \sqrt{\rho} \tag{38}$$

Thus it results in

$$\left[U_\mu - \frac{\lambda}{2} (d\tau)^{(2/D_F - 1)} \partial_\mu \right] \partial^\mu U_\alpha = \lambda^2 (d\tau)^{(4/D_F - 2)} \left(\partial^\mu \ln \sqrt{\rho} \partial_\mu \partial_\alpha \ln \sqrt{\rho} + \frac{1}{2} \partial^\mu \partial_\mu \partial_\alpha \ln \sqrt{\rho} \right) \tag{39}$$

Since the identities from (33) work in variable $\ln \sqrt{\rho}$, Eq. (39) becomes

$$\begin{aligned} \left[U_\mu - \frac{\lambda}{2} (d\tau)^{(2/D_F - 1)} \partial_\mu \right] \partial^\mu U_\alpha &= \frac{\lambda^2}{2} (d\tau)^{(4/D_F - 2)} \partial_\alpha \left(\partial^\mu \ln \sqrt{\rho} \partial_\mu \ln \sqrt{\rho} + \partial^\mu \partial_\mu \ln \sqrt{\rho} \right) \\ &= \lambda^2 (d\tau)^{(4/D_F - 2)} \partial_\alpha \left(\frac{\partial^\mu \partial_\mu \sqrt{\rho}}{\sqrt{\rho}} \right) \end{aligned} \tag{40}$$

which implies through the specific non-differentiable potential

$$\begin{aligned} Q &= \frac{\lambda^2}{2} (d\tau)^{(4/D_F - 2)} \frac{\partial^\mu \partial_\mu \sqrt{\rho}}{\sqrt{\rho}} = \frac{1}{2} U^\mu U_\mu - \lambda (d\tau)^{(2/D_F - 1)} \partial^\mu U_\mu = \\ &= \lambda^2 (d\tau)^{(4/D_F - 2)} \partial_\alpha \left(\frac{\partial^\mu \partial_\mu \sqrt{\rho}}{\sqrt{\rho}} \right) \end{aligned} \tag{41}$$

the specific non-differentiable force

$$F_\alpha = \frac{\lambda^2}{2} (d\tau)^{(4/D_F - 2)} \partial_\alpha \left(\frac{\partial^\mu \partial_\mu \sqrt{\rho}}{\sqrt{\rho}} \right) = \left[U^\mu - \frac{\lambda}{2} (d\tau)^{(2/D_F - 1)} \partial^\mu \right] \partial^\mu U_\alpha \tag{42}$$

Thus, the first equation (28) takes the form

$$\frac{\hat{d}V_\alpha}{d\tau} = \partial_\tau V_\alpha + V^\mu \partial_\mu V_\alpha = \frac{\lambda^2}{2}(d\tau)^{(4/D_F-2)}\partial_\alpha\left(\frac{\partial^\mu \partial_\mu \sqrt{\rho}}{\sqrt{\rho}}\right) \tag{43}$$

If

$$V_\alpha = \lambda(d\tau)^{(2/D_F-1)}\partial_\alpha S \tag{44}$$

which implies

$$V^\nu \partial_\nu V_\alpha = V^\nu \partial_\alpha V_\nu \tag{45}$$

the relation (43) becomes

$$\frac{\hat{d}V_\alpha}{d\tau} = \partial_\tau V_\alpha + V^\nu \partial_\nu V_\alpha - \frac{\lambda^2}{2}(d\tau)^{(4/D_F-2)}\partial_\alpha\left(\frac{\partial^\nu \partial_\nu \sqrt{\rho}}{\sqrt{\rho}}\right) \tag{46}$$

and more, for $\partial_\tau V_\alpha = 0$:

$$\partial_\alpha\left[V^\nu V_\nu - \frac{\lambda^2}{2}(d\tau)^{(4/D_F-2)}\frac{\partial^\mu \partial_\mu \sqrt{\rho}}{\sqrt{\rho}}\right] = 0 \tag{47}$$

Now, by a suitable choice of the constant integration and knowing that [2]:

$$V^\nu V_\nu = \left(\frac{E}{m_0 V_0}\right)^2 - \left(\frac{\boldsymbol{p}}{m_0}\right)^2 = V_0^2 + \lambda^2(d\tau)^{(4/D_F-2)}\frac{\Box\sqrt{\rho}}{\sqrt{\rho}}$$

we obtain the non-differentiable energy expression in the form

$$E = \pm V_0\left[(m_0 V_0)^2 + \boldsymbol{p}^2 + (m_0\lambda)^2(d\tau)^{(4/D_F-2)}\frac{\Box\sqrt{\rho}}{\sqrt{\rho}}\right]^{1/2} \tag{48}$$

where

$$\Box = -\frac{\partial^2}{\partial x^2} - \frac{\partial^2}{\partial y^2} - \frac{\partial^2}{\partial z^2} + V_0^{-2}\frac{\partial^2}{\partial t^2}$$

For relativistic motions on Peano curves, D_F = 2 at Compton scale, $\overline{\Lambda}_0 = \lambda/V_0 \equiv /(m_0 c), \lambda = /m_0, V_0 = c$, the fractal energy (48) is reduced to the de Broglie's relation:

$$E = \pm c\left[(m_0 c)^2 + \boldsymbol{p}^2 + \hbar^2\frac{\Box\sqrt{\rho}}{\sqrt{\rho}}\right]^{1/2} \tag{49}$$

Relation (48) specifies the following: (i) information propagates with a limit speed V_0 which differs from one biophysical structure to another; (ii) energy, through $\Box\sqrt{\rho}/\sqrt{\rho}$, depends on the state of the biophysical structure; and (iii) the non-differentiable mass

$$M = \pm m_0\left[1 + \frac{\boldsymbol{p}^2}{(m_0 V_0)^2} + \left(\frac{\lambda}{V_0}\right)^2(d\tau)^{(4/D_F-2)}\frac{\Box\sqrt{\rho}}{\sqrt{\rho}}\right]^{1/2} \tag{50}$$

depends also on the state of the biophysical structure, through $\square\sqrt{\rho}/\sqrt{\rho}$.

1.8 Non-differentiable state density conservation law

Let us consider Eq. (35) and its complex conjugate:

$$\lambda^2(d\tau)^{(4/D_F-2)}\partial_\mu\partial^\mu\overline{\Psi} - i\lambda(d\tau)^{(2/D_F-1)}\partial_\tau\overline{\Psi} + F^2(\tau)\overline{\Psi} = 0 \tag{51}$$

Multiplying (35) by $(i\lambda)^{-1}(d\tau)^{1-2/D_F}\overline{\Psi}$, (51) by $(i\lambda)^{-1}(d\tau)^{1-2/D_F}\Psi$ and subtracting the results, one obtains the state density conservation law:

$$\partial_\tau\rho + \partial_\mu j^\mu = 0 \tag{52}$$

where

$$\rho = \Psi\overline{\Psi}, j^\mu = i\lambda(d\tau)^{(2/D_F-1)}\left(\Psi\partial^\mu\overline{\Psi} - \overline{\Psi}\partial^\mu\Psi\right) \tag{53}$$

In the above relations, ρ defines the state density, while j^μ defines the state density 4-current. If Ψ does not depend on τ, which implies $\partial_\tau\rho \equiv 0$, then for relativistic motions on Peano curves, $D_F = 2$ at Compton scale $\overline{\Lambda}_0 = /(m_0c)$, and relation (52) reduces to the state density standard conservation law:

$$\partial_\mu j^\mu = 0 \tag{54}$$

2. Applications of the mathematical model

2.1 Stationary dynamics of the cholesterol at fractal/multifractal scale resolutions

Since cholesterol in any of its forms (principally LDL and HDL) is a fundamental component of blood, its dynamics will be dictated by those of the blood at fractal/multifractal scale resolutions having in view the average dimensions of the cholesterol particles (9–10 nm for HDL and 20–27 nm for LDL [9–12]).

In such a framework, nonrelativistic equations of the non-differentiable hydrodynamics at fractal/multifractal scale resolutions for the stationary case write like

$$f^i = \left(U^i + \lambda(dt)^{(2/D_F-1)}\partial_l\right)\partial^l U^i = 0 \tag{55}$$

$$\partial_l U^l = 0, i = 1, 2, 3 \tag{56}$$

results obtained from Eq. (28) under the conditions $V^i \equiv 0$ and $|U^i| << V_0$.

The first of these equations corresponds to the canceling of specific multifractal force at a differentiable scale resolution, while the second equation corresponds to the incompressibility of the blood at non-differentiable scale.

Generally, it is difficult to obtain an analytical solution for our previous equation system, taking into account its nonlinear nature (induced both by means of non-differentiable convection $U^l\partial_l U^i$ and by the non-differentiable dissipation $\lambda(dt)^{2/D_F-1}\partial_l\partial^l U^i$).

We can still obtain an analytic solution in the case of a plane symmetry (in x, y coordinates) of the dynamics of the blood. For this purpose, let us consider the equation system (55) and (56) in the form:

$$u\partial_x u + v\partial_y u = \nu\partial^2_{yy}u \tag{57}$$

$$\partial_x u + \partial_y v = 0 \tag{58}$$

where we substituted

$$U_x = u(x,y), U_y = v(x,y), \nu = \lambda(dt)^{2/D_F - 1} \tag{59}$$

Using the similarities method given in [6, 7] to solve the equation system (57) and (58) with limit conditions

$$\lim_{y\to 0} v(x,y) = 0, \lim_{y\to 0} \frac{\partial u}{\partial y} = 0, \lim_{y\to\infty} u(x,y) = 0 \tag{60}$$

and a constant flux momentum per unit of depth,

$$q = \rho \int_{-\infty}^{+\infty} u^2 dy = const., \tag{61}$$

we obtain the field of velocities as solutions of the equation system (57) and (58) in the form:

$$u = \frac{1.5\left(\frac{q}{6\rho}\right)^{\frac{2}{3}}}{(\nu x)^{\frac{1}{3}}} sech^2\left[\frac{0.5y\left(\frac{q}{6\rho}\right)^{\frac{1}{3}}}{(\nu x)^{\frac{2}{3}}}\right] \tag{62}$$

$$v = \frac{1.9\left(\frac{q}{6\rho}\right)^{\frac{2}{3}}}{(\nu x)^{\frac{1}{3}}}\left\{\frac{y\left(\frac{q}{6\rho}\right)^{\frac{1}{3}}}{(\nu x)^{\frac{2}{3}}} sech^2\left[\frac{0.5y\left(\frac{q}{6\rho}\right)^{\frac{1}{3}}}{(\nu x)^{\frac{2}{3}}}\right] - tanh\left[\frac{0.5y\left(\frac{q}{6\rho}\right)^{\frac{1}{3}}}{(\nu x)^{\frac{2}{3}}}\right]\right\} \tag{63}$$

The above equations are simplified greatly if we introduce both non-dimensional variables and non-dimensional parameters:

$$X = \frac{x}{x_0}, Y = \frac{y}{y_0}, U = \frac{u}{w_0}, V = \frac{v}{w_0}, \tag{64}$$

$$\xi = \frac{\nu}{\nu_0}, \nu_0 = \frac{y_0^{\frac{3}{2}}}{x_0}\left(\frac{q}{6\rho}\right)^{\frac{1}{2}}, w_0 = \frac{1}{(y_0)^{\frac{1}{2}}}\left(\frac{q}{6\rho}\right)^{\frac{1}{2}}, \tag{65}$$

where x_0, y_0, w_0, and ν_0 are lengths, velocity, and "multifractality degree" specific to the blood. The normalized velocity field is obtained:

$$U = \frac{1.5}{(\xi X)^{\frac{1}{3}}} sech^2\left[\frac{0.5Y}{(\xi X)^{\frac{2}{3}}}\right], \tag{66}$$

$$V = \frac{1.9}{(\xi X)^{\frac{1}{3}}}\left\{\frac{Y}{(\xi X)^{\frac{2}{3}}} sech^2\left[\frac{0.5Y}{(\xi X)^{\frac{2}{3}}}\right] - tanh\left[\frac{0.5Y}{(\xi X)^{\frac{2}{3}}}\right]\right\}, \tag{67}$$

Any of Eqs. (62)–(65) specifies the nonlinearity of the velocity fields: a multifractal soliton for the velocity field across the Ox axis, respectively, "mixtures" of multifractal soliton-multifractal kink for the velocity fields across the Oy axis.

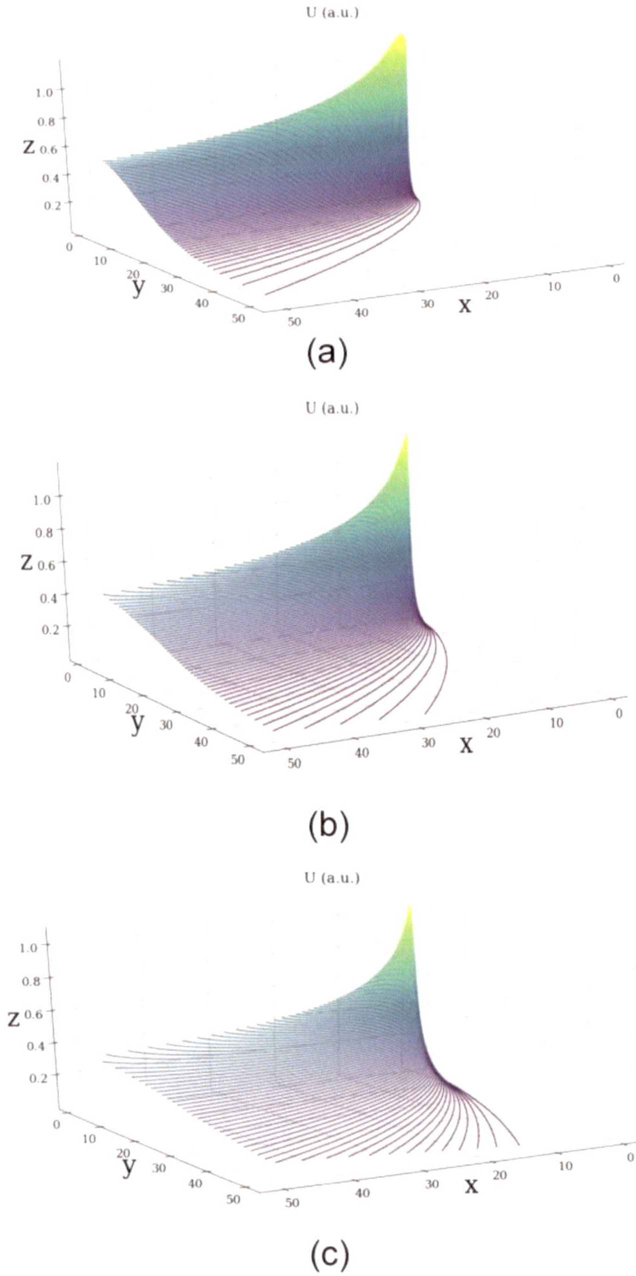

Figure 1.
Normalized velocity field U for various fractal degrees: (a) ξ = 0.4; (b) ξ = 1.0; (c) ξ = 1.9.

The multifractality of the system is "explained" through its dependence from scale resolutions [**Figures 1a–c** and **2a–c**].

The velocity fields (66) and (67) induce the multifractal minimal vortex (**Figure 3a–c**).

$$\Omega = \left(\frac{\partial U}{\partial Y} - \frac{\partial V}{\partial Y}\right) = \frac{0.57Y}{(\xi X)^2} + \frac{0.63\xi}{(\xi X)^{\frac{4}{3}}} tanh\left[\frac{0.5Y}{(\xi X)^{\frac{2}{3}}}\right] + \frac{1.9Y}{(\xi X)^2} sech^2\left[\frac{0.5Y}{(\xi X)^{\frac{2}{3}}}\right] -$$

$$-\frac{0.57Y}{(\xi X)^2} tanh^2\left[\frac{0.5Y}{(\xi X)^{\frac{2}{3}}}\right] - \left[\frac{1.5}{\xi X} + \frac{1.4Y^2}{X(\xi X)^{\frac{5}{3}}}\right] sech^2\left[\frac{0.5Y}{(\xi X)^{\frac{2}{3}}}\right] tanh\left[\frac{0.5Y}{(\xi X)^{\frac{2}{3}}}\right], \quad (68)$$

Since the fractal degree depends on the dimensions of the cholesterol particle (the bigger, the lower the fractal degree), from the analysis of both the velocity field and the vortex field, it results that the LDL particles will deposit at the wall, while the HDL particles will not deposit themselves at the wall.

2.2 On the chameleonic behavior of cholesterol

Cholesterol fractions, especially LDL and HDL cholesterol, are frequently analyzed biomarkers in clinical laboratories [9]. Observational studies have shown that LDL and HDL have opposing associations with the risk of myocardial infarction, with LDL cholesterol being a positive factor and HDL cholesterol being a negative (protective) factor [10]. Observational studies cannot separate the causal role in the pathological process from the role of a marker of the underlying pathophysiology. The results of both randomized trials of LDL-cholesterol-lowering treatments [11] and from human Mendelian diseases [12] are suggesting that plasma LDL

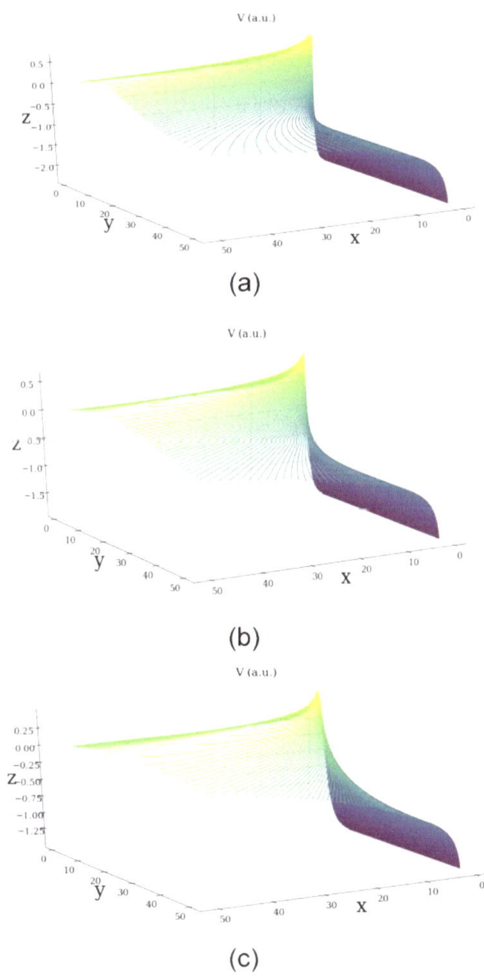

Figure 2.
Normalized velocity field V *for various fractal degrees: (a)* $\xi = 0.4$; *(b)* $\xi = 1.0$; *(c)* $\xi = 1.9$.

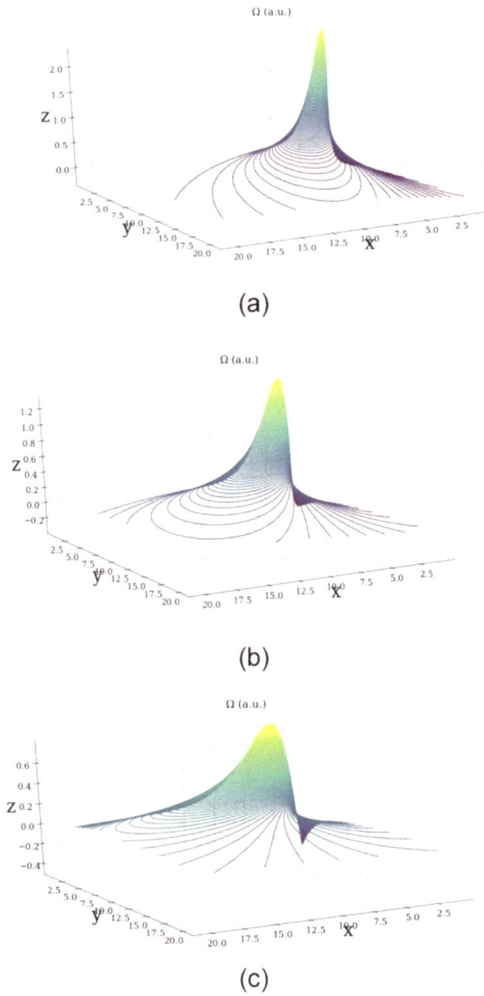

Figure 3.
Multifractal minimal normalized vortex field Ω for various fractal degrees: (a) $\xi = 0.4$*; (b)* $\xi = 1.0$*; (c)* $\xi = 1.9$*.*

cholesterol is related to the risk of myocardial infarction. However, few proofs are available for the causal relevance of HDL cholesterol from randomized trials or Mendelian diseases, and the existing ones are inconsistent [10, 11]. Moreover, more and more studies are starting to oppose the idea that raising plasma HDL cholesterol will surely translate into a risk reduction of myocardial infarction [9–12]. Therefore both LDL and HDL cholesterol can constitute risk factors for myocardial infarction. Such a behavior has been called by experts in the field the "chameleonic effect" of cholesterol [9–12]. In the present paragraph, using our previous mathematical model, LDL and HDL cholesterol dynamics is proposed. In such a context, a fractal/multifractal tunneling effect for biostructures with spontaneous symmetry breaking is analyzed. If the spontaneous symmetry breaking is assimilated to an inflammation (in the form of a specific scalar potential), then two fractal/multifractal states can be observed. In these conditions, these two states, which have been associated with biostructures such as LDL and HDL, transfer their states through a fractal/multifractal tunneling effect. As a result, in our opinion, the widely used notions of "good" and "bad" cholesterol must be redefined as two different states of the same biostructure named "cholesterol," such as in nuclear physics the neutron and proton are two different states of the same particle named nucleon.

With this aim in view, let us reconsider the differential equation (35) with $F^2(\tau) = 0$ subjected to an external constraint independent on τ given as a scalar potential U. One gets

$$\lambda^2 (d\tau)^{(4/D_F - 2)} \partial_\mu \partial^\mu \Psi + i\lambda (d\tau)^{(2/D_F - 1)} \partial_\tau \Psi - \frac{U}{2} \Psi = 0 \quad (69)$$

For nonrelativistic dynamics, Eq. (69) in the one-dimensional case admits the fractal/multifractal stationary solution:

$$\psi(z,t) = \theta(z) \exp\left[-\frac{i}{m_0 \lambda (dt)^{(2/D_F) - 1}} Et \right] \quad (70)$$

where E is the fractal/multifractal energy of the fractal/multifractal stationary cholesterol state $\theta(x)$ and m_0 is the rest mass of the cholesterol particle. Then $\theta(x)$ becomes a fractal/multifractal solution of the fractal/multifractal space equation:

$$\partial_{zz} \theta(z) + \frac{1}{m_0 \lambda^2 (dt)^{(4/D_F) - 2}} (E - U) \theta(z) = 0 \quad (71)$$

If, in such a context, we suppose that the state transfer between LDL and HDL cholesterol implies spontaneous symmetry breaking [13], then $U = V(z)$ from (71) must have the form of an effective potential, as shown in **Figure 4**.

In these conditions, the stationary fractal/multifractal equation becomes

$$\frac{d^2 \theta_\alpha}{dz^2} + \frac{1}{m_0 \lambda^2 (dt)^{(4/D_F) - 2}} [E - V_\alpha] \theta_\alpha = 0, \alpha = \overline{1,3} \quad (72)$$

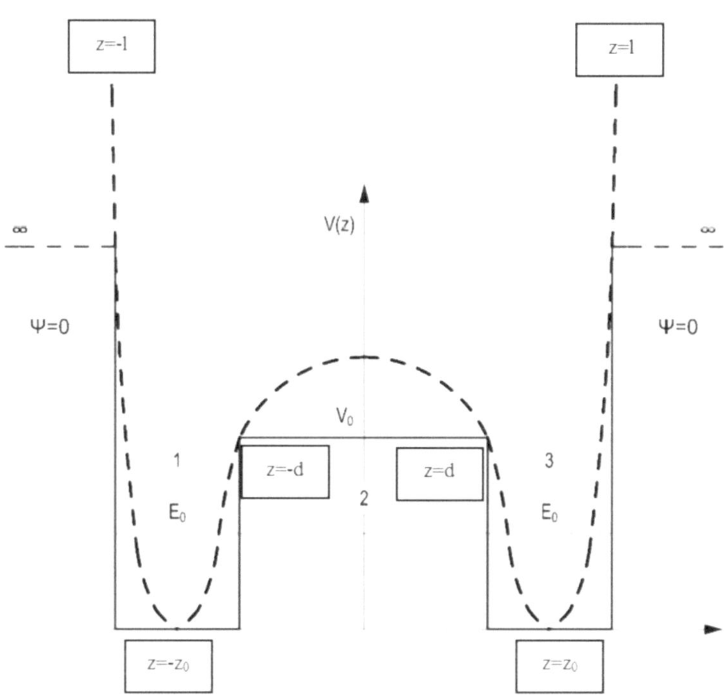

Figure 4.
The effective potential of a fractal/multifractal tunneling effect in the dynamics of biostructures with spontaneous symmetry breaking.

For each of the three regions, the solutions of the equations are

$$\begin{aligned} \theta_1(z) &= C_+ e^{ikz} + C_- e^{-ikz} \\ \theta_2(z) &= Be^{qz} + Ce^{-qz} \\ \theta_3(z) &= D_+ e^{ikz} + D_- e^{-ikz} \end{aligned} \tag{73}$$

with

$$\begin{aligned} k &= \left[\frac{E}{m_0\lambda^2(dt)^{\left(4/D_F\right)-2}}\right]^{1/2} \\ q &= \left[\frac{V_0 - E}{m_0\lambda^2(dt)^{\left(4/D_F\right)-2}}\right]^{1/2} \end{aligned} \tag{74}$$

and

$$C_+, C_-, B, C, D_+, D_-$$

integration constants.

Due to the infinite potential in the two extreme regions, $|z|>l$, the fractal/multifractal state function continuity in $z = \pm l$ implies

$$\begin{aligned} \theta_2(-l) &= C_+ e^{-ikl} + C_- e^{ikl} = 0 \\ \theta_3(l) &= D_+ e^{ikl} + D_- e^{-ikl} = 0 \end{aligned} \tag{75}$$

Since the state density $|\Psi|^2$ is not altered by the multiplication of the fractal/multifractal state function in the form of a constant phase factor, the two equations for $C_\pm$ and $D_\pm$ can be immediately solved by imposing the forms:

$$\begin{aligned} C_+ &= \frac{A}{2i}e^{ikl}, C_- = -\frac{A}{2i}e^{-ikl} \\ D_+ &= \frac{D}{2i}e^{-ikl}, D_- = -\frac{D}{2i}e^{ikl} \end{aligned} \tag{76}$$

so that $\theta_{1,3}$ are given through simple expressions:

$$\begin{aligned} \theta_1(z) &= A\sin[k(z+l)] \\ \theta_3(z) &= D\sin[k(z-l)] \end{aligned} \tag{77}$$

These, along with θ_2, lead to the concrete form of "alignment conditions" in $z = \pm d$

$$\begin{aligned} &\theta_1(-d) = \theta_2(-d), \theta_2(d) = \theta_3(d) \\ &\frac{d\theta_1}{dz}(-d) = \frac{d\theta_2}{dz}(-d), \frac{d\theta_2}{dz}(d) = \frac{d\theta_3}{dz}(d) \end{aligned} \tag{78}$$

namely

$$\begin{aligned} &e^{-qd}B + e^{qd}C = A\sin[k(l-d)] \\ &qe^{-qd}B - qe^{qd}C = kA\cos[k(l-d)] \text{ in } z = -d \\ &e^{qd}B + e^{-qd}C = -D\sin[k(l-d)] \\ &qe^{qd}B - qe^{-qd}C = kD\cos[k(l-d)] \text{ in } z = d \end{aligned} \tag{79}$$

Due to the algebraic form of the two equation pairs, in order to establish the actual expression of the "secular equation" (for eigenvalues E of the energy), $\Delta[E] = 0$, we avoid calculating the 4th order determinant, $\Delta[k(E), q(E)]$, formed with the fractal/multifractal amplitude coefficients A, B, C, D, by employing the following: we note with ρ the ratio C/B, and we divide the first equation to the second one, for each pair. It results in

$$\begin{aligned}\frac{e^{2qd}\rho+1}{e^{2qd}\rho-1}&=-\frac{q}{k}\tan\left[k(l-d)\right]\\ \frac{e^{-2qd}\rho+1}{e^{-2qd}\rho-1}&=\frac{q}{k}\tan\left[k(l-d)\right]\end{aligned}\tag{80}$$

which leads to the equation for ρ:

$$\frac{e^{2qd}\rho+1}{e^{2qd}\rho-1}+\frac{e^{-2qd}\rho+1}{e^{-2qd}\rho-1}=0\tag{81}$$

We find

$$\rho^2=1$$

which implies

$$\rho_-=-1,\rho_+=1\tag{82}$$

For $\rho_+ = 1$, the amplitude function, $\theta_2(z)\cong\coth(qz)$, is symmetric just as the fractal/multifractal states of cholesterol with regard to the (spatial) reflectivity against the origin. Then the permitted value equation of the energy of these states, E_s, has the actual form:

$$\tan\left[k_S(l-d)\right]=-\frac{\coth\left(q_Sd\right)}{q_S}k_S\tag{83}$$

where

$$\begin{aligned}k_S&=\left[\frac{E_S}{m_0\lambda^2(dt)^{\left(4/D_F\right)-2}}\right]^{1/2}\\ q_S&=\left[\frac{V_0-E_S}{m_0\lambda^2(dt)^{\left(4/D_F\right)-2}}\right]^{1/2}\end{aligned}\tag{84}$$

For $\rho_- = -1$, the amplitude function $\theta_2(z)\cong\sinh(qz)$, so that the states will be antisymmetric and permitted values equation, E_A, becomes

$$\tan\left[k_A(l-d)\right]=-\frac{\tanh\left(q_Ad\right)}{q_A}k_A\tag{85}$$

where

$$\begin{aligned}k_A&=\left[\frac{E_A}{m_0\lambda^2(dt)^{\left(4/D_F\right)-2}}\right]^{1/2}\\ q_A&=\left[\frac{V_0-E_A}{m_0\lambda^2(dt)^{\left(4/D_F\right)-2}}\right]^{1/2}\end{aligned}\tag{86}$$

It results in, for now, at least qualitatively that the presence of the barrier (of finite height V_0) between $-d$ and d leads to the splitting of the fundamental level E_0 into two sublevels E_s and E_A accounting for the two types of fractal/multifractal states, symmetric and antisymmetric, respectively, in which the system can be found (both states can be associated to LDL and HDL). Because both eigenvalue equations are strongly transcendent, a direct estimation of solutions $E_{s,A}$ could be possible only by means of numerical methods, which in our case is not necessary. More precisely, we can see here a process of coupling between two different fractal/multifractal (LDL and HDL) states, made possible through a fractal/multifractal tunneling effect.

Taking the above into account, we can thus state that LDL and HDL are two different states of the same biostructure, like in the case of neutron and proton which are two different states of the same particle, named nucleon. The state transfer between LDL and HDL occurs by means of a fractal/multifractal tunneling effect (**Figure 5**).

The fact presented above is in accordance with the latest study results. Thus, we can unequivocally state that the role of cholesterol fractions must be clearly reconsidered. Our model could offer an explanation of why high values of HDL cholesterol can be "toxic" or why, in certain conditions, LDL cholesterol can be a protective factor. We can practically discuss about different states of the same entity, HDL and LDL being expressions of a unique entity—cholesterol—with a pro- or antiatherogenic effect modeled by the instant state and the alternation between the two possible sides. As a consequence, as long as cholesterol fractions maintain a continuous "fluidity," the maximum benefit will be attained if the total cholesterol, in absolute value, is decreased. Our mathematical model only enforces the recent medical findings in the field, which are more and more frequent. At the same time, in our opinion, the present mathematical model confirms and explains the apparent paradoxes from clinical studies.

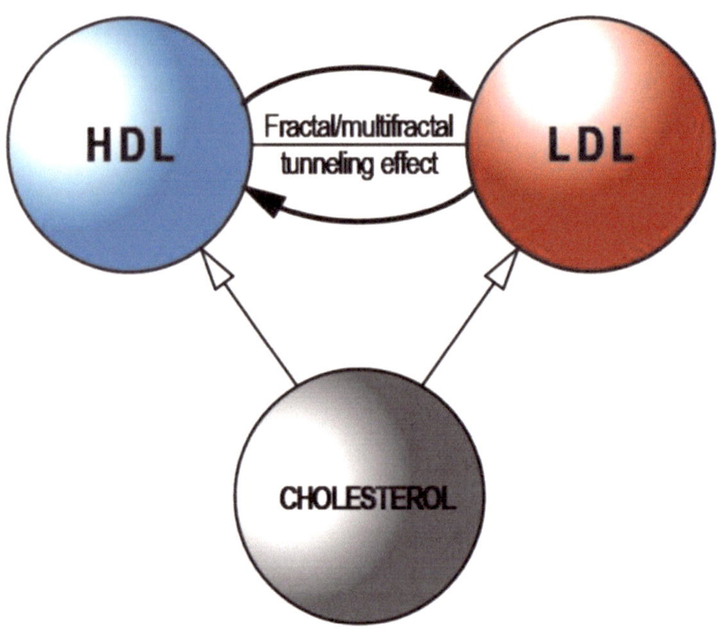

Figure 5.
Schematic representation of the chameleonic behavior of cholesterol.

The mathematical model developed here allows also some numerical evaluations on both the time of transfer between the LDL and HDL states and on the probability of achieving such a transfer. Thus, having in view the nonrelativistic relations,

$$E = 2m_0\lambda(dt)^{(2/D_F-1)}\frac{1}{\tau} \tag{87}$$

$$E = \frac{m_0 v_0^2}{2}$$

one gets through $\lambda = \alpha v_0$, in the case of motion on Peano curves of the cholesterol particles, a time of transfer τ, of the state, of the form

$$\tau = \frac{4\alpha}{v_0} \tag{88}$$

In the relations (87) and (88), α is the dimension of the cholesterol particle and v_0 the blood flow speed, e.g., knowing that in the arteries the average speed of the blood flow is $v_0 \approx 12$ cm/s [10, 11] and the average dimensions of the cholesterol particles are $\alpha_{HDL} \approx 9$ nm and $\alpha_{LDL} \approx 25$ nm, then through (88) we get $\tau_{HDL} \approx 0.189$ μs and $\tau_{LDL} \approx 0.526$ μs. Accordingly, the HDL → LDL transition is faster than the inverse one.

3. Conclusions

The main conclusions of the present work are as follows: (i) we develop a dynamics of the biological systems on a fractal space-time manifold. In such a context, we build the motion operator and the equations of geodesics for rotational and irrotational motions on non-differentiable curves induced by Markov and non-Markov type *stochasticities*, and we establish correlations with known theories of motion (relativity theory, de Broglie relativistic model, etc.). (ii) In the two-dimensional relativistic case, we determine both the velocity field and the vortex one of the cholesterol type biological structure. Based on these we show that the process of wall deposition of the LDL cholesterol is much more accentuated than the HDL cholesterol; (iii) using a multifractal Schrödinger-type equation, we show that by spontaneous symmetry breaking HDL transforms into LDL and vice versa by means of a fractal tunneling effect. We calculate the time transfer probability HDL ↔ LDL, and we show that the HDL → LDL process is more probable than the inverse one.

Author details

Maricel Agop[1,2], Calin Buzea[3*], Decebal Vasincu[4] and Daniel Timofte[5]

1 Technical University "Gheorghe Asachi", Iasi, Romania

2 Academy of Romanian Scientists, București, Romania

3 National Institute of Research and Development for Technical Physics, Iasi, Romania

4 Biophysics Department, University of Medicine and Pharmacy "Grigore T. Popa", Iasi, Romania

5 Surgical Department, University of Medicine and Pharmacy "Grigore T. Popa", Iasi, Romania

*Address all correspondence to: calinb2003@yahoo.com

References

[1] Nottale L. Fractal Space-Time and Microphysics: Towards a Theory of Scale Relativity. Singapore: World Scientific; 1993

[2] Nottale L. Scale Relativity and Fractal Space-Time—A New Approach to Unifying Relativity and Quantum Mechanics. London: Imperial College Press; 2011

[3] Nottale L. Scale relativity: A fractal matrix for organization in nature. Electronic Journal of Theoretical Physics. 2007;**4**(16(III)):15-102

[4] Mandelbrot B. The Fractal Geometry of Nature (Updated and Augm. Ed.). New York: W.H. Freeman; 1983

[5] Merches I, Agop M. Differentiability and Fractality in Dynamics of Physical Systems. Singapore: World Scientific; 2016

[6] Agop M, Paun VP. On the New Perspectives of Fractal Theory. Fundaments and Applications. Bucuresti: Romanian Academy Publishing House; 2017

[7] Agop M, Merches I. Operation Procedures Describing Physical Systems. Boca Raton, Florida, United States: CRC Press, Taylor and Francis Group; 2019

[8] Cresson J. Scale calculus and the Schrödinger equation. Journal of Mathematical Physics. 2003;**44**(11): 4907-4938

[9] Fogelman AM. Further evidence that high-density lipoprotein is a chameleon-like lipoprotein. European Heart Journal. 2015;**36**(43):3017-3019

[10] Zewinger S, Drechler C, Kleber ME, Dressel A, Riffel J, Triem S, et al. Serum amyloid A (SAA): High-density lipoproteins (HDL) interaction and cardiovascular risk. European Heart Journal. 2015;**36**:3007-3016

[11] Kathiresan S, Melander O, Anevski D, et al. Polymorphisms associated with cholesterol and risk of cardiovascular events. The New England Journal of Medicine. 2008;**358**: 1240-1249

[12] Ravnskov U, Diamond DM, Hama R, et al. Lack of an association or an inverse association between low-density-lipoprotein cholesterol and mortality in the elderly: a systematic review. BMJ Open 2016;**6**:e010401. DOI: 10.1136/bmjopen-2015-010401

[13] Chaichian M, Nelipa NF. Introduction to Gauge Field Theories. Berlin, Heidelberg: Springer-Verlag; 1984

Chapter 9

Stueckelberg-Horwitz-Piron Canonical Quantum Theory in General Relativity and Bekenstein-Sanders Gauge Fields for TeVeS

Lawrence P. Horwitz

Abstract

A consistent (off-shell) canonical classical and quantum dynamics in the framework of special relativity was formulated by Stueckelberg in 1941 and generalized to many-body theory by Horwitz and Piron in 1973 (SHP). This theory has been embedded into the framework of general relativity (GR), here denoted by SHPGR. The canonical Poisson brackets of the SHP theory remain valid (invariant under local coordinate transformations) on the manifold of GR and provide the basis for formulating a canonical quantum theory. The relation between representations based on coordinates and momenta is given by Fourier transform; a proof is given here for this functional relation on a manifold. The potential which may occur in the SHP theory emerges as a spacetime scalar mass distribution in GR. Gauge fields, both Abelian and non-Abelian on the quantum mechanical SHPGR Hilbert space in both the single particle and many-body theory, may be generated by phase transformations. Application to the construction of Bekenstein and Sanders in their solution to the lensing problem in TeVeS is discussed.

Keywords: relativistic dynamics, general relativity, quantum theory on curved space, non-Abelian gauge fields, Bekenstein-Sanders field, TeVeS

1. Introduction

The relativistic canonical Hamiltonian dynamics of Stueckelberg, Horwitz, and Piron (SHP) [1] with scalar potential and gauge field interactions for single- and many-body theories can, by local coordinate transformation, be embedded into the framework of general relativity (GR). This embedding provides a basis for the work of Horwitz et al. [2, 3] in their discussion of the origin of the field introduced by Bekenstein and Sanders [4] to explain gravitational lensing in the TeVeS formulation of modified Newtonian dynamic (MOND) theories [5–10].

The theory was originally formulated for a single particle by Stueckelberg in [11–13]. Stueckelberg envisaged the motion of a particle along a world line in spacetime that can curve and turn to flow backward in time, resulting in the phenomenon of pair annihilation in classical dynamics. The world line was then described by an invariant monotonic parameter τ. The theory was generalized by Horwitz and Piron in [14] (see also [15, 16]) to be applicable to many-body systems by assuming that the parameter τ is universal (as for Newtonian time, enabling them to solve the two-body problem classically, and later, a quantum solution was found by Arshansky and Horwitz [17–19], both for bound states and scattering theory).

Performing a coordinate transformation to general coordinates, along with the corresponding transformation of the momenta (the cotangent space of the original Minkowski manifold), one obtains [20] the SHP theory in a curved space of general coordinates and momenta with a canonical Hamilton-Lagrange (symplectic) structure. We shall refer to this generalization as SHPGR. We discuss the extension of the Abelian gauge theory described in Ref. [20] to the non-Abelian gauge discussed in [2, 3].

The invariance of the Poisson bracket under local coordinate transformations provides a basis for the canonical quantization of the theory, for which the evolution under τ is determined by the covariant form of the Stueckelberg-Schrödinger equation [1].

In this chapter, we assume a τ-independent background gravitational field; the local coordinate transformations from the flat Minkowski space to the curved space are taken to be independent of τ, consistently with an energy momentum tensor that is τ independent. In a more dynamical setting, when the energy momentum tensor depends on τ, the spacetime is evolved nontrivially [20, 21].

2. Embedding of single particle dynamics with external potential in GR

We write the SHP Hamiltonian [1, 11–13] as

$$K = \frac{1}{2M}\eta^{\mu\nu}\pi_\mu\pi_\nu + V(\xi) \tag{1}$$

where $\eta^{\mu\nu}$ is the flat Minkowski metric $(-+++)$ and π_μ, ξ^μ are the spacetime canonical momenta and coordinates in the local tangent space of a general manifold, following Einstein's use of the equivalence principle.

The existence of a potential term (which we assume to be a Lorentz scalar), representing nongravitational forces, implies that the "free fall" condition is replaced by a local dynamics carried along by the free falling system (an additional force acting on the particle within the "elevator" according to the coordinates in the tangent space).

The canonical equations are

$$\dot{\xi}^\mu = \frac{\partial K}{\partial \pi_\mu} \qquad \dot{\pi}_\mu = -\frac{\partial K}{\partial \xi^\mu} = -\frac{\partial V}{\partial \xi^\mu}, \tag{2}$$

where the dot here indicates $\frac{d}{d\tau}$, with τ the invariant universal "world time." Since then

$$\begin{aligned} \dot{\xi}^\mu &= \frac{1}{M}\eta^{\mu\nu}\pi_\nu, \\ \text{or} \quad \pi_\nu &= \eta_{\nu\mu}M\dot{\xi}^\mu, \end{aligned} \tag{3}$$

the Hamiltonian can then be written as

$$K = \frac{M}{2}\eta_{\mu\nu}\dot{\xi}^\mu\dot{\xi}^\nu + V(\xi). \tag{4}$$

It is important to note that, as clear from (3), that $\dot{\xi}^0 = \frac{dt}{d\tau}$ has a sign *opposite* to π_0 which lies in the cotangent space of the manifold, as we shall see in the Poisson bracket relations below. The energy of the particle for a normal timelike particle should be positive (negative energy would correspond to an antiparticle [1, 11–13]). The *physical momenta and energy* therefore correspond to the mapping

$$\pi^\mu = \eta^{\mu\nu}\pi_\mu, \tag{5}$$

back to the tangent space. Thus, equivalently, from (2), $\dot{\xi}^\mu = (1/M)\pi^\mu$. This simple observation will be important in the discussion below of the dynamics of a particle in the framework of general relativity, for which the metric tensor is nontrivial.

We now transform the local coordinates (contravariantly) according to the diffeomorphism

$$d\xi^\mu = \frac{\partial \xi^\mu}{\partial x^\lambda} dx^\lambda \tag{6}$$

to attach small changes in ξ to the corresponding small changes in the coordinates x on the curved space, so that

$$\dot{\xi}^\mu = \frac{\partial \xi^\mu}{\partial x^\lambda} \dot{x}^\lambda. \tag{7}$$

The Hamiltonian then becomes

$$K = \frac{M}{2} g_{\mu\nu} \dot{x}^\mu \dot{x}^\nu + V(x), \tag{8}$$

where $V(x)$ is the potential at the point ξ corresponding to the point x (a function of ξ in a small neighborhood of the point x) and

$$g_{\mu\nu} = \eta_{\lambda\sigma} \frac{\partial \xi^\lambda}{\partial x^\mu} \frac{\partial \xi^\sigma}{\partial x^\nu} \tag{9}$$

Since V has significance as the source of a force in the local frame only through its derivatives, we can make this pointwise correspondence with a globally defined scalar function $V(x)$.[1]

The corresponding Lagrangian is then

$$L = \frac{M}{2} g_{\mu\nu} \dot{x}^\mu \dot{x}^\nu - V(x), \tag{10}$$

In the locally flat coordinates in the neighborhood of x^μ, the symplectic structure of Hamiltonian mechanics implies that the momentum[2] π_μ, lying in the cotangent space of the manifold $\{\xi^\mu\}$, transforms covariantly under the local transformation (5), that is, as does $\frac{\partial}{\partial \xi^\mu}$, so that we may define

---

[1] Since $V(x)$ has the dimension of mass, one can think of this function as a scalar mass field, reflecting forces acting in the local tangent space at each point. It may play the role of "dark energy" [2, 3]. If $V = 0$, our discussion reduces to that of the usual general relativity, but with a well-defined canonical momentum variable.

[2] We shall call the quantity π_μ in the cotangent space as *canonical momentum*, although it must be understood that its map back to the tangent space π^μ corresponds to the actual physically measureable momentum.

$$p_\mu = \frac{\partial \xi^\lambda}{\partial x^\mu} \pi_\lambda. \tag{11}$$

This definition is consistent with the transformation properties of the momentum defined by the Lagrangian (10):

$$p_\mu = \frac{\partial L(x, \dot{x})}{\partial \dot{x}^\mu}, \tag{12}$$

yielding

$$p_\mu = M g_{\mu\nu} \dot{x}^\nu. \tag{13}$$

The second factor in the definition (9) of $g_{\mu\nu}$ in (13) acts on $\dot{x}^\nu$; with (7) we then have (as in (11))

$$\begin{aligned} p_\mu &= M \eta_{\lambda\sigma} \frac{\partial \xi^\lambda}{\partial x^\mu} \dot{\xi}^\sigma \\ &= \frac{\partial \xi^\lambda}{\partial x^\mu} \pi_\lambda. \end{aligned} \tag{14}$$

As we have remarked above for the locally flat space in (5), the *physical* energy and momenta are given, according to the mapping,

$$p^\mu = g^{\mu\nu} p_\nu = M \dot{x}^\nu \tag{15}$$

back to the tangent space of the manifold, which also follows directly from the local coordinate transformation of (3) and (5).

It is therefore evident from (15) that

$$\dot{p}^\mu = M \ddot{x}^\mu. \tag{16}$$

We see that $\dot{p}^\mu$, which should be interpreted as the force acting on the particle, is proportional to the *acceleration along the orbit of motion* (a covariant derivative plus a gradient of the potential), as described by the geodesic-type relation. This Newtonian-type relation in the general manifold reduces in the limit of a flat Minkowski space to the corresponding SHP dynamics and in the nonrelativistic limit, to the classical Newton law. We remark that this result does not require taking a post-Newtonian limit, the usual method of obtaining Newton's law from GR.

We now discuss the geodesic equation obtained by studying the condition

$$\ddot{\xi}^\mu = -\frac{1}{M} \dot{\pi}_\mu = -\frac{1}{M} \eta^{\mu\nu} \frac{\partial V(\xi)}{\partial \xi^\nu}. \tag{17}$$

To do this, we compute

$$\begin{aligned} \ddot{\xi}^\mu = \frac{d}{d\tau}\left(\frac{\partial \xi^\mu}{\partial x^\lambda} \dot{x}^\lambda\right) &= \frac{\partial^2 \xi^\mu}{\partial x^\lambda \partial x^\gamma} \dot{x}^\gamma \dot{x}^\lambda \\ + \frac{\partial \xi^\mu}{\partial x^\lambda} \ddot{x}^\lambda &= -\frac{1}{M} \eta^{\mu\nu} \frac{\partial x^\lambda}{\partial \xi^\nu} \frac{\partial V(x)}{\partial x^\lambda}, \end{aligned} \tag{18}$$

so that, after multiplying by $\frac{\partial x^\sigma}{\partial \xi^\mu}$ and summing over μ, we obtain

$$\begin{aligned}\ddot{x}^\sigma &= -\frac{\partial x^\sigma}{\partial \xi^\mu}\frac{\partial^2 \xi^\mu}{\partial x^\lambda \partial x^\gamma}\dot{x}^\gamma \dot{x}^\lambda \\ &\quad -\frac{1}{M}\eta^{\mu\nu}\frac{\partial x^\lambda}{\partial \xi^\nu}\frac{\partial x^\sigma}{\partial \xi^\mu}\frac{\partial V(x)}{\partial x^\lambda}.\end{aligned} \tag{19}$$

Finally, with (9) and the usual definition of the connection

$$\Gamma^\sigma{}_{\lambda\gamma} = \frac{\partial x^\sigma}{\partial \xi^\mu}\frac{\partial^2 \xi^\mu}{\partial x^\lambda \partial x^\gamma} \tag{20}$$

we obtain the modified geodesic-type equation

$$\ddot{x}^\sigma = -\Gamma^\sigma{}_{\lambda\gamma}\dot{x}^\gamma \dot{x}^\lambda - \frac{1}{M}g^{\sigma\lambda}\frac{\partial V(x)}{\partial x^\lambda}, \tag{21}$$

from which we see that the derivative of the potential $V(\xi)$ is mapped, under this coordinate transformation into a force resulting in a modification of the acceleration along the geodesic-like curves, that is, (16) now reads

$$\dot{p}^\mu = M\ddot{x}^\nu = -M\Gamma^\sigma{}_{\lambda\gamma}\dot{x}^\gamma \dot{x}^\lambda - g^{\sigma\lambda}\frac{\partial V(x)}{\partial x^\lambda} \tag{22}$$

The procedure that we have carried out here provides a canonical dynamical structure for motion in the curvilinear coordinates. We now remark that the Poisson bracket remains valid for the coordinates $\{x, p\}$. In the local coordinates $\{\xi, \pi\}$, the τ derivative of a function $F(\xi, \pi)$ is

$$\begin{aligned}\frac{dF(\xi,\pi)}{d\tau} &= \frac{\partial F(\xi,\pi)}{\partial \xi^\mu}\dot{\xi}^\mu + \frac{\partial F(\xi,\pi)}{\partial \pi_\nu}\dot{\pi}_\mu \\ &= \frac{\partial F(\xi,\pi)}{\partial \xi^\mu}\frac{\partial K}{\partial \pi_\mu} - \frac{\partial F(\xi,\pi)}{\partial \pi_\mu}\frac{\partial K}{\partial \xi^\nu} \\ &\equiv [F, K]_{PB}(\xi, \pi).\end{aligned} \tag{23}$$

If we replace in this formula

$$\begin{aligned}\frac{\partial}{\partial \xi^\mu} &= \frac{\partial x^\lambda}{\partial \xi^\mu}\frac{\partial}{\partial x^\lambda} \\ \frac{\partial}{\partial \pi_\mu} &= \frac{\partial \xi^\mu}{\partial x^\lambda}\frac{\partial}{\partial p_\lambda},\end{aligned} \tag{24}$$

we immediately (as assured by the invariance of the Poisson bracket under local coordinate transformations) obtain

$$\frac{dF(\xi,\pi)}{d\tau} = \frac{\partial F}{\partial x^\mu}\frac{\partial K}{\partial p_\mu} - \frac{\partial F}{\partial p_\mu}\frac{\partial K}{\partial x^\nu} \equiv [F, K]_{PB}(x, p) \tag{25}$$

In this definition of Poisson bracket, we have, as for the ξ^μ, π_ν relation,

$$[x^\mu, p_\nu]_{PB}(x, p) = \delta^\mu{}_\nu. \tag{26}$$

The Poisson bracket of x^μ with the (physical energy momentum) tangent space variable p^μ has then the tensor form

$$[x^\mu, p^\nu]_{PB}(x,p) = g^{\mu\nu}. \tag{27}$$

In the flat space limit, this relation reduces to the SHP bracket,

$$[\xi^\mu, \pi^\nu]_{PB}(\xi, \pi) = \eta^{\mu\nu}. \tag{28}$$

Continuing our analysis with p_μ (we drop the (x,p) label on the Poisson bracket henceforth),

$$\left[p_\mu, F(x)\right]_{PB} = -\frac{\partial F}{\partial x^\mu}, \tag{29}$$

so that p_μ acts infinitesimally as the *generator of translation* along the coordinate curves and

$$[x^\mu, F(p)]_{PB} = \frac{\partial F(p)}{\partial p_\mu}, \tag{30}$$

so that x^μ is the generator of translations in p_μ. In the classical case, if $F(p)$ is a general function of p^μ, we can write at some point x,[3]

$$[x^\mu, F(p)]_{PB} = g^{\mu\nu}(x)\frac{\partial F(p)}{\partial p^\nu}. \tag{31}$$

This structure clearly provides a phase space which could serve as the basis for the construction of a canonical quantum theory on the curved spacetime.

We now turn to a discussion of the dynamics introduced into the curved space by the procedure outlined above.

We may also write (22) in terms of the full connection form by noting that with (9),

$$\frac{\partial g_{\lambda\gamma}}{\partial x^\mu} = \eta_{\alpha\beta}\left(\frac{\partial^2 \xi^\alpha}{\partial x^\lambda \partial x^\mu}\frac{\partial \xi^\beta}{\partial x^\gamma} + \frac{\partial \xi^\alpha}{\partial x^\lambda}\frac{\partial^2 \xi^\beta}{\partial x^\gamma \partial x^\mu}\right). \tag{32}$$

Multiplying by $\dot{x}^\gamma \dot{x}^\lambda$, the two terms combine to give a factor of two. We then return to the original definition of Γ in (20) in the form

$$\frac{\partial^2 \xi^\alpha}{\partial x^\lambda \partial x^\mu} = \frac{\partial \xi^\alpha}{\partial x^\sigma}\Gamma^\sigma{}_{\lambda\mu}, \tag{33}$$

so we can write

$$\begin{aligned} \frac{\partial g_{\lambda\gamma}}{\partial x^\mu}\dot{x}^\gamma \dot{x}^\lambda &= 2\eta_{\alpha\beta}\frac{\partial \xi^\alpha}{\partial x^\sigma}\frac{\partial \xi^\beta}{\partial x^\gamma}\Gamma^\sigma{}_{\lambda\mu}\dot{x}^\gamma \dot{x}^\lambda \\ &= 2g_{\sigma\gamma}\Gamma^\sigma{}_{\lambda\mu}\dot{x}^\gamma \dot{x}^\lambda. \end{aligned} \tag{34}$$

---

[3] In the quantized form, the factor $g^{\mu\nu}(x)$ cannot be factored out from polynomials, so, as for Dirac's quantization procedure [22–25], some care is required.

We therefore have

$$\dot{p}_\mu = -\frac{\partial V(x)}{\partial x^\mu} + M g_{\sigma\gamma} \Gamma^\sigma{}_{\lambda\mu} \dot{x}^\gamma \dot{x}^\lambda. \tag{35}$$

3. Quantum theory on the curved space

The Poisson bracket formulas (25) and (26) can be considered as a basis for defining a quantum theory with canonical commutation relations

$$\left[x^\mu, p_\nu\right] = i\hbar \delta^\mu{}_\nu, \tag{36}$$

so that

$$\left[p_\mu, F(x)\right] = -i\hbar \frac{\partial F}{\partial x^\mu}, \tag{37}$$

and

$$[x^\mu, F(p)] = i\hbar \frac{\partial F(p)}{\partial p_\mu}. \tag{38}$$

The transcription of the Stueckelberg-Schrödinger equation for a wave function $\psi_\tau(x)$ can be taken to be (see also [26–28])

$$i \frac{\partial}{\partial \tau} \psi_\tau(x) = K \psi_\tau(x), \tag{39}$$

where the operator valued Hamiltonian can be taken to be the Hermitian form (42), written below, on a Hilbert space defined with scalar product (with invariant measure; we write $g = -\det\{g^{\mu\nu}\}$),

$$(\psi, \chi) = \int d^4x \sqrt{g} \psi^*{}_\tau(x) \chi_\tau(x). \tag{40}$$

To construct a Hermitian Hamiltonian, we first study the properties of the canonical momentum in coordinate representation. Clearly, in coordinate representation, $-i\frac{\partial}{\partial x^\mu}$ is not Hermitian due to the presence of the factor $\sqrt{g}$ in the integrand of the scalar product. The problem is somewhat analogous to that of Newton and Wigner [29] in their treatment of the Klein-Gordon equation in momentum space. It is easily seen that the operator

$$p_\mu = -i \frac{\partial}{\partial x^\mu} - \frac{i}{2} \frac{1}{\sqrt{g(x)}} \frac{\partial}{\partial x^\mu} \sqrt{g(x)} \tag{41}$$

is essentially self-adjoint in the scalar product (40), satisfying as well as the commutation relations (36).[4]

---

[4] The physically observable momentum can be defined, as in (15), as $\frac{1}{2}\{g^{\mu\nu}, p_\nu\}$, with commutation relations of the form (27). This operator can be transformed, as for the Newton-Wigner operator [29], to the form $-i\frac{\partial}{\partial x^\mu}$ by the Foldy-Wouthuysen transformation [30] $(g(x))^{\frac{1}{4}} p_\mu (g(x))^{-\frac{1}{4}}$.

Since p_μ is Hermitian in the scalar product (41), we can write the Hermitian Hamiltonian as

$$K = \frac{1}{2M} p_\mu g^{\mu\nu} p_\nu + V(x), \tag{42}$$

consistent with the local coordinate transformation of (1). The integration (40) must be considered as a total volume sum with invariant measure on the whole space, consistent with the notion of Lebesgue measure and the idea that the norm is the sum of probability measures on every subset contained. We return to this point in our discussion of the Fourier transform below.

4. Canonical quantum theory and the Fourier transform

To complete the construction of a canonical quantum theory on the curved space of GR, we discuss first the formulation of the Fourier transform $f(x) \to \tilde{f}(p)$ for a scalar function $f(x)$ (we shall use x^μ and the canonically conjugate p_μ in this discussion). Let us define ($g \equiv -\det g_{\mu\nu}$)

$$\tilde{f}(p) = \int d^4x \sqrt{g(x)} e^{ip_\mu x^\mu} f(x). \tag{43}$$

The inverse is given by

$$\int e^{-ip_\mu x^\mu} \tilde{f}(p) d^4p = \int d^4p e^{-ip_\mu (x^\mu - x'^\mu)} f(x') \sqrt{g(x')} d^4x' = (2\pi)^4 f(x) \sqrt{g(x)} \tag{44}$$

so that

$$\tilde{f}'(p) = \frac{1}{(2\pi)^4 \sqrt{g(x)}} \int e^{-ip_\mu x^\mu} \tilde{f}(p) d^4p. \tag{45}$$

One sees immediately that under diffeomorphisms, for which with the scalar property $f(x) = f'(x'), \tilde{f}(p) \to \tilde{f}'(p)$. The Fourier transform of $f'(x')$ is

$$\tilde{f}'(p) = \int d^4x' \sqrt{g(x')} e^{ip_\mu x'^\mu} f'(x'), \tag{46}$$

By change of integration variables, we have

$$\tilde{f}'(p) = \int d^4x \sqrt{g(x)} e^{ip_\mu x^\mu} f'(x), \tag{47}$$

In Dirac notation,

$$f(x) = \langle x|f\rangle, \tag{48}$$

and we write as well

$$\tilde{f}(p) = \langle p|f\rangle. \tag{49}$$

For

$$
\begin{aligned}
<x|p> &= \frac{1}{(2\pi)^4\sqrt{g(x)}} e^{-ip_\mu x^\mu} \\
<p|x> &= \sqrt{g(x)} e^{ip_\mu x^\mu},
\end{aligned}
\tag{50}
$$

we have, for example, the usual action of transformation functions

$$
\int <x|p><p|f> d^4p = <x|f>,
\tag{51}
$$

where we have used

$$
\begin{aligned}
\int <x|p><p|x'> d^4p &= \frac{1}{(2\pi)^4\sqrt{g(x)}} \int d^4p e^{-ip_\mu x^\mu} e^{ip_\mu x'^\mu} \sqrt{g(x')} \\
&= \delta^4(x - x').
\end{aligned}
\tag{52}
$$

Note that the transformation functions $<x|p>$ and $<p|x>$ are not simple complex conjugates of each other, but require nontrivial factors of $\sqrt{g(x)}$ and its inverse to satisfy the necessary transformation laws on the manifold. Conversely (the factors $\sqrt{g(x)}$ and its inverse cancel), we should obtain

$$
\int <p'|x><x|p> d^4x = \delta^4(p' - p).
\tag{53}
$$

The validity of (53) is not obvious on a curved space. We therefore provide a simple, but not trivial, proof of (53). For

$$
\int d^4p e^{ip_\mu(x^\mu - x'^\mu)} = (2\pi)^4 \frac{\delta^4(x - x')}{\sqrt{g}}
\tag{54}
$$

we must have

$$
\tilde{f}(p) = \frac{1}{(2\pi)^4} \int d^4x \int d^4p' e^{i(p_\mu - p'_\mu)x^m u} \tilde{f}(p'),
\tag{55}
$$

that is, exchanging the order of integrations, on the set $\{\tilde{f}(p)\}$,

$$
\Delta(p - p') = \frac{1}{(2\pi)^4} \int d^4x e^{i(p_\mu - p'_\mu)x^\mu} = \delta^4(p - p').
\tag{56}
$$

We now represent the integral as a sum over small boxes around the set of points $\{x_B\}$ that cover the space and eventually take the limit as for a Riemann integral.[5] In each small box, the coordinatization arises from an invertible transformation from the local tangent space in that neighborhood. We write

$$
x^\mu = x_B{}^\mu + \eta^\mu \quad \in \text{box B}
\tag{57}
$$

where

---

[5] We follow here essentially the method discussed in Reed and Simon [31] in their discussion of the Lebesgue integral.

$$\eta^\mu = \frac{\partial x^\mu}{\partial \xi^\lambda} \xi^\lambda \tag{58}$$

and ξ^λ is in the flat local tangent space at x_B.

We now write the integral (56) as

$$\begin{aligned}\Delta(p-p') &= \frac{1}{(2\pi)^4} \Sigma_B \int_B d^4\eta e^{i(p_\mu - p'_\mu)(x_B{}^\mu + \eta^\mu)} \\ &= \frac{1}{(2\pi)^4} \Sigma_B e^{i(p_\mu - p'_\mu)x_B{}^\mu} \int_B d^4\eta e^{i(p_\mu - p'_\mu)\eta^\mu}.\end{aligned} \tag{59}$$

Let us call

$$I_B = \int_B d^4\eta e^{i(p_\mu - p'_\mu)\eta^\mu}. \tag{60}$$

In this neighborhood, call

$$\frac{\partial x^\mu}{\partial \xi^\lambda} = a^\mu{}_\lambda(B), \tag{61}$$

which we assume a constant matrix (Lorentz transformation) in each box. In (60), we then have

$$I_B = \int_B \frac{d^4\xi}{\sqrt{\det a}} e^{i(p_\mu - p'_\mu)a^\mu{}_\lambda(B)\xi^\lambda}. \tag{62}$$

However, we can make a change of variables; we are left with

$$I_{B'} = \int_B d^4\xi e^{i(p_\mu - p'_\mu)\xi^\mu}. \tag{63}$$

in each box.

However, the transformations $a^\mu{}_\lambda(B)$ in the neighborhood of each point B may be different, and therefore the set of transformed boxes may not cover (boundary deficits) the full domain of spacetime coordinates (one can easily estimate that the deficit from an arbitrarily selected set can be infinite in the limit).

We may avoid this problem by assuming geodesic completeness of the manifold and taking the covering set of boxes, constructed of parallel transported edges, along geodesic curves. Parallel transport of the tangent space boxes then fills the space in the neighborhood of the geodesic curve we are following, and each infinitesimal box may carry an invariant volume (Liouville-type flow) transported along a geodesic curve. For successive boxes along the geodesic curve, since the boundaries are determined by parallel transport (rectilinear shift in the succession of local tangent spaces), there is no volume deficit between adjacent boxes.

We may furthermore translate a geodesic curve to an adjacent geodesic by the mechanism discussed in [32], so that boxes associated with adjacent geodesics are also related by parallel transport. In this way, we may fill the entire geodesically accessible spacetime volume.

Let us assign a measure to each point B:

$$\Delta\mu(B, p - p') \equiv I_B. \tag{64}$$

We may then write (59) as

$$\Delta(p-p') = \frac{1}{(2\pi)^4}\Sigma_B e^{i(p_\mu - p'_\mu)x_B{}^\mu}\Delta\mu(B, p-p'), \tag{65}$$

Our construction has so far been based on elements constructed in the tangent space in the neighborhood of each point B. Relating all points along a geodesic to the corresponding tangent spaces and putting each patch in correspondence by continuity, we may consider the set $\{x_B\}$ to be in correspondence with an extended flat space $\{\xi\}$, for which $x_B \sim \xi_B$ to obtain[6]

$$\Delta(p-p') = \frac{1}{(2\pi)^4}\Sigma_B e^{i(p_\mu - p'_\mu)\xi_B{}^\mu}\Delta\mu(\xi_B, p-p'), \tag{66}$$

In the limit of vanishing spacetime box volume, this approaches the Lebesgue-type integral on a flat space:

$$\Delta(p-p') = \frac{1}{(2\pi)^4}\int e^{i(p_\mu - p'_\mu)\xi^\mu} d\mu(\xi, p-p'). \tag{67}$$

If the measure is differentiable, we could write

$$d\mu(\xi, p-p') = m(\xi, p-p')d^4\xi. \tag{68}$$

Since the kernel $\Delta(p-p')$ is to act on elements of a Hilbert space $\left\{\tilde{f}(p)\right\}$, the support for $p' \to \infty$ vanishes, so that $p-p'$ is essentially bounded, as we discuss below. In the small box, say, size ϵ,

$$\int_{-\epsilon/2}^{\epsilon/2} d\xi^0 d\xi^1 d\xi^2 d\xi^3 e^{i(p_\mu - p'_\mu)\xi^\mu} = (2i)^4 \Pi_{j=0}^{j=3}\frac{\sin\left(p_j - p'_j\right)\frac{\epsilon}{2}}{\left(p_j - p'_j\right)} \to \epsilon^4 \sim d^4\xi, \tag{69}$$

so that $m(\xi, p-p') = 1$, and we have

$$\Delta(p-p') = \frac{1}{(2\pi)^4}\int e^{i(p_\mu - p'_\mu)\xi^\mu} d^4\xi, \tag{70}$$

or[7]

$$\Delta(p-p') = \delta^4(p-p'). \tag{71}$$

It is clear that the assertion (69) requires some discussion. For $\epsilon \to 0$ we must be sure that p' does not become too large, so that our local measure is equivalent to $d^4\xi$. In one of the dimensions, what we want to find are conditions for which

$$\frac{sinp\epsilon}{p} \to \epsilon \tag{72}$$

[6] Similar to the method followed in the simpler case of constant curvature by Georgiev [33].

[7] Note that Abraham et al. [34] apply the formal Fourier transform on a manifold in three dimensions without proof.

for $\epsilon \to 0$, where we have written p for $p - p'$. As a distribution, on functions $g(p)$, the left member of (72) acts as

$$G(\epsilon) \equiv \int_{-\infty}^{\infty} \frac{sinp\epsilon}{p} g(p). \tag{73}$$

The function $G(\epsilon)$ is analytic in the neighborhood of $\epsilon = 0$ if $p^n g(p)$ has a Fourier transform for all n and the series is convergent in this neighborhood, since $G(0)$ is identically zero and successive derivatives correspond to the Fourier transforms of $p^n g(p)$ (differentiating under the integral). This implies that if the (usual) Fourier transform of $g(p)$ is a C^∞ function (as a simple sufficient condition) in the local tangent space $\{\xi\}$ and we have appropriate convergence properties, we can reliably use the first order term in the Taylor expansion;

$$\frac{d}{d\epsilon} G(\epsilon)\Big|_{\epsilon=0} = \int \cos \epsilon p \; g(p)\Big|_{\epsilon=0} \tag{74}$$

so that, for $\epsilon \to 0$,

$$G(\epsilon) \to \epsilon \tilde{g}(0), \tag{75}$$

where $\tilde{g}(\xi)$ is the Fourier transform of $g(p)$. As a distribution on such functions $g(p)$, the assertion (3.39) then follows.

5. Application to the Bekenstein-Sanders fields

We have discussed the construction of a canonical quantum theory in terms of an embedding of the SHP relativistic classical and quantum theory into general relativity. We show in this section that this systematic embedding provides a framework for the method developed by Bekenstein and Milgrom for understanding the MOND [5–10] that appeared necessary to explain the galactic rotation curves [35].

The remarkable development of observational equipment and power of computation has resulted in the discovery that Newtonian gravitational physics leads to a prediction for the dynamics of stars in galaxies that is not consistent with observation. It was proposed that there should be a matter present which does not radiate light which would resolve this difficulty, but so far no firm evidence of the existence of such matter has emerged. Milgrom [5–10] proposed a modification of Newton's law (MOND) which could resolve the problem. However, since Newton's law of gravitation emerges in the "post-Newtonian approximation" to the geodesic motion in Einstein's theory of gravity [35], the modification of Newton's law must involve a modification of Einstein's theory.[8] Such a modification was proposed by Bekenstein and Milgrom [5–10] in terms of a conformal factor multiplying the usual Einstein metric.

The origin of such a conformal factor can be found in the potential term of the special relativistic SHP theory. The embedding of this theory in GR [20] brings this potential term as a world scalar. The Hamiltonian for the general relativistic case then has the form (8). It was shown by Horwitz et al. [37] that a very sensitive test

[8] Yahalom [36] has proposed an alternative view involving the retardation effects associated with gravitational waves, presently being tested and developed. We do not discuss this approach further here.

by geodesic deviation can be formulated by to study stability by transforming a standard nonrelativistic Hamiltonian of the form

$$H = \frac{\mathbf{p}^2}{2M} + V(\mathbf{r}) \tag{76}$$

to the form

$$H = \frac{1}{2M} p_i g^{ij}(\mathbf{r}) p_j, \tag{77}$$

with

$$g^{ij}(\mathbf{r}) = \phi(\mathbf{r}) \frac{E}{E - V} \delta^{ij}, \tag{78}$$

that is, a conformal factor on the original metric. Applying the same idea to the Hamiltonian (8), with the $g^{\mu\nu}(x)$ of Einstein replaced by the conformal form

$$\tilde{g}\mu\nu(x) = \phi(x) g^{\mu\nu}(x) \tag{79}$$

where

$$\phi(x) = \frac{k}{k - V(x)}, \tag{80}$$

with k a point in the spectrum of K, so that

$$H = \frac{1}{2M} p_\mu \tilde{g}\mu\nu(x) p_\nu. \tag{81}$$

We see that we can in this way achieve the structure proposed by Bekenstein and Milgrom [5–10] systematically. Moreover, in addition to providing a mechanism for achieving a realization of the MOND theory, in the original form (8), the world scalar term $V(x)$ could represent the so-called dark energy [2, 3], establishing a relation between the MOND picture and the anomalous expansion of the universe, a question presently under study.

The theory proposed by Bekenstein and Milgrom [5–10] did not, however, account for the lensing of light observed when light passes a galaxy. To solve this problem, Bekenstein and Sanders [4] proposed the introduction of a *vector field* $n^\mu(x)$, satisfying the normalization constraint

$$n^\mu n_\mu = -1, \tag{82}$$

so that the vector is timelike.

This vector field can then be used to construct a modified meric of the form

$$\tilde{g}\mu\nu_T = \phi(g^{\mu\nu}(x) + n^\mu(x) n^\nu(x)) + \phi^{-1} n^\mu(x) n^\nu(x). \tag{83}$$

With this modification, Bekenstein and Sanders [4] could explain the lensing effect. In the following, we show that this new field may arise from a *non-Abelian* gauge transformation [38, 39] on the quantum theory that we have discussed in Section 3. Although Contaldi et al. [40] point out that a gauge field in this context can have caustic singularities due to the presence of a massive system, Horwitz et al. [2, 3] show that in the limit in which the gauge field approaches the Abelian limit, as required by Bekenstein and Sanders [4], there is a residual term that can cancel the caustic singularities.

To preserve the normalization condition (83), it is clear that we have the possibility of moving the n field on a hyperbola with a Lorentz transformation, which we can perform by a gauge transformation.

A Lorentz transformation on n^μ is noncommutative, and therefore the gauge field is non-Abelian [21].

An analogy can be drawn to the usual Yang-Mills gauge on $SU(2)$, where there is a two-valued index for the wave function $\psi_\alpha(x)$. The gauge transformation is a two-by-two matrix function of x and acts only on the indices α. The condition of invariant absolute square (probability) is

$$\sum_\alpha \left| \sum_\beta U_{\alpha\beta}\psi_\beta \right|^2 = \sum |\psi_\alpha|^2 \tag{84}$$

Generalizing this structure, one can take the indices α to be infinite dimensional, and even continuous, so that (84) becomes (in the spectral representation for n^μ)

$$\int (dn) \left| \int (dn')U(n,n')\psi(n',x) \right|^2 = \int (dn)|\psi(n,x)|^2, \tag{85}$$

implying that $U(n,n')$ (at each point x) is a unitary operator on a Hilbert space $L^2(dn)$. Since we are assuming that n^μ lies on a hyperbola determined by (83), the measure is

$$(dn) = \frac{d^3n}{n^0}, \tag{86}$$

that is, a three-dimensional Lorentz invariant integration measure.

We now examine the gauge condition:

$$(p^\mu - \epsilon n'^\mu)U\psi = U(p^\mu - \epsilon n^\mu)\psi \tag{87}$$

Since the Hermitian operator p_μ acts as a derivative under commutation relations, we obtain

$$n'_\mu = Un_\mu U^{-1} - \frac{i}{\epsilon}\frac{\partial U}{\partial x^\mu}U^{-1}, \tag{88}$$

in the same form as the Yang-Mills theory [38, 39]. It is evident in the Yang-Mills theory, due to the operator nature of the second term, the field will be algebra-valued, and thus we have the usual structure of the Yang-Mills non-Abelian gauge theory. Here, if the transformation U is a Lorentz transformation, the numerical-valued field n_μ would be carried, at least in the first term, to a new value on a hyperbola. However, the second term is operator valued on $L^2(dn)$, and thus, as in the Yang-Mills theory, n'_μ would become operator valued. Therefore, in general, the gauge field n^μ is operator valued.

It follows from (87) that the "field strengths"

$$f^{\mu\nu} = \frac{\partial n^\mu}{\partial x_\nu} - \frac{\partial n^\nu}{\partial x_\mu} + i\epsilon[n^\mu, n^\nu]. \tag{89}$$

Under a gauge transformation $n^\mu \to n'^\mu$, the new fields create field strengths in the transformed form

$$f'^{\mu\nu} = \frac{\partial n'^\mu}{\partial x_\nu} - \frac{\partial n'^\nu}{\partial x_\mu} + i\epsilon[n'^\mu, n'^\nu] \tag{90}$$

according to

$$f'^{\mu\nu}(x) = Uf^{\mu\nu}(x)U^{-1}, \tag{91}$$

just as in the finite-dimensional Yang-Mills theories.

For

$$U \cong 1 + iG, \tag{92}$$

where G is infinitesimal, (87) becomes

$$n'^\mu = n^\mu + i[G, n^\mu] + \frac{1}{\epsilon}\frac{\partial G}{\partial x_\mu} + O(G^2). \tag{93}$$

Then,

$$\begin{aligned} n'^\mu n'_\mu &\cong n^\mu n_\mu + i\left(n^\mu [G, n_\mu] + [G, n^\mu]\right)n_\mu \\ &+ \frac{1}{\epsilon}\left(\frac{\partial G}{\partial x_\mu} n_\mu + n^\mu \frac{\partial G}{\partial x^\mu}\right). \end{aligned} \tag{94}$$

Let us take

$$\begin{aligned} G &= -\frac{i\epsilon}{2}\sum\left\{\omega_{\lambda\gamma}(n, x), \left(n^\lambda \frac{\partial}{\partial n_\gamma} - n^\gamma \frac{\partial}{\partial n_\lambda}\right)\right\} \\ &\equiv \frac{\epsilon}{2}\sum\left\{\omega_{\lambda\gamma}(n, x), N^{\lambda\gamma}\right\} \end{aligned} \tag{95}$$

where symmetrization is required since $\omega_{\lambda\gamma}$ is a function of n as well as x and

$$N^{\lambda\gamma} = -i\left(n^\lambda \frac{\partial}{\partial n_\gamma} - n^\gamma \frac{\partial}{\partial n^\lambda}\right). \tag{96}$$

Our investigation in the following will be concerned with a study of the infinitesimal gauge neighborhood of the Abelian limit, where the components of n^μ do not commute and therefore still constitute a Yang-Mills-type field. We shall show in the limit that the corresponding field equations acquire nonlinear terms and may therefore nullify the difficulty found by Contaldi et al. [40] demonstrating a dynamical instability for an Abelian vector-type TeVeS gauge field. They found that nonlinear terms associated with a non-Maxwellian-type action, such as $(div\mathbf{n})^2$, could nullify this caustic singularity, so that the nonlinear terms we find as a residue of the Yang-Mills structure induced by our gauge transformation might achieve this effect in a natural way.

Now, the second term of (94), which is the commutator of G with $n^\mu n_\mu$, vanishes, since this product is Lorentz invariant (the symmetrization in G does not affect this result).

We now consider the third term in (94).

$$\frac{1}{\epsilon}\frac{\partial G}{\partial x_\mu}n_\mu + n^\mu \frac{\partial G}{\partial x^\mu} = \frac{1}{2}\left\{\frac{\partial \omega_{\lambda\gamma}}{\partial x_\mu}, N^{\lambda\gamma}\right\} n_\mu + n_\mu\left\{\frac{\partial \omega_{\lambda\gamma}}{\partial x_\mu}, N^{\lambda\gamma}\right\}$$
$$= \frac{1}{2}N^{\lambda\gamma}\frac{\partial \omega_{\lambda\gamma}}{\partial x_\mu}n_\mu + \frac{\partial \omega_{\lambda\gamma}}{\partial x_\mu}N^{\lambda\gamma}n_\mu + n_\mu N^{\lambda\gamma}\frac{\partial \omega_{\lambda\gamma}}{\partial x_\mu} + n_\mu \frac{\partial \omega_{\lambda\gamma}}{\partial x_\mu}N^{\lambda\gamma} \tag{97}$$

There are two terms proportional to

$$\frac{\partial \omega_{\lambda\gamma}}{\partial x_\mu}n_\mu.$$

If we take

$$\omega_{\lambda\gamma}(n,x) = \omega_{\lambda\gamma}(k^\nu x_\nu), \tag{98}$$

where $k^\nu n_\nu = 0$, then

$$\frac{\partial \omega_{\lambda\gamma}}{\partial x_\mu}n_\mu = k^\mu n_\mu \omega'_{\lambda\gamma} = 0. \tag{99}$$

For the remaining two terms,

$$\begin{aligned} n_\mu N^{\lambda\gamma}\frac{\partial \omega_{\lambda\gamma}}{\partial x_\mu} + \frac{\partial \omega_{\lambda\gamma}}{\partial x_\mu}N^{\lambda\gamma}n_\mu \\ = N^{\lambda\gamma}n^\mu \frac{\partial \omega_{\lambda\gamma}}{\partial x_\mu} \\ + \left[n_\mu, N^{\lambda\gamma}\right]\frac{\partial \omega_{\lambda\gamma}}{\partial x_\mu} + \frac{\partial \omega_{\lambda\gamma}}{\partial x_\mu}n_\mu N^{\lambda\gamma} \\ + \frac{\partial \omega_{\lambda\gamma}}{\partial x_\mu}\left[N^{\lambda\gamma}, n_\mu\right]. \end{aligned} \tag{100}$$

The commutators contain only terms linear in n^μ and they cancel; the remaining terms are zero, and therefore the condition $n^\mu n_\mu = -1$ is invariant under this gauge transformation. It involves the coefficient $\omega_{\lambda\gamma}$ which is a function of the projection of x^μ onto a hyperplane orthogonal the n^μ. The vector k^μ of course depends on n^μ. We take, for definiteness, $k^\mu = n^\mu(n\cdot b) + b^\mu$, for some $b^\mu \neq 0$.

We now consider the derivation of field equations from a Lagrangian constructed with the ψ_s and $f^{\mu\nu}f_{\mu\nu}$. We take the Lagrangian to be of the form

$$\mathcal{L} = \mathcal{L}_f + \mathcal{L}_m, \tag{101}$$

where

$$\mathcal{L}_f = -\frac{1}{4}f^{\mu\nu}f_{\mu\nu} \tag{102}$$

and

$$\mathcal{L}_m = \psi^*\left(i\frac{\partial}{\partial\tau} - \frac{1}{2M}(p^\mu - \epsilon n^\mu)\left(p_\mu - \epsilon n_\mu\right) - \Phi\right)\psi + \text{c.c.} \tag{103}$$

In carrying out the variation of $\mathcal{L}_m$, the contributions of varying the ψs with respect to n vanish due to the field equations (Stueckelberg-Schrödinger equation) obtained by varying ψ^* (or ψ), and therefore in the variation with respect to n, only the explicit presence of n in (103) need be taken into account.

Note that for the general case of n generally operator valued, we can write

$$\psi^*(p^\mu - \epsilon n^\mu)\left(p_\mu - \epsilon n_\mu\right)\psi = ((p^\mu - \epsilon n^\mu)\psi)^*\left(p_\mu - \epsilon n_\mu\right)\psi, \tag{104}$$

since the Lagrangian density (108) contains an integration over $(dn')(dn'')$ (in spectral representation, considered in lowest order, as well as an integration over $(dx)\sqrt{g(x)}$ in the action). In the limit in which n is evaluated in the spectral representation, and noting that p_μ is represented by an imaginary differential operator, we can write this as

$$\psi^*(p^\mu - \epsilon n^\mu)\left(p_\mu - \epsilon n_\mu\right)\psi = -(p^\mu + \epsilon n^\mu))\psi^*\left(p_\mu - \epsilon n_\mu\right)\psi, \tag{105}$$

that is, replacing explicitly p_μ by $-i(\partial/\partial x^\mu) \equiv -i\partial_\mu$ (since it acts by commutator with the fields); we have

$$\delta_n\mathcal{L}_m = \frac{-i\epsilon}{2M}\{\psi^*(\partial_\mu - i\epsilon n_\mu)\psi - ((\partial_\mu + i\epsilon n_\mu)\psi^*)\psi\}\delta n^\mu, \tag{106}$$

or

$$\delta_n\mathcal{L}_m = j_\mu(n,x)\delta n^\mu, \tag{107}$$

where j_μ has the usual form of a gauge invariant current.

For the calculation of the variation of $\mathcal{L}_f$, we note that the commutator term in (89) is, in lowest order, a c-number function.

Calling

$$\omega'_\lambda{}^\mu n^\lambda \equiv v^\mu, \tag{108}$$

we compute the variation of

$$[n'^\mu, n'^\nu] = 2i(k^\nu v^\mu - k^\mu v^\nu) \tag{109}$$

Then, for

$$\delta_n[n'^\mu, n'^\nu] = \delta_{n^\gamma}\frac{\partial}{\partial n^\gamma}[n'^\mu, n'^\nu], \tag{110}$$

we compute

$$\frac{\partial}{\partial n^\gamma}[n'^\mu, n'^\nu] = 2i\left(\frac{\partial k^\nu}{\partial n^\gamma}v^\mu + k^\nu\frac{\partial v^\mu}{\partial n^\gamma} - (\mu \leftrightarrow \nu)\right). \tag{111}$$

With our choice of $k^\nu = n^\nu(n\cdot b) + b^\nu$,

$$\frac{\partial k^\nu}{\partial n^\gamma} = \delta^\nu{}_\gamma(n\cdot b) + n^\nu b_\gamma, \tag{112}$$

so that

$$\frac{\partial}{\partial n^\gamma}[n'^\mu, n'^\nu] = 2i\big(\big(\delta^\nu{}_\gamma(n\cdot b) + n^\nu b_\gamma\big)v^\mu + k^\nu \frac{\partial v^\mu}{\partial n^\gamma} - (\mu \leftrightarrow \nu)\big). \tag{113}$$

Here,

$$\frac{\partial v^\mu}{\partial n^\gamma} = \omega\prime^\mu{}_\gamma + \omega''{}_\lambda{}^{\mu}{}^{\lambda} n^\lambda \frac{\partial k_\sigma}{\partial n^\gamma} x^\sigma,$$

so we see that

$$\frac{\partial}{\partial n^\gamma}[n'^\mu, n'^\nu] \equiv \mathcal{O}_\gamma{}^{\mu\nu}, \tag{114}$$

where the quantity $\mathcal{O}_\gamma{}^{\mu\nu}\delta n^\gamma$ depends on the first and second derivatives of ω'^μ_λ, in general, nonlinear in n^μ. We therefore have

$$\delta_n[n'^\mu, n'^\nu] = \mathcal{O}_\gamma{}^{\mu\nu}\delta n^\gamma \tag{115}$$

In the limit that $\omega \to 0$, its derivative and higher derivatives which appear in $\mathcal{O}_\gamma{}^{\mu\nu}$ may not vanish (somewhat analogous to the case in gravitational theory when the connection form vanishes, but the curvature does not), so that this term can contribute in the limit of the an Abelian gauge.

Returning to the variation of $\mathcal{L}_f$, we see that

$$\begin{aligned}\delta\mathcal{L}_f &= -\frac{1}{4}\big(\big(\partial^\mu \delta n^\nu - \partial^\nu \delta n_\mu + i\epsilon\delta[n^\mu, n^\mu]\big)f_{\mu\nu} \\ &\quad + f^{\mu\nu}\big(\partial_\mu \delta n_\nu - \partial_\nu \delta n_\mu + i\epsilon\delta[n_\mu, n_\mu]\big)\big) \\ &= -\partial^\nu f_{\mu\nu}\delta n^\mu + 2if_{\mu\nu}\delta[n^\mu, n^\nu],\end{aligned} \tag{116}$$

where we have taken into account the fact that $[n_\mu, n_\mu]$ is a c-number function and integrated by parts the derivatives of δn. We then obtain

$$\delta\mathcal{L}_f = -\partial^\nu f_{\mu\nu}\delta n^\mu + 2i\epsilon f_{\lambda\sigma}\mathcal{O}^{\lambda\sigma}{}_\mu \delta n^\mu \tag{117}$$

Since the coefficient of δn^μ must vanish, we obtain the Yang-Mills equations for the fields given the source currents:

$$\partial^\nu f_{\mu\nu} = j_\mu - 2i\epsilon f_{\lambda\sigma}\mathcal{O}^{\lambda\sigma}{}_\mu, \tag{118}$$

which is nonlinear in the fields n^μ, as we have seen, even in the Abelian limit, where, from (106),

$$j_\mu = -i\frac{\epsilon}{2M}\{\psi^*(\partial_\mu - i\epsilon n_\mu)\psi - ((\partial_\mu + i\epsilon n_\mu)\psi^*)\psi\}. \tag{119}$$

6. Summary

In this chapter, we have shown that the formulation of MOND theory by Bekenstein and Milgrom [5–10] can have a systematic origin within the framework of the embedding of the SHP [1] theory into general relativity [20]. The SHP formalism admits a scalar potential term that appears both in the conformal factor giving rise to the MOND functions in the galaxy and, in the original form of the Hamiltonian, to a possible candidate for "dark energy." The solution of the lensing problem by Bekenstein and Sanders [4] by introduction of a local vector field was also shown to arise in a natural way in terms of a non-Abelian gauge field, for which, in the Abelian limit, there is a residual term that can cancel the caustic singularity found by Contaldi et al. [40] which can arise in a purely Abelian gauge theory.

Author details

Lawrence P. Horwitz[1,2,3]

1 School of Physics and Astronomy, Tel Aviv University, Ramat Aviv, Israel

2 Department of Physics, Bar Ilan University, Ramat Gan, Israel

3 Department of Physics, Ariel University, Ariel, Israel

*Address all correspondence to: larry@tauex.tau.ac.il

References

[1] Horwitz L. Relativistic Quantum Mechanics, Fundamental Theories of Physics. Vol. 180. Dordrecht: Springer; 2015

[2] Horwitz LP, Gershon A, Schiffer M. Map to conformal modification of spacetime metric: Kaluza Klein and TeVeS. Foundations of Physics. 2010;**41**: 141

[3] Gershon A, Horwitz LP. Kaluyza-Klein theory as a dynamics in a dual geometry. Journal of Mathematical Physics. 2009;**50**:102704

[4] Bekenstein JD, Sanders RH. Gravitational lenses and unconventional gravity theories. The Astrophysical Journal. 1994;**429**:480

[5] Milgrom M. A modification of the dynamics as a possible alternative to the hidden mass hypothesis. The Astrophysical Journal. 1983;**270**:365

[6] Milgrom M. A modification of the Newtonian dynamics-Implications for galaxies I. The Astrophysical Journal. 1983;**270**:371

[7] Milgrom M. A modification of the Newtonian dynamics-Implications for galaxies II. The Astrophysical Journal. 1983;**270**:384

[8] Bekenstein JD, Milgrom M. Does the missing mass problem signal the breakdown of Newtonian gravity?. The Astrophysical Journal. 1984;**286**:7

[9] Bekenstein JD. Relativistic gravitation theory for the modified Newtonian dynamics. Physical Review D. 2004;**70**:083509

[10] Bekenstein JD. The modified newtonian dynamics -MOND, and its implications for new physics. Contemporary Physics. 2006;**47**:387

[11] Stueckelberg ECG. Pair annihilation in classical physics. Helvetica Physica Acta. 1941;**14**:372

[12] Stueckelberg ECG. Le significance de temps propre en mecanique ondulatoire. Helvetica Physica Acta. 1941;**14**:588

[13] Stueckelberg ECG. Helvetica Physica Acta. 1942;**15**:23

[14] Horwitz LP, Piron C. Relativistic dynamics. Helvetica Physica Acta. 1973; **66**:316

[15] Collins RE, Fanchi JR. Relativistic quantum mechanics: A space-time formalism for spin-zero particles. Il Nuovo Cimento. 1978;**48A**:314

[16] Fanchi JR. Parametrized Relativistic Quantum Theory. Dordrecht: Kluwer; 1993

[17] Arshansky RI, Horwitz LP. The relativistic two body bound state I: The spectrum. Journal of Mathematical Physics. 1989;**30**:66

[18] Arshansky RI, Horwitz LP. The relativistic two body bound state II: The induced representation of SL(2,C). Journal of Mathematical Physics. 1989; **30**:380

[19] Arshansky RI, Horwitz LP. Relativistic potential scattering and phase shift analysis. Journal of Mathematical Physics. 1989;**30**:213

[20] Horwitz LP. An elementary canonical classical and quantum dynamics for general relativity. The European Physical Journal Plus. To be published. arXiv: 1810.09248

[21] Land M. Journal of Physics: Speeds of light in Stueckelberg-Horwitz-Piron

Theory. Conference Series - IOPscience. 2017;**845**:012024

[22] Dirac PAM. Quantum Mechanics. 1st ed. London: Oxford University Press; 1930

[23] Dirac PAM. Quantum Mechanics. 3rd ed. London: Oxford University Press; 1947

[24] van Hove L. Sur certaines representations unitaires d'un groupe infini de transformations. Proceedings of the Royal Academy of Science, Letters and Fine Arts of Belgium. 1951; **26**:1

[25] Groenwold HJ. On the principles of elementary quantum Mechanics. Physica. 1946;**12**:405

[26] Schwinger J. On gauge invarianca dn vascuum polarization. Physical Review. 1951;**82**:664

[27] DeWitt BS. Quantum field theory in curved spacetime. Physics Reports. 1975;**19**:295

[28] DeWitt BS. The Global Approach to Quantum Field Theory. Oxford: Oxford University Press; 2002

[29] Newton TD, Wigner E. Localized states for elementary systems. Reviews of Modern Physics. 1949;**21**:400

[30] Foldy LL, Wouthuysen SA. On the dirac theory of spin 1/2 particles and its non-relativistic limit. Physics Review. 1950;**78**:29

[31] Reed M, Simon B. Methods of Modern Mathematical Physics I, Functional Analysis. New York: Academic Press; 1979

[32] Strauss Y, Horwitz LP, Levitan J, Yahalom A. Quantum field theory of hamiltonian chaos. Journal of Mathematical Physics. 2015;**56**:072701

[33] Georgiev V. Chapter 8: Fourier transform on manifolds with constant negative curvature. In: Semilinear Hyperbolic Equations. Japan: Tokyo Mathematical Society; 2005. p. 126

[34] Abraham R, Marsden JE, Ratiu T. Manifolds, tensor analysis and applications. In: Applied Mathematical Sciences. Vol. 75. New York: Springer-Verlag; 1988

[35] Weinberg S. Gravitation and Cosmology. New York: John Wiley and Sons; 1972

[36] Yahalom A. The effect of retardation on galactic rotation curves. Journal of Physics: Conference Series - IOPscience. 2019;**1239**:012006

[37] Horwitz LP, Ben Zion Y, Lewkowicz M, Schiffer M, Levitan J. Geometry of hamiltonian chaos. Physical Review Letters. 2007;**98**:234301

[38] Yang CN. Magnetic monopoles, fiber bundles and gauge fields. Annals of the New York Academy of Sciences. 1977;**294**:86

[39] Yang CN, Mills R. Conservation of isotopic spin and isotopic gauge invariance. Physics Review. 1954;**96**:191

[40] Contaldi CR, Wiseman T, Withers B. TeVeS gets caught on caustics. Physical Review D. 2008;**78**:044034

Chapter 10

Fast Indicators for Orbital Stability: A Survey on Lyapunov and Reversibility Errors

Giorgio Turchetti and Federico Panichi

Abstract

We present a survey on the recently introduced fast indicators for Hamiltonian systems, which measure the sensitivity of orbits to small initial displacements, Lyapunov error (LE), and to a small additive noise, reversibility error (RE). The LE and RE are based on variational methods and require the computation of the tangent flow or map. The modified reversibility error method (REM) measures the effect of roundoff and is computed by iterating a symplectic map forward and backward the same number of times. The smoothest indicator is RE since it damps the oscillations of LE. It can be proven that LE and RE grow following a power law for regular orbits and an exponential law for chaotic orbits. There is a numerical evidence that the growth of RE and REM follows the same law. The application to models of celestial and beam dynamics has shown the reliability of these indicators.

Keywords: variational principles, reversibility error, additive noise, roundoff

1. Introduction

The global stability properties of Hamiltonian systems and symplectic maps have a solid theoretical foundation [1, 2]. Nevertheless, the determination of the orbital stability by computing the maximum Lyapunov exponent is a procedure difficult to implement numerically, because of the $t \to \infty$ limit. For this reason a variety of fast indicators has been developed during the last two decades [3–7]. The variational methods mentioned above measure the sensitivity to initial conditions of the orbit computed for finite times. The spectral methods [8, 9] relate the stability to the behavior of the Fourier spectrum of the orbit computed for finite times.

In the framework of the variational methods, we have proposed two indicators [10–12] the Lyapunov error (LE) and the reversibility error (RE) introducing also the modified reversibility error method (REM). The LE is due to a small displacement of the initial condition, the RE is due to an additive noise, and REM is due to roundoff. The reversibility error due to the roundoff or noise is more convenient with respect to the error occurring in the forward evolution of the map.[1]

[1] The forward error (FE) due to additive noise in the forward evolution of a map can be defined and written in terms of the tangent map. However, RE is very simply related to LE, whereas FE is not. In addition the error due to roundoff in the forward evolution requires in principle the evaluation of the exact orbit or, in practice, its evaluation with a much higher accuracy.

IntechOpen

In the limit of a vanishing amplitude of the initial displacement or of the random displacement, the LE and RE are defined by using the tangent map along the orbit. Furthermore, RE is related to LE by a very simple formula. A reversibility error is always present in numerical computations due to roundoff even when no additive noise is introduced. We compute REM by iterating n times the map M, then its inverse n times, and dividing the norm of the displacement from the initial point, by the roundoff amplitude. The procedure is extremely simple and does not require the knowledge of the tangent map. Though the effect of roundoff on a single iteration is not equivalent to a random displacement, after many iterations the cumulative result is comparable if the computational complexity of the map is sufficiently high. The main difference is that for an additive noise, the error is defined as the root mean square deviation of the noisy orbit with respect to the exact one, obtained by averaging over all possible realizations of the noise, whereas for the roundoff a unique realization is available. As a consequence REM fluctuates with the iteration number n, whereas RE does not. A statistical analysis of roundoff compared to a random noise was previously performed using the fidelity method [13, 14], and a comparison of REM with other fast indicators was initially carried out for the standard map [15]. The growth of errors, for REM-, RE-, and Lyapunov-based indicators, is governed by the tangent map. For LE a small initial displacement is propagated and amplified along the orbit. For RE or REM, a random or pseudorandom displacement is introduced at any (forward or backward) iteration of the map and is propagated and amplified along the orbit. The final random displacement is the sum of the global displacements triggered by the local displacements (due to noise or roundoff) occurring at any iteration. Therefore, it is not surprising that the square of RE is twice the sum of the squares of LE computed along the orbit and that all the numerical experiments suggest that REM exhibits a similar behavior even though with larger fluctuations.

For an integrable map, the growth of LE and RE follows asymptotically a power law n^α, and the exact analytical result is known. This result can be extended to quasi integrable maps by using the normal forms theory. For uniformly chaotic maps (hyperbolic automorphisms of the torus), the LE and RE have an exponential growth $e^{\lambda n}$. For generic maps, the asymptotic growth of LE and RE follows a power law in the regions of regular motion and an exponential law in the regions of chaotic motion, and the same behavior is observed for REM. For an integrable or quasi integrable map, LE has an asymptotic linear growth $\alpha = 1$ with oscillations, whereas RE has an asymptotic power law growth with $\alpha = 3/2$ without oscillations, since they are rapidly damped. The oscillations of LE disappear when the map is written in normal coordinates. For a linear map conjugated to a rotation, the power law exponents are $\alpha = 0$ for LE and $\alpha = 1/2$ for RE. For REM the power law exponent α varies between 0 and 1, its value depending on the computational complexity of the map and therefore on the choice of coordinates.

The definition of LE we propose differs from fast Lyapunov indicator (FLI) [3] or orthogonal fast Lyapunov indicator (OFLI) [4], which are based on the growth along the orbit of the norm of a given initial displacement vector. Indeed, we compute the growth of the vectors of an orthogonal basis, which amounts to defining LE, which we denote as e_n^L, as the square root of $\mathrm{Tr}\left((DM^n(\mathbf{x}_0))^T DM^n(\mathbf{x}_0)\right)$ where $M(\mathbf{x})$ is the map, $DM(\mathbf{x})$ denotes the tangent map, and $\mathbf{x}_0$ is the initial condition. This definition has the obvious advantage of insuring the correct asymptotic growth.

Indeed the anomalies in the behavior of FLI [16], due to the choice of the initial vector, are not met. The use of exponential growth factor of nearby orbits (MEGNO) [17] allows to filter the oscillations which are still present in LE. The RE is obtained from the covariance matrix which is computed from the tangent map. We denote this error by e_n^R, which has a very simple relation with LE given by the square root of $\left(e_0^L\right)^2 + 2\left(e_1^L\right)^2 + \ldots + 2\left(e_{n-1}^L\right)^2 + \left(e_n^L\right)^2$. We first analyze the case of linear maps to explore the behavior of REM. A systematic comparison of LE, RE, and REM is presented for two basic models: the standard map and the Hénon map. The asymptotic power law exponents are computed by using the MEGNO filter. For nonlinear two-dimensional maps, the behavior of the errors has been compared moving along a one-dimensional grid in the phase plane: crossing of islands has a clear signature, the chaotic regions are very neatly distinguished, and good agreement with the theoretical predictions is found.

A rectangular region of phase plane has been examined by choosing a grid and using a color, logarithmic scale for the errors at each point. Also in this case, a good correspondence with the phase space portrait is found. On the basis of the analysis presented here and the experience gained in investigating more complex models from celestial mechanics [18] and beam dynamics [19], we suggest to compare RE and REM with LE, possibly, filtered with MEGNO, to damp the oscillations[2]. For maps of dimension 4 or higher, a direct geometric inspection of the orbits is not possible since the Poincaré section requires an interpolation Hamiltonian. As a consequence the use of fast indicators is the only practical approach to analyze the orbital stability. Hamiltonian systems have a continuous time flow, and the errors LE and RE denoted by $e^L(t)$ and $e^R(t)$, respectively, are computed by using the fundamental matrix $\mathrm{L}(t)$ of the tangent flow. In this case $e_L(t) = \left(\mathrm{Tr}\left(\mathrm{L}^T(t)\,\mathrm{L}(t)\right)\right)^{1/2}$ and $e^R(t)$ are given by the square root of $2\int_0^t ds\,\left(e^L(s)\right)^2$, whose trapezoidal rule approximation gives the relation found for the maps [12]. Standard procedures allow to approximate the orbit $\mathbf{x}(t)$ by the iterates $M^n(\mathbf{x}_0)$ of a symplectic integrator map M (see [20]) and the fundamental matrix $\mathrm{L}(t)$ by $DM^n(\mathbf{x}_0)$ (see [21]). The paper has the following structure. In Section 2, we recall the definitions for LE and RE and obtain their mutual relation. In Section 3, we present the analytical results on LE and RE together with the numerical results on REM for integrable maps. In Section 4, the key features of two prototype models, the standard map and the Hénon map, are summarized. In Section 5, we present a detailed numerical analysis of LE, RE, and REM for the standard map. In Section 6, the same analysis is presented for the Hénon map. In Section 7, the summary and conclusions are presented.

2. Definition of errors

Given a symplectic map $M(\mathbf{x})$ where $\mathbf{x} \in \mathbb{R}^{2d}$, we consider the orbits $\mathbf{x}_n = M^n(\mathbf{x}_0)$ and $\mathbf{y}_n = M^n(\mathbf{y}_0)$ for two initial points $\mathbf{x}_0$ and $\mathbf{y}_0 = \mathbf{x}_0 + \epsilon\boldsymbol{\eta}_0$, respectively, where $\boldsymbol{\eta}$ is a unit vector. We consider the normalized displacement $\boldsymbol{\eta}_n$ at iteration n defined by

$$\boldsymbol{\eta}_n = \lim_{\epsilon\to 0} \frac{\mathbf{y}_n - \mathbf{x}_n}{\epsilon} = \lim_{\epsilon\to 0} \frac{M(\mathbf{y}_{n-1}) - M(\mathbf{x}_{n-1})}{\epsilon} \tag{1}$$

[2] The application of MEGNO to RE is not necessary due to the absence of oscillations, whereas its application to REM is not recommended because the fluctuations are not filtered and the computational cost is quadratic rather than linear in the iteration order.

which satisfies the linear recurrence

$$\boldsymbol{\eta}_n = DM(\mathbf{x}_{n-1})\boldsymbol{\eta}_{n-1} \qquad (DM)_{ij} = \frac{\partial M_i}{\partial x_j} \tag{2}$$

where DM is the tangent map. For any finite ϵ, we have $\mathbf{y}_n = \mathbf{x}_n + \epsilon\boldsymbol{\eta}_n + O(\epsilon^2)$. We might define the error as the norm of $\boldsymbol{\eta}_n$ which is closely related to the fast Lyapunov indicator FLI (see [3]) as

$$e_n(\boldsymbol{\eta}_0) = \|\boldsymbol{\eta}_n\| = \|DM^n(\mathbf{x}_0)\boldsymbol{\eta}_0\| \qquad (FLI)_n = \max_{1\le k\le n} \log e_k(\boldsymbol{\eta}_0) \tag{3}$$

and to its variants such as OFLI [4]. The mean exponential growth factor of nearby orbits, MEGNO [17], denoted by $Y_n = Y(e_n)$ is the double average of the slope, and we denote it as Δe_n^2. When n is a continuous variable, then $\Delta e_n^2 = d\log e_n^2/d\log n$. When n is an integer, the standard definition is

$$\Delta e_n^2 = n\left(\log e_n^2 - \log e_{n-1}^2\right) \qquad Y_n = \langle\langle \Delta e_n^2 \rangle\rangle. \tag{4}$$

2.1 Lyapunov error

We propose a definition of the Lyapunov error which is independent from the choice of the initial vector:

$$\begin{aligned} e_n^L &= \left(\mathrm{Tr}\left(\mathrm{A}_n^T \mathrm{A}_n\right)\right)^{1/2} \\ \mathrm{A}_n &= DM^n(\mathbf{x}_0) = DM(\mathbf{x}_{n-1})\mathrm{A}_{n-1} \qquad \mathrm{A}_0 = \mathrm{I} \end{aligned} \tag{5}$$

It is immediate to verify that given an orthonormal basis $\boldsymbol{\eta}_{0\,k}$, we have

$$e_n^L = \left(\sum_{k=1}^{2d} e_n^2(\boldsymbol{\eta}_{0\,k})\right)^{1/2} \tag{6}$$

and obviously the result does not depend on the choice of the basis. The computational cost of $e_n(\boldsymbol{\eta}_0)$ is $2d$ times higher with respect to e_n^L, but this difference is negligible with respect to the computational cost of the matrix $DM(\mathbf{x}_{n-1})$, which recursively gives $\boldsymbol{\eta}_n$ and A_n. A similar definition is proposed in the case of Hamiltonian flows (see the last Subsection 2.6 and [12] for more details). An advantage of the proposed definition is that it takes into account the error growth on all possible directions of the initial displacement vector. As a consequence, no spurious effects due to the choice of the initial vector have to be faced (see [16]).

2.2 Forward error

When an additive noise of amplitude ϵ is introduced, the reference orbit $\mathbf{x}_n$ is replaced by the noisy one $(\mathbf{y}_n)$ having the same initial condition:

$$\mathbf{y}_n = M(\mathbf{y}_{n-1}) + \epsilon\boldsymbol{\xi}_n \qquad \mathbf{y}_0 = \mathbf{x}_0 \tag{7}$$

where $\boldsymbol{\xi}_n$ are independent random vectors satisfying

$$\langle \boldsymbol{\xi}_n \rangle = 0 \qquad \langle \boldsymbol{\xi}_n \boldsymbol{\xi}_m^T \rangle = \mathrm{I}\,\delta_{nm} \tag{8}$$

The global stochastic displacement satisfies a linear nonhomogeneous equation and is defined by

$$\Xi_n = \lim_{\epsilon \to 0} \frac{\mathbf{y}_n - \mathbf{x}_n}{\epsilon} \qquad \Xi_n = DM(\mathbf{x}_{n-1})\Xi_{n-1} + \xi_n \tag{9}$$

with initial condition $\Xi_0 = 0$. Letting $\Sigma_n^{2F} = \langle \Xi_n \Xi_n^T \rangle$ be the covariance matrix the forward error is defined by

$$e_n^F = \langle \Xi_n \cdot \Xi_n \rangle^{1/2} = \left(\mathrm{Tr}\left(\Sigma_n^{2F}\right)\right)^{1/2} \tag{10}$$

The explicit solution for Ξ_n is given by

$$\Xi_n = \sum_{k=1}^{n} DM^{n-k}(\mathbf{x}_k)\xi_k = \sum_{k=0}^{n-1} \mathrm{B}_k \xi_{n-k} \qquad \mathrm{B}_k = DM^k(\mathbf{x}_{n-k}) \tag{11}$$

where B_k can be evaluated recursively as

$$\mathrm{B}_k = \mathrm{B}_{k-1} DM(\mathbf{x}_{n-k}) \qquad \mathrm{B}_0 = \mathrm{I}. \tag{12}$$

The expression for the forward error finally reads

$$e_n^F = \left(\sum_{k=0}^{n-1} \mathrm{Tr}\left(\mathrm{B}_k^T \mathrm{B}_k\right)\right)^{1/2}. \tag{13}$$

The computation cost of B_n is negligible, once we have evaluated the tangent map, but the storage of the tangent map along the orbit up to n is required.

2.3 Reversibility error

We have just defined the forward error, but it will not be used, because it is only an intermediate step toward the definition of the reversibility error. Consider the backward evolution $\mathbf{y}_{n,\,-k}$, given by the inverse map M^{-1}, with initial point $\mathbf{y}_{n,\,0} = \mathbf{y}_n$. The point $\mathbf{y}_n$ is reached by iterating n times the map M with a random displacement of amplitude ϵ at each step, starting from $\mathbf{y}_0 = \mathbf{x}_0$ (see the previous subsection). The orbit $\mathbf{y}_{n,\,-k}$ is obtained by iterating k times the map M^{-1} with a random displacement of the same amplitude at each step:

$$\mathbf{y}_{n,\,-k} = M^{-1}\left(\mathbf{y}_{n,\,-k+1}\right) + \epsilon \boldsymbol{\xi}_{-k} \qquad k = 1, \ldots, n \tag{14}$$

The random backward displacements $\boldsymbol{\xi}_{-k}$ are independent from the forward displacements $\xi_{k'}$, namely, $\langle \xi_{-k}\xi_{k'} \rangle = 0$ and $\langle \xi_{-k}\xi_{-k'} \rangle = \mathrm{I}\,\delta_{k\,k'}$, for any $k, k'>0$. We consider then the stochastic process $\Xi_{n,\,-k}$ defined by

$$\Xi_{n,\,-k} = \lim_{\epsilon \to 0} \frac{\mathbf{y}_{n,\,-k} - \mathbf{x}_{n-k}}{\epsilon} \qquad \Xi_{n,\,-k} = DM^{-1}(\mathbf{x}_{n-k+1})\Xi_{n,\,-k+1} + \xi_{-k} \tag{15}$$

with initial condition $\Xi_{n,\,0} = \Xi_n$. The solution of the recurrence reads

$$\Xi_{n,\,-k} = DM^{-k}(\mathbf{x}_n)\Xi_n + \sum_{j=1}^{k} DM^{-(k-j)}(\mathbf{x}_{n-j})\xi_{-j} \tag{16}$$

For $k = n$, we obtain the global normalized displacement $\Xi_n^R = \Xi_{n,\,-n}$ after n forward and n backward iterations with noise of vanishing amplitude:

$$\Xi_n^R = DM^{-n}(\mathbf{x}_n)\,\Xi_n + \sum_{j=1}^{n} DM^{-(n-j)}\left(\mathbf{x}_{n-j}\right)\boldsymbol{\xi}_{-j} \tag{17}$$

Letting $\Sigma_n^{2R} = \left\langle \Xi_n^R \left(\Xi_n^R\right)^T \right\rangle$ be the covariance matrix, the reversibility error (RE) is defined by

$$e_n^R = \left\langle \Xi_n^R \cdot \Xi_n^R \right\rangle^{1/2} = \left(\mathrm{Tr}\,\Sigma_n^{2R}\right)^2 \tag{18}$$

and using Eqs. (17) and (8), an explicit expression involving only the tangent maps is obtained. Indeed the global stochastic displacement reads

$$\Xi_n^R = \sum_{k=1}^{n} DM^{-n}(\mathbf{x}_n)\,DM^{n-k}(\mathbf{x}_k)\,\boldsymbol{\xi}_k + \sum_{k=1}^{n} DM^{-(n-k)}(\mathbf{x}_{n-k})\,\boldsymbol{\xi}_{-k}. \tag{19}$$

Taking into account the independence of $\boldsymbol{\xi}_k$ and $\boldsymbol{\xi}_{-k'}$, the expression for the reversibility error $e_n^R = \left\langle \Xi_n^R \cdot \Xi_n^R \right\rangle^{1/2}$ is immediately obtained.

2.4 Analytical relation between RE and LE indicators

The RE can be obtained from LE in a very simple way. We first notice that

$$DM^{-n}(\mathbf{x}_n)\,DM^{n-k}(\mathbf{x}_k) = DM^{-k}(\mathbf{x}_k) \tag{20}$$

We prove this relation by writing $M^{-n}\left(M^{n-k}(\mathbf{x})\right) = M^{-k}(\mathbf{x})$, computing the tangent map $DM^{-n}\left(M^{n-k}(\mathbf{x})\right)DM^{n-k}(\mathbf{x}) = DM^{-k}(\mathbf{x})$, and evaluating it for $\mathbf{x} = \mathbf{x}_k$. As a consequence the expression for the reversibility error becomes

$$\begin{aligned}
\left(e_n^R\right)^2 &= \mathrm{Tr}\left\langle \Xi_n^R \left(\Xi_n^R\right)^T \right\rangle = \sum_{k=1}^{n}\Bigg(\mathrm{Tr}\left[\left(DM^{-k}(\mathbf{x}_k)\right)^T \left(DM^{-k}(\mathbf{x}_k)\right)\right] + \\
&\quad + \mathrm{Tr}\left[\left(DM^{-(n-k)}(\mathbf{x}_{n-k})\right)^T DM^{-(n-k)}(\mathbf{x}_{n-k})\right]\Bigg) = \\
&= 2\sum_{k=1}^{n-1} \mathrm{Tr}\left[\left(DM^{-k}(\mathbf{x}_k)\right)^T \left(DM^{-k}(\mathbf{x}_k)\right)\right] + \\
&\quad + \mathrm{Tr}\left[\left(DM^{-n}(\mathbf{x}_n)\right)^T DM^{-n}(\mathbf{x}_n)\right] + \mathrm{Tr}(\mathrm{I})
\end{aligned} \tag{21}$$

Starting from $M^{-k}\left(M^k(\mathbf{x})\right) = \mathbf{x}$, computing the tangent map, and evaluating it at $\mathbf{x} = \mathbf{x}_0$, it follows that

$$DM^{-k}(\mathbf{x}_k) = \left(DM^k(\mathbf{x}_0)\right)^{-1} \tag{22}$$

Given any symplectic matrix L[3], we can prove that

[3] A symplectic matrix L is defined by $\mathrm{LJL}^T = \mathrm{J}$ where J is antisymmetric and $\mathrm{J}^2 = -\mathrm{I}$. As a consequence $\mathrm{L}^{-1} = -\mathrm{JL}^T\mathrm{J}$, and $\left(\mathrm{L}^{-1}\right)^T = -\mathrm{JLJ}$ so that $\mathrm{Tr}\left(\mathrm{L}^{-1T}\mathrm{L}^{-1}\right) = \mathrm{Tr}\left(\mathrm{JLJ}^2\mathrm{L}^T\mathrm{J}\right) = \mathrm{Tr}\left(\mathrm{LL}^T\right)$.

$$\mathrm{Tr}\left(\left(\mathrm{L}^{-1}\right)^T \mathrm{L}^{-1}\right) = \mathrm{Tr}\left(\mathrm{L}^T \mathrm{L}\right). \tag{23}$$

As a consequence in Eq. (21), we can use the following relation:

$$\mathrm{Tr}\left[\left(DM^{-k}(\mathbf{x}_k)\right)^T \left(DM^{-k}(\mathbf{x}_k)\right)\right] = \mathrm{Tr}\left[\left(DM^{k}(\mathbf{x}_0)\right)^T \left(DM^{k}(\mathbf{x}_0)\right)\right] = \left(e_k^L\right)^2. \tag{24}$$

Finally, the relation between LE and RE is given by

$$\left(e_n^R\right)^2 = \sum_{k=1}^{n} \left(\left(e_k^L\right)^2 + \left(e_{n-k}^L\right)^2\right) = 2\left(\sum_{k=1}^{n-1} \left(e_k^L\right)^2 + \frac{1}{2}\left(e_0^L\right)^2 + \frac{1}{2}\left(e_n^L\right)^2\right). \tag{25}$$

This relation clearly shows how the error due to random kicks along the orbit is related to the error due to initial orthogonal kicks.

2.5 Roundoff-induced reversibility error

The reversibility error method (REM) is a very simple procedure based on n iterations of the map M followed by n iterations of the inverse map. The distance from the initial point normalized by the roundoff amplitude ϵ defines the REM error. Denoting with M_ϵ the map with roundoff, we have

$$e_n^{REM} = \left(\frac{\|M_\epsilon^{-n} \circ M_\epsilon^n(\mathbf{x}_0) - \mathbf{x}_0\|}{\epsilon}\right) \tag{26}$$

where ϵ is the roundoff amplitude. For the eight-byte representation of reals, we choose $\epsilon = 10^{-17}$. If the map has a sufficiently high computational complexity, the displacement $\boldsymbol{\xi}$ defined by $M_\epsilon(\mathbf{x}) - M(\mathbf{x}) = \epsilon\boldsymbol{\xi}$ is almost random, but a unique realization is available. (For a discussion on the roundoff error, see [22]). As a consequence, the e_n^{REM} has large fluctuations, whereas e_n^R has a smooth dependence on n since it is defined by an average overall possible realizations of the stochastic displacements occurring at each iteration.

2.6 Errors for Hamiltonian flows

For Hamiltonian flows, we define the Lyapunov error $e_L(t)$ according to

$$e_L(t) = \left(\mathrm{Tr}\left(\mathrm{L}^T(t)\,\mathrm{L}(t)\right)\right)^{1/2} \tag{27}$$

where $\mathrm{L}(t)$ is the fundamental matrix for the tangent flow, which satisfies the linear equation $d\mathrm{L}/dt = \mathrm{JHL}$, H denotes the Hessian of the Hamiltonian computed along the orbit $\mathrm{H}_{ij} = \partial^2 H/\partial x_i \partial x_j$, and the initial condition for the matrix $\mathrm{L}(t)$ is $\mathrm{L}(0) = \mathrm{I}$. The relation with the standard fast indicators is the same as for the symplectic maps. Let $\Xi_R(t)$ be the stochastic displacement from $\mathbf{x}_0$ after a noisy evolution up to time t and backward to $t = 0$, divided by the noise amplitude ε in the limit $\varepsilon \to 0$. It has been proven [12] that $\Xi_R(t)$ satisfies a linear Langevin equation whose solution reads

$$\Xi^R(t) = \int_0^t \mathrm{L}^{-1}(s)\left(\boldsymbol{\xi}(s) - \boldsymbol{\xi}(s-t)\right)ds. \tag{28}$$

The reversibility error in this case is defined by the mean square deviation of the random displacement $e^R(t) = \left\langle \Xi^R(t) \cdot \Xi^R(t) \right\rangle^{1/2}$. As shown in [12] and from Eq. (28), we immediately obtain

$$e^R(t) = \left(2 \int_0^t \left(e^L(s)\right)^2 ds \right)^{1/2}. \tag{29}$$

If the continuous time t is replaced by an integer n and we approximate the integral with the trapezoidal rule, we recover the relation in Eq. (25) obtained for a symplectic map.

3. Integrable maps

We evaluate the errors for integrable maps with an elliptic fixed point at the origin, whose normal form is a rotation $\mathrm{R}(\Omega)$ with a frequency Ω depending on the distance from the origin. The LE asymptotic growth is linear, and oscillations are present unless the coordinates are normal. The RE asymptotic growth follows a power law n^α with exponent $\alpha = 3/2$. If the map is linear, its asymptotic growth follows a power law with $\alpha = 0$ for LE and $\alpha = 1/2$ for RE. The oscillations reflect the loss of rotational symmetry when generic coordinates are used. The roundoff induced reversibility error REM is also sensitive to the choice of coordinates, and a comparison between RE and REM is presented in the next sections.

3.1 Change of coordinate system

In generic coordinates an integrable map $M(\mathbf{x})$ is conjugated to its normal form $N(\mathbf{X})$ by a symplectic coordinate transformation $\mathbf{x} = \Phi(\mathbf{X})$; as a consequence the conjugation equation and its iterates read

$$M(\mathbf{x}) = \Phi \circ N \circ \Phi^{-1}(\mathbf{x}) \qquad\qquad M^n(\mathbf{x}) = \Phi \circ N^n \circ \Phi^{-1}(\mathbf{x}) \tag{30}$$

which imply that the orbits $\mathbf{x}_n = M^n(\mathbf{x}_0)$ and $\mathbf{X}_n = N^n(\mathbf{X}_0)$ are related by $\mathbf{x}_n = \Phi(\mathbf{X}_n)$. The tangent maps are given by

$$DM^n(\mathbf{x}_0) = D\Phi(\mathbf{X}_n) DN^n(\mathbf{X}_0) D\Phi^{-1}(\mathbf{x}_0) = D\Phi(\mathbf{X}_n) DN^n(\mathbf{X}_0) \left(D\Phi(\mathbf{X}_0)\right)^{-1} \tag{31}$$

where we used $D\Phi(\mathbf{X}) D\Phi^{-1}(\mathbf{x}) = \mathrm{I}$, a relation which is proved to hold by computing the Jacobian of $\Phi \circ \Phi^{-1}(\mathbf{x}) = \mathbf{x}$. As a consequence, the expression for the Lyapunov error in both coordinate systems is

$$\begin{aligned} \left(e_n^L(\mathbf{X}_0)\right)^2 &= \mathrm{Tr}\left[\left(DN^n(\mathbf{X}_0)\right)^T \left(DN^n(\mathbf{X}_0)\right.\right] \\ \left(e_n^L(\mathbf{x}_0)\right)^2 &= \mathrm{Tr}\left[\left(DM^n(\mathbf{x}_0)\right)^T \left(DM^n(\mathbf{x}_0)\right.\right] \end{aligned} \tag{32}$$

Taking Eq. (31) into account, the last equation can be written as

$$\begin{aligned} \left(e_n^L(\mathbf{x}_0)\right)^2 &= \mathrm{Tr}\left[\mathrm{V}^{-1}(\mathbf{X}_0) \left(DN^n(\mathbf{X}_0)\right)^T \mathrm{V}(\mathbf{X}_n) DN^n(\mathbf{X}_0) \right] \\ \mathrm{V}(\mathbf{X}) &= \left(D\Phi(\mathbf{X})\right)^T D\Phi(\mathbf{X}) \end{aligned} \tag{33}$$

Notice that V is a positive-defined matrix and that its determinant is equal to 1 if Φ is symplectic. For a two-dimensional map, we can write

$$\mathrm{V} \equiv \begin{pmatrix} a & b \\ b & c \end{pmatrix} \qquad \mathrm{V}^{-1} = \begin{pmatrix} c & -b \\ -b & a \end{pmatrix} \tag{34}$$

where $ac - b^2 = 1$.

3.2 Isochronous rotations: oscillations in LE and RE

If the given map is linear and two-dimensional and $M(\mathbf{x}) = \mathrm{L}\mathbf{x}$ with $|\mathrm{Tr\,L}| < 2$, then the map is conjugated to a rotation $\mathrm{R}(\omega)$:

$$\mathrm{L} = \mathrm{T}\mathrm{R}(\omega)\mathrm{T}^{-1} \qquad \mathrm{R}(\omega) = \begin{pmatrix} \cos(\omega) & \sin\omega \\ -\sin(\omega) & \cos(\omega) \end{pmatrix} \tag{35}$$

Letting $\mathbf{x}_n = \mathrm{L}^n \mathbf{x}_0$ and $\mathbf{X}_n = \mathrm{R}^n \mathbf{X}_0$, the orbits in the coordinate $\mathbf{x}$ and the normal coordinate $\mathbf{X} = \mathrm{T}^{-1}\mathbf{x}$ and the Lyapunov errors are given by

$$\begin{aligned} \left(e_n^L(\mathbf{X}_0)\right)^2 &= \mathrm{Tr}\left[\mathrm{R}(-n\omega)\mathrm{R}(n\omega)\right] = 2 \\ \left(e_n^L(\mathbf{x}_0)\right)^2 &= \mathrm{Tr}\left[\mathrm{V}^{-1}\mathrm{R}(-n\omega)\mathrm{V}\mathrm{R}(n\omega)\right] = 2\cos^2(n\omega) + \left(a^2 + c^2 + 2b^2\right)\sin^2(n\omega) = \\ &= \frac{(a+c)^2}{2} + \left(2 - \frac{(a+c)^2}{2}\right)\cos(2n\omega) \end{aligned} \tag{36}$$

where $\mathrm{V} = \mathrm{T}^T\mathrm{T}$, and we have used the representation given by Eq. (34) where $D\Phi = \mathrm{T}$. The error is constant in normal coordinate $\mathbf{X}$ and oscillates between 2 and $(a+c)^2 - 2 = a^2 + c^2 + 2b^2$ in the coordinate $\mathbf{x}$. The geometric interpretation is obvious since the orbits of the map belong to an ellipse rather than a circle. The result for the reversibility error is given by

$$\begin{aligned} \left(e_n^R(\mathbf{X}_0)\right)^2 &- 4n \\ \left(e_n^R(\mathbf{x}_0)\right)^2 &= 2n\,\frac{(a+c)^2}{2} + \left(2 - \frac{(a+c)^2}{2}\right)f(n) \\ f(n) = \sum_{k=1}^{n}\left(\cos(2k\omega) + \cos(2(n-k)\omega)\right) &= \cos(2n\omega) + \frac{\cos(2(n-1)\omega) - \cos(2n\omega)}{1 - \cos(2\omega)} \end{aligned} \tag{37}$$

We shall first consider the dependence of the errors on the iteration order n from $n = 1$ up to a maximum value N. Then, we shall consider the dependence on the initial condition $\mathbf{x}_0$ when it is varied on a one-dimensional grid crossing the origin for the value N of the iteration number. We choose the linear map L which depends on a single parameter λ, and its relation with the rotation frequency is

$$\mathrm{L} = \begin{pmatrix} 1-\lambda & 1 \\ -\lambda & 1 \end{pmatrix} \qquad \sin\frac{\omega}{2} = \frac{\sqrt{\lambda}}{2} \qquad 0 \le \lambda \le 4 \tag{38}$$

The rotation $\mathrm{R}(\omega)\mathbf{x}$ is the linear part for the Hénon map, whereas $\mathrm{L}\mathbf{x}$ is the linear part of the standard map that will be discussed in the next sections. The behavior of

LE and RE for these maps is provided by Eqs. (36) and (37). The error growth follows a power law with exponent $\alpha = 0$ for LE and $\alpha = 1/2$ for RE. Oscillations are present when the coordinates are not normal.

For a generic map such as L defined by Eq. (38), the growth of REM follows a power law exponent $\alpha = 1/2$ as RE, for almost any value of λ, as shown by **Figure 1**, right panel, where the plot of MEGNO corresponding to 2α is shown. The result for the map $R(\omega)$ in normal coordinates is shown in the left panel of the same figure, and the exponent is $\alpha = 1$ for almost all the values of ω.

Letting $\mathbf{X} = (X, P)^T$ and (ϕ, J) be the action angle coordinates defined by $X = (2J)^{1/2} \cos\phi$ and $P = -(2J)^{1/2} \sin\phi$, the rotation in the $\mathbf{X}$ plane becomes a translation on the cylinder:

$$\phi_n = \phi_{n-1} + \omega \mod 2\pi \qquad J_n = J_{n-1} \tag{39}$$

and in this case REM vanishes. These results show that REM strongly depends on the computational complexity of the map. The error growth always follows a power law, but, depending on the choice of the coordinates, the exponent α varies in the range $[0, 1]$. Unlike RE, we observe that REM depends linearly on the distance of the initial condition $\mathbf{x}_0$ from the origin. In **Figure 2**, we plot e_N^{REM} as a function of the initial condition when it varies on a one-dimensional grid issued from the origin for

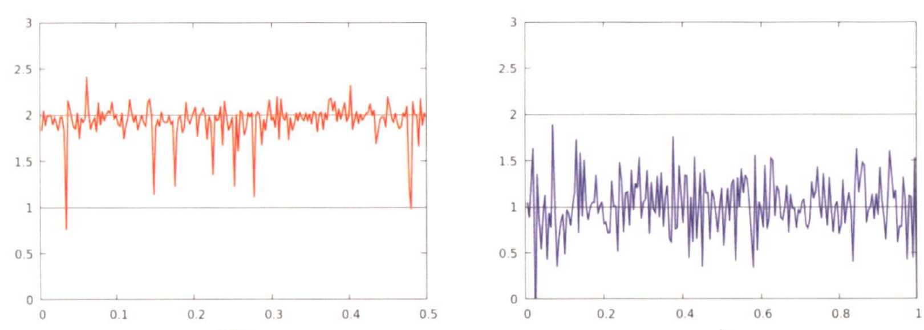

Figure 1.
Left frame: twice the asymptotic power law exponent provided by the MEGNO filter Y_N with $N = 1000$ applied to REM for a rotation $R(\omega)$ where $\omega/2\pi$ varies in the interval $[0, 1/2]$. The initial condition is $x_0 = 0.1, p_0 = 0$. Right frame: twice the asymptotic power law exponent provided by the MEGNO filter Y_N with $N = 1000$ applied to REM for a linear map L given by Eq. (38) whose parameter λ varies in $[0, 1]$. Initial condition $x_0 = 0.1, p_0 = 0$.

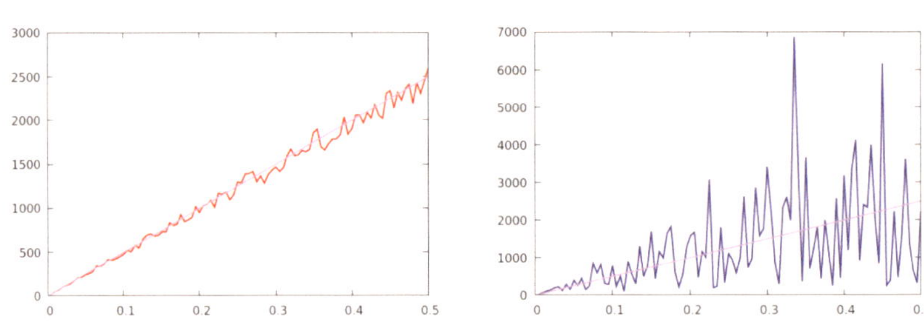

Figure 2.
Left frame: reversibility error due to roundoff e_N^{REM} for a rotation $R(\omega)$ with $\omega = 2\pi(\sqrt{2} - 1)$ and $N = 1000$ when the initial condition is varied. We choose $x_0 \in [0, 0.5], p_0 = 0$. The dependence on x_0 is evident and a linear fit $f(x_0) = 5000 x_0$ is shown, purple line. Right frame: computation of the error for the linear map with $\lambda = 4\sin^2(\omega/2)$ where ω has the same value. The linear fit is the same.

the rotation $\mathrm{R}(\omega)$ and the linear map L. The linear dependence is evident in both cases, even though the fluctuations are large for the linear map.

3.3 Anisochronous rotations

An integrable map M in normal coordinates and the tangent map DM^n read

$$M(\mathbf{x}) = \mathrm{R}(\Omega(J))\mathbf{x} \qquad DM^n(\mathbf{x}) = \mathrm{R}(n\Omega) + n\Omega'\mathrm{R}'(n\Omega)\mathbf{x}\mathbf{x}^T \tag{40}$$

where $J = \|\mathbf{x}\|^2/2$ is the action. The square of the Lyapunov error[4] reads

$$\left(e_n^L\right)^2 = \mathrm{Tr}\left(\left(DM^n\right)^T DM^n\right) = 2 + n^2(2J\Omega')^2 \tag{41}$$

and the square of the reversibility error is given by

$$\begin{aligned}\left(e_n^R\right)^2 = \sum_{k=1}^{n}\left(\left(e_k^L\right)^2 + \left(e_{n-k}^L\right)^2\right) &= 4n + (2J\Omega')^2\left(\sum_{k=1}^{n}(n-k)^2 + \sum_{k=1}^{n}k^2\right)\\ &= 4n + (2J\Omega')^2\,2\left(\frac{n^3}{3} + \frac{n}{6}\right)\end{aligned} \tag{42}$$

For a fixed value of the invariant J, the slope of $\left(e_n^R\right)^2$, whose asymptotic value is 2α, is defined as $d\log\left(e_n^R\right)^2/d\log n$, and its double average is given by MEGNO $Y_n \equiv Y(e_n)$. The range of variation is $[1, 3]$. One can prove that for a given initial condition, the intermediate value $Y_n = 2$ is reached for

$$n = \frac{14.5}{2J|\Omega'|} = \frac{14.5}{\left(x_0^2 + p_0^2\right)|\Omega'|} \tag{43}$$

In **Figure 3**, we show the variation with $n \in [1, 1000]$ of Y_n computed for RE given by Eq. (42), corresponding to the map presented in Eq. (40), where $\Omega(J)$ is a linear function of J, and find a perfect agreement with the analytical estimate of the value of n for which the value $Y_n = 2$ is reached. In **Figure 4**, we show the variation of e_N^R and the corresponding MEGNO filter Y_N with the initial condition chosen on a one-dimensional grid crossing the origin for $N = 1000$. The integrable map is given by Eq. (40), where Ω' is constant. The error reaches a minimum value at the origin, and a similar behavior for Y_N is observed. Also e_N^{REM} decreases by approaching the origin so that the behavior is similar even though in this case the fluctuations are large. We notice that MEGNO does not eliminate the fluctuations of REM. In order to compute Y_n, one needs the sequence e_m^{REM} for $m = 1, \ldots, n$ whose computational cost is of the order of n^2. This can be avoided by storing the sequence $\mathbf{x}_{\varepsilon,m}$ and computing $\hat{e}_m{}^{REM} = \|M_\varepsilon^{-m}(\mathbf{x}_{\varepsilon,n}) - \mathbf{x}_{n-m}\|$ which turns out to be comparable with e_m^{REM}.

---

[4] The standard definition for an initial displacement along the unit vector $\boldsymbol{\eta}_0$ is $e_n^L(\boldsymbol{\eta}_0) = \|DM^n\boldsymbol{\eta}_0\|$ where $\|DM^n\boldsymbol{\eta}_0\|^2 = 1 + (n\Omega')^2\|\mathbf{x}\|^2(\boldsymbol{\eta}_0\cdot\mathbf{x})^2 + 2n\Omega'(\boldsymbol{\eta}_0\cdot\mathbf{x})(\boldsymbol{\eta}_0\cdot \mathrm{J}\mathbf{x})$ and $\mathrm{J} = \begin{pmatrix} 0 & 1\\ -1 & 0\end{pmatrix}$. The sum $\left\|DM^n\begin{pmatrix}1\\0\end{pmatrix}\right\|^2 + \left\|DM^n\begin{pmatrix}0\\1\end{pmatrix}\right\|^2$ gives the Lyapunov error $\left(e_n^L\right)^2$.

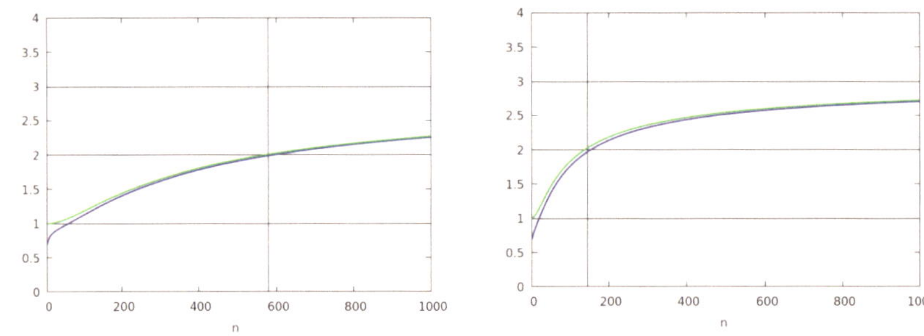

Figure 3.
Left frame: plot of MEGNO Y_n *on the error* e_N^R *for* $1 \le n \le N$ *with* $N = 1000$ *for the integrable map with* $\Omega' = 0.1$ *and initial condition* $x_0 = 0.5, p_0 = 0$ *(blue line). The green line refers to a modified definition* YM_n*, where* $n\left(\log e_n^2 - \log e_{n-1}^2\right)$ *is replaced with* $\left(\log e_n^2 - \log e_{n-1}^2\right)/(\log n - \log(n-1))$*, which, applied to the sequence* $e_n = n^\alpha$*, gives* 2α *for any n. The vertical line gives the theoretical estimate of the value of n for which* $Y_n = 2$ *(see Eq. (43)) Right frame: the same for* $x_0 = 1, p_0 = 0$*.*

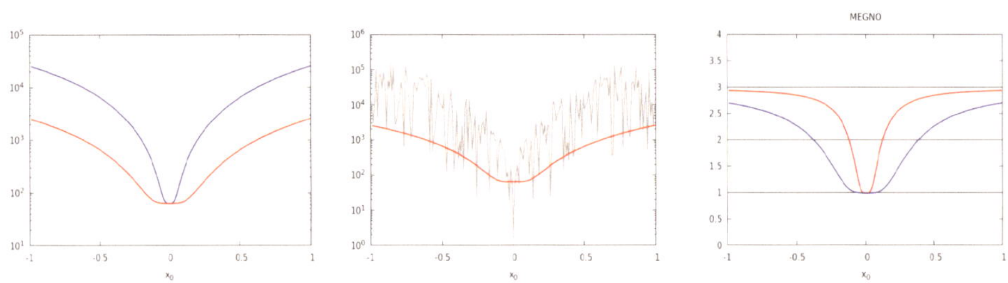

Figure 4.
Left frame: plot or the error e_N^R *with* $N = 1000$ *for the integrable map as a function of the initial condition* $x_0, p_0 = 0$ *with* $\Omega'(J) = 0.1$ *(red line) and* $\Omega'(J) = 1$ *(blue line). Center frame: same plot with* e_N^{REM} *for* $\omega = 2\pi\left(\sqrt{2} - 1\right)$ *and* $\Omega'(J) = 0.1$*, gray line, compared with* e_n^R*, red line. Right frame: plot of* Y_N *for the integrable map with* $\Omega' = 0.1$ *(red line) and* $\Omega'(J) = 1$*(blue line).*

If the coordinates are not normal, which is usually the case for a quasi integrable map, the correspondence between RE and REM is better, and it is confirmed by comparing the results for MEGNO. Just a shift of 1/2 in the exponent of the power law n^α occurs close to the origin, if the linear part is a rotation R, as for the Hénon map. If the linear part is L as for the standard map, there is no shift. The better correspondence is not surprising since the computational complexity of the map is higher when the coordinates are not normal.

4. Non-integrable maps

We examine here the behavior of the proposed dynamical indicators for two basic models, the standard map and the Hénon map.

The **standard map** is defined as a map on the torus $\mathbb{T}^2$ and reads

$$p_{n+1} = p_n - \frac{\lambda}{2\pi} \sin\left(2\pi x_n\right) \mod 1 \qquad x_{n+1} = x_n + p_{n+1} \mod 1 \tag{44}$$

where x, p belong to the interval $[-1/2, 1/2]$ whose ends are identified. For $\lambda \ll 1$ and $|p| \gg \sqrt{\lambda}$, it is just a weakly perturbed rotator, and x, p are action angle coordinates. The origin is an elliptic fixed and very close to it; the map is approximated by a linear map

$$x_{n+1} = (1-\lambda)x_n + p_n \qquad p_{n+1} = -\lambda x_n + p_n \tag{45}$$

This map is conjugated to a rotation $R(\omega)$ for $0 < \lambda < 2$ where $\sin(\omega/2) = \sqrt{\lambda}/2$. The point $x = \pm 1/2, p = 0$ is hyperbolic, and for $\lambda \ll 1$ the corresponding orbit is approximated by the separatrix of the Hamiltonian:

$$H = \frac{p^2}{2} - \frac{\lambda}{(2\pi)^2}\cos(2\pi x) \tag{46}$$

which is the interpolating Hamiltonian of the map when $\lambda \to 0$. We observe that the frequency for small oscillations is $\omega = \sqrt{\lambda}$ (see Eq. (38)) when $\lambda \to 0$. Since the time scale of the Hamiltonian H is $T = 2\pi/\sqrt{\lambda} \gg 1$, the symplectic integrator in Eq. (44), obtained for a time step $\Delta t = 1$, provides a good approximation to the orbit. Conversely, the Hamiltonian provides a good interpolation to the orbit of the map. The equation for the separatrix of H is given by

$$p = \pm\frac{\sqrt{\lambda}}{\pi}\cos(\pi x). \tag{47}$$

As a consequence, for λ small the width of the separatrix is $2\sqrt{\lambda}/\pi$. When λ increases, non-integrable features appear, such as chains of islands corresponding to resonances and a chaotic region near the separatrix due to homoclinic intersections.

The **Hénon map** is defined by

$$\begin{pmatrix} x_{n+1} \\ p_{n+1} \end{pmatrix} = R(\omega)\begin{pmatrix} x_n \\ p_n + x_n^2 \end{pmatrix} \qquad R(\omega) = \begin{pmatrix} \cos\omega & \sin\omega \\ -\sin\omega & \cos\omega \end{pmatrix} \tag{48}$$

Close to the origin, this is just a rotation with frequency ω. For $\omega \to 0$ this is a good symplectic integrator of the Hamiltonian:

$$H = \omega\frac{p^2 + x^2}{2} - \frac{x^3}{3} \tag{49}$$

with time step $\Delta t = 1$. The approximation is good since the characteristic time is the period of the linear rotation $T = 2\pi/\omega$. The motion is bounded by the orbit issued from the hyperbolic fixed point of the map $(x = 2\tan(\omega/2), p = -2\tan^2(\omega/2))$ which corresponds to the critical point $(x = \omega, p = 0)$ of the Hamiltonian. The stability boundary is approximated by $H(x,p) = \omega^3/6$ whose orbit explicitly reads $p = \pm(\omega - x)\sqrt{(\omega + 2x)/3}$. The Birkhoff normal forms provide an integrable approximation to the map and the corresponding interpolating Hamiltonian, from which the errors may be analytically computed.

5. The standard map

We have analyzed the errors e_n for a fixed initial condition by varying n up to a maximum value N, by varying the initial condition on a one-dimensional grid for $n = N$ and by choosing a grid in the phase plane for $n = N$. The LE shows oscillations with n, RE grows without oscillations, and the behavior of RE is similar to RE although with large fluctuations. The results obtained by filtering the errors with

MEGNO confirm this observation. In **Figure 5**, we plot the errors e_n for $\lambda = 0.1$ and $Y(e_n)$ by varying n. The fast oscillations of LE and the large fluctuations of REM are clearly visible.

When the orbit is chaotic, the growth of all errors is exponential. However, LE and RE can grow until the overflow is reached, whereas REM can grow only up to $1/\epsilon$ where ϵ is the machine accuracy. Typically in double precision, the overflow corresponds to 10^{300} where $\epsilon^{-1} \sim 10^{17}$. The same limitation is met when the Lyapunov error is computed using the shadow orbit method without renormalization rather than with the variational method. In **Figure 6**, we show the errors for a chaotic orbit when $\lambda = 0.8$. Both LE and RE exhibit an exponential growth after an initial transitory phase. The behavior of REM is very similar until $n \leq 300$. For higher values the saturation to 10^{17} is evident, and REM ceases to grow exponentially.

5.1 Initial conditions on a one-dimensional grid

Figure 7 shows the variation of LE, RE, and REM for $\lambda = 0.1$, with the initial condition chosen on a regular grid in the vertical axis p for a fixed order N. The LE oscillates when the initial condition varies, RE does not oscillate, and REM fluctuates. When the MEGNO filter is applied, LE and RE are equally smooth, whereas REM still fluctuates.

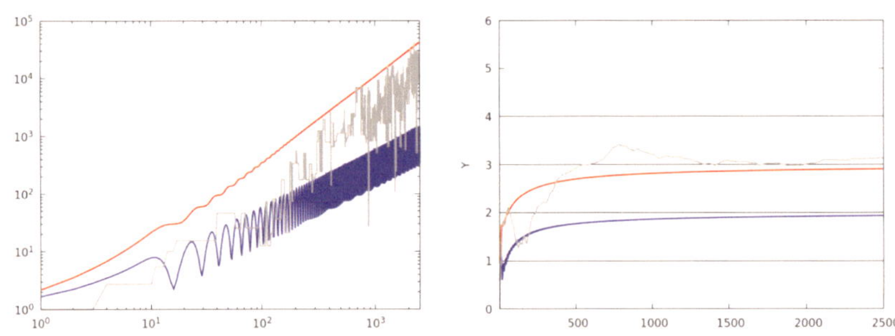

Figure 5.
Left frame: plot of the errors for the standard map with $\lambda = 0.1$ and initial condition $x_0 = 0, p_0 = 0.075$. Lyapunov error e_n^L (blue line), reversibility errors e_n^R (red line), and e_n^{REM} (gray line), for $1 \leq n \leq 2500$. Right frame: plots for the MEGNO filter Y_n for the same errors.

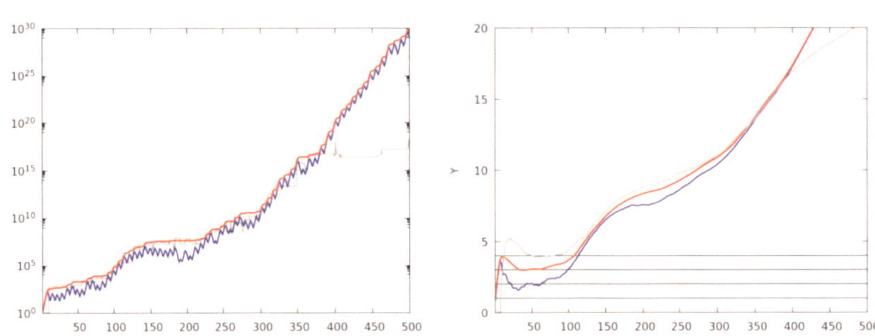

Figure 6.
Left frame: plot of the errors for the standard map with $\lambda = 0.8$ and initial condition $x_0 = 0, p_0 = 0.26$ corresponding to a chaotic orbit. Lyapunov error e_n^L (blue line), reversibility errors e_n^R (red line), and e_n^{REM} (gray line), for $n \leq 500$. Right frame: plots for the MEGNO filter Y_n for the same errors.

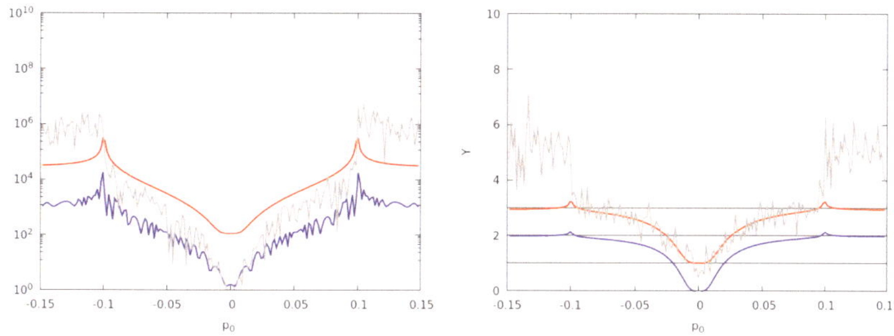

Figure 7.
Left frame: variation for the standard map with $\lambda = 0.1$ of the errors LE (blue line), RE (red line), REM (gray line) computed at $N = 1000$ with the initial condition $x_0 = 0, p_0 \in [-0.15, 1, 0.15]$. Right frame: the same for MEGNO Y_N.

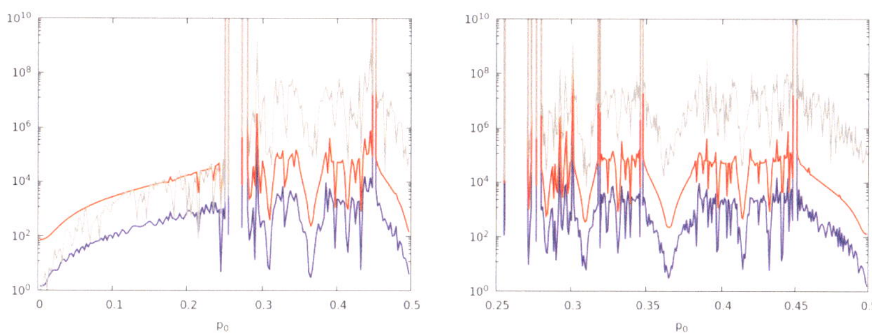

Figure 8.
Left frame: variation for the standard map with $\lambda = 0.8$ of the errors LE (blue line), RE (red line), and REM (gray line), computed at $N = 1000$ with the initial condition $x_0 = 0, p_0 \in [0, 0.5]$. Right frame: magnification in the interval $p_0 \in [0.25, 0.5]$.

In **Figure 8**, the same results are shown for a higher value of the parameter $\lambda = 0.8$ at which the dynamical structure is rich due to the presence of many resonances and small chaotic regions. The effectiveness in detecting the resonances is evident.

5.2 Initial conditions on a two-dimensional domain

We compare here LE, RE, and REM when the initial conditions are chosen in a two-dimensional phase space domain and the iteration number has a fixed value N. The most effective way of analyzing the results is to plot the errors using a logarithmic, color scale. Following the conclusions of our previous section, we show LE, RE, and REM, in a logarithmic color scale. We choose a regular two-dimensional grid in a square (or rectangular) domain of phase space with $N_g \times N_g$ points, where we compute the errors and show the result using a color scale. In order to analyze the details, smaller squares may be chosen eventually increasing the iteration number. In **Figure 9**, we show for $N = 500$ and $N_g = 200$ the color plots for the errors of the standard map with $\lambda = 0.8$ and in **Figure 10** for $\lambda = 1.5$. In the first case, the measure of chaotic orbits is small with respect to the regular ones. We observe that LE has some weak structures within the main regular component surrounding the origin, visible when the figure is sufficiently magnified. Such structures of LE, related to the oscillating growth with n, disappear when MEGNO is computed and

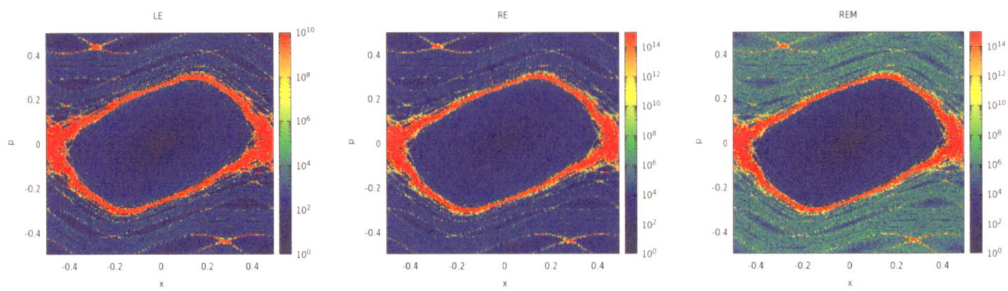

Figure 9.
Left frame: standard map with $\lambda = 0.8$ color plot of LE in a logarithmic scale for $N = 500$ and a grid with $N_g = 200$. Center frame: standard map with $\lambda = 0.8$ color plot of RE in a logarithmic scale for $N = 500$ and a grid with $N_g = 200$. Right frame: color plot of REM in a logarithmic scale.

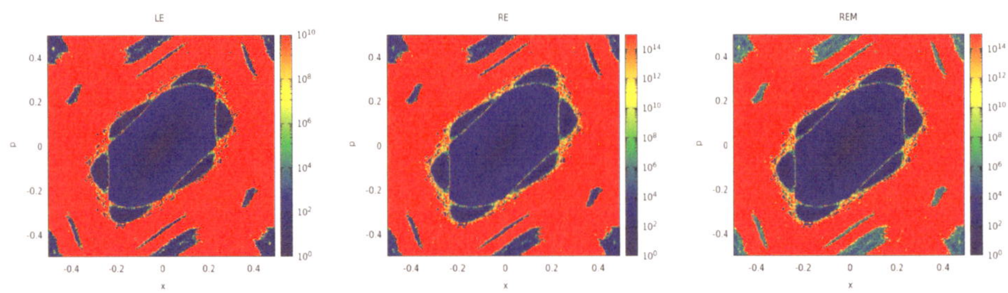

Figure 10.
Left frame: standard map with $\lambda = 0.8$ color plot of LE in a logarithmic scale for $N = 500$ and a grid with $N_g = 200$. Center frame: standard map with $\lambda = 1.5$ color plot of RE in a logarithmic scale for $N = 500$ and a grid with $N_g = 200$. Right frame: color plot of REM in a logarithmic scale.

are not present in the RE and REM plots. The spurious structures observed in FLI, which depend on the choice of the initial vector, are not present in LE, because in our definition the error does not depend on the choice of an initial displacement vector. Notice that the chosen scales have maximum equal to 10^{10} for LE and 10^{15} for RE and REM. This choice is suggested by the asymptotic behavior n^{α} of the error for regular orbits where $\alpha = 1$ for LE and $\alpha = 3/2$ for RE.

6. The Hénon map

We briefly report in this section the numerical results on the errors computed on domains of dimensions 1 and 2 in phase space. Close to the origin, the linear map in this case is a rotation $R(\omega)$. As a consequence the power law exponent of REM varies from 1 to 2, whereas the exponent for RE varies from 1/2 to 3/2. Within the main island, the variation range of the exponent for RE and REM is the same $[1/2, 3/2]$. The behavior of LE and RE close to the origin is analytically obtained by using the normal forms. The frequency $\Omega(J)$, from normal forms at the lowest order, reads

$$\Omega \simeq \omega + J\Omega_2 \qquad \Omega_2 = -\frac{1}{8}\left(3\cot\left(\frac{\omega}{2}\right) + \cot\left(\frac{3\omega}{2}\right)\right) \tag{50}$$

a formula valid for frequencies $\omega/2\pi$ not approaching the unstable resonances 0 and 1/3 where Ω_2 diverges.

In the normal coordinates (X, P), the behavior of errors is given by Eqs. (41) and (42). In the original coordinates (x, p), the error could be evaluated using Eq. (33).

In normal coordinates, the errors grow as $2J|\Omega_2|n^\alpha$ where $\alpha = 1$ for LE and $\alpha = 3/2$ for RE. When the frequency attains a low resonant value, a chain of islands appears. Close to the separatrix $J = J_s$, the frequency vanishes as $\Omega \sim 1/\log(J_s - J)$ and consequently $\Omega'(J) \sim (J_s - J)^{-1}$ as J_s is approached, up to a logarithmic correction. The errors diverge by approaching the separatrix even though the power law growth does not change except on the separatrix itself. As a consequence, LE and RE can detect the separatrix. If the remainder in the normal form interpolating Hamiltonian is taken into account, then the separatrix becomes a thin chaotic region where the errors have an exponential growth and MEGNO rises linearly with n. The REM behaves as RE neglecting its fluctuations. The Hénon map, we consider here, has a linear frequency $\omega/2\pi = 0.21$ which is close to the resonance $1/5$. As a consequence a chain of five islands appears before reaching the dynamic aperture, namely, the boundary of the stability region, beyond which the orbits escape to infinity.

In **Figure 11**, we show the variation of LE, RE, and REM computed at a fixed order N and after filtering them with MEGNO, when the initial conditions are chosen on a one-dimensional grid. The resonance $1/5$ is met, as shown by the appearance of a large chain of islands, since $\Omega(J)$ is monotonically decreasing. The chaotic layer at the border of the islands chain is very thin so that LE and RE grow by approaching it, as for an integrable map with a separatrix.

In **Figure 12**, we show the color plots of LE, RE, and REM in a square domain of phase space. The weakly chaotic separatrix is detectable in LE and is more clearly visible in RE. The REM plot differs from RE for the up-shit 1/2 of the power law exponent before the thin chaotic separatrix and for the presence of fluctuations.

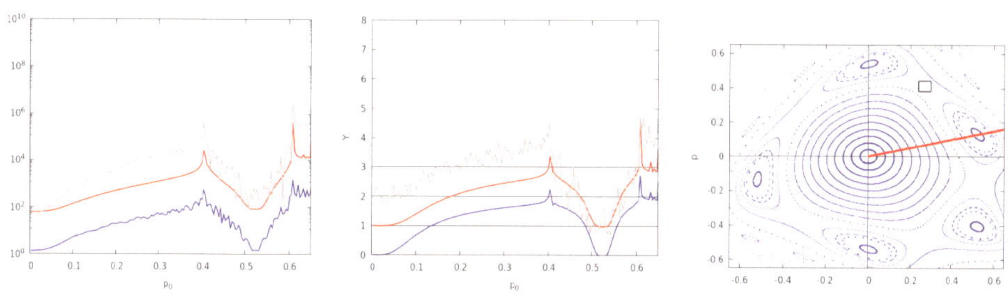

Figure 11.
Left frame: errors for the Hénon with $\omega = 0.21 \times (2\pi)$: LE (blue line), RE (red line), and REM (gray line) computed at iteration number $N = 1000$ along the line $x = r\cos\alpha$, $p = r\sin\alpha$ with $\alpha = 14^o$ joining the origin with the center of the first of five islands. Center frame: computation of MEGNO with $N = 1000$ for LE (blue line), RE (red line), and REM (gray line). Right frame: phase portrait of the Hénon map. The initial conditions for the errors in the left and right frames are chosen on the red segment.

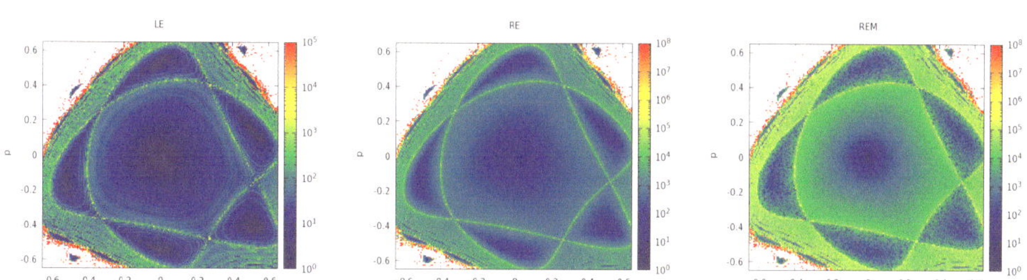

Figure 12.
Left frame: Hénon map with $\omega = 0.21 \times (2\pi)$ color plot of LE in a logarithmic scale for $N = 500$ and a grid with $N_g = 200$. The white points belong to the unstable region beyond the dynamic aperture. Center frame: color plot of RE in a logarithmic scale. Right frame: color plot of REM in a logarithmic scale.

7. Conclusions

We have presented a detailed analysis of the stability indicators LE, RE, and REM recently proposed. Defining the square of LE as the trace of the tangent map times, its transpose renders this indicator independent from the choice of an initial vector, which can introduce spurious structures. The RE is the reversibility error due to additive random noise, whereas REM is the reversibility error due to the roundoff. A very simple relation is found between RE and LE. The oscillations, which affect the fast Lyapunov indicator, can be filtered with MEGNO. Since RE has a smooth behavior and does not exhibit oscillations, filtering it by MEGNO is not necessary. The asymptotic behavior of REM is similar to RE even though it exhibits large fluctuations. The displacements caused by roundoff are almost random vectors, if the map has a high computational complexity, but since just a single realization of the process is available, the fluctuations cannot be averaged.

We have first examined the behavior of LE and RE for linear maps and for integrable maps. If the fixed point is elliptic, then the asymptotic growth follows a power law n^{α}, and the exponent does not depend on the chosen coordinates for LE and RE. Conversely, the presence of oscillations and their amplitude depends on the choice of coordinates. The growth of REM also follows a power law, but the choice of coordinates affects the exponent itself.

For a generic map which has regular and chaotic components, the error growth follows a power law and an exponential law, respectively. For the standard map and the Hénon map, the behavior of LE, RE, and REM has been compared first by varying the iteration order n up to a some value N, for a selected initial condition. Then the errors for $n = N$ have been compared when the initial point moves on a line. The theoretical predictions concerning the power law growth in the regular regions and the exponential growth in the chaotic ones are confirmed. For two-dimensional maps, the error plots for initial conditions in a rectangular domain of phase space are very similar, and the correspondence with the phase space portraits is excellent. Moreover, the different plots allow a quantitative comparison of the orbital sensitivity to initial displacements, noise, and roundoff. For maps of dimension 4 or higher, the proposed error plots on selected phase planes allow to inspect the orbital stability. Hamiltonian flows must be approximated with a high accuracy by symplectic maps, with algorithms which provide simultaneously the corresponding tangent maps [20, 21], in order to compute the errors discussed so far. A special care is required in comparing RE with REM when the chosen phase plane is invariant. Indeed given an initial point in the invariant plane, the noise brings the orbit out of it, whereas the roundoff usually does not. In this case a random kick before reversing the orbit is sufficient to bring the orbit out of the invariant plane and to restore the correspondence between REM and RE. The satisfactory results obtained so far, not only in the simple models presented here but also in high dimensional models of celestial mechanics, prove that the method we propose has a wide range of applicability.

DOI: http://dx.doi.org/10.5772/intechopen.88085

Author details

Giorgio Turchetti[1*] and Federico Panichi[2]

1 Department of Physics and Astronomy, Alma Mater Studiorum—University of Bologna, Bologna, Italy

2 Faculty of Mathematics and Physics, Institute of Physics, University of Szczecin, Szczecin, Poland

*Address all correspondence to: giorgio.turchetti@unibo.it

References

[1] Arnold VI. The stability of the equilibrium position of a Hamiltonian system of ordinary differential equations in the general elliptic case. Soviet Mathematics Doklady. 1961;**2**: 247-249

[2] Oseledec I. A multiplicative ergodic theorem. Lyapunov characteristic numbers for dynamical systems. Transactions of the Moscow Mathematical Society. 1968;**19**:197

[3] Froeschlé C, Lega E. On the structure of symplectic mappings: The fast Lyapunov indicator: A very sensitive tool. Celestial Mechanics and Dynamical Astronomy. 2000;**78**:167

[4] Fouchard M, Lega E, Froeschlé C. On the relationship between fast Lyapunov indicator and periodic orbits for continuous flows. Celestial Mechanics and Dynamical Astronomy. 2002; **83, 205**

[5] Barrio R. Sensitivity tools vs. Poincaré sections. Chaos, Solitons & Fractals. 2005;**25**:711

[6] Skokos C. Alignment indices: A new simple method to determine the ordered or chaotic nature of orbits. Journal of Physics A: Mathematical and General. 2001;**34**:10029

[7] Skokos Ch, Antonopoulos Ch, Bountis TC, Vrahatis MN. Detecting order and chaos in Hamiltonian systems by the SALI method. Journal of Physics A: Mathematical and General. 2005;**37**: 6269-6284. Available from: https://arxiv.org/pdf/nlin/0404058.pdf

[8] Laskar J. The chaotic motion of the solar system—A numerical estimate of the size of the chaotic zones. Icarus. 1990;**88**:266

[9] Laskar J, Froeschlé C, Celletti A. The measure of chaos by numerical analysis of fundamental frequencies: Application to the standard map. Physica D. 1992; **56**:253

[10] Panichi F, Ciotti L, Turchetti G. Fidelity and reversibility in the restricted 3 body problem. Communications in Nonlinear Science and Numerical Simulation. 2015;**35**: 53-68

[11] Turchetti G, Panichi F, Sinigardi S, Vaienti S. Errors, correlations, fidelity for noisy Hamiltonian flows. Theory and numerical examples. Journal of Physics A: Mathematical and Theoretical. 2017; **50**:064001. Available from: https://arxiv.org/pdf/1509.07738.pdf

[12] Panichi F, Turchetti G. Lyapunov and reversibility errors for Hamiltonian flows. Chaos, Solitons & Fractals. 2018; **112**:83

[13] Turchetti G, Vaienti S, Zanlungo F. Asymptotic distribution of global errors in the numerical computation of dynamical systems. Physica A: Statistical Mechanics and its Applications. 2010;**389**:4994-5006

[14] Turchetti G, Vaienti S, Zanlungo F. Relaxation to the asymptotic distribution of global errors due to round off. Europhysics Letters. 2010;**89**: 40006-40010

[15] Faranda D, Mestre FM, Turchetti G. Analysis of round-off errors with reversibility test as a dynamical indicator. International Journal of Bifurcation and Chaos. 2012;**22**(09): 1-14

[16] Barrio R, Borczyk W, Breiter S. Spurious structures in chaos indicators maps. Chaos, Solitons & Fractals. 2009; **40**:1697

[17] Cincotta PM, Simo C. Simple tools to study the global dynamics in non

axisymmetric galactic potentials. Astronomy and Astrophysics Supplement Series. 2000;**147**:205

[18] Panichi F, Godziewski K, Turchetti G. The reversibility error method (REM): A new, dynamical fast indicator for planetary dynamics. Monthly Notices of the Royal Astronomical Society. 2017;**468**:469-491 https://arxiv.org/abs/1703.10596

[19] Panichi F, Turrchetti G. Birkhoff normal forms and stability indicators for betatronic motion. In: Di Mitri S, editor. NOCE Conference Proceedings. World Scientific; 2019

[20] Yoshida H. Construction of higher order symplectic integrators. Physics Letters A. 1990;**150**:262

[21] Skokos Ch, Gerlach E. Numerical integration of variational equations. Physical Review E, Statistical, Nonlinear, and Soft Matter Physics. 2010;**82**:1-19

[22] Hairer E, Lubich C, Gerhard Geometric W. Numerical Integration: Structure-Preserving Algorithms for Ordinary Differential Equations. 2nd ed. Berlin: Springer; 2006. Available from: https://cds.cern.ch/record/1250576

Chapter 11

BH M87: Beyond the Gates of Hell

Pawel Gusin, Andy T. Augousti and Andrzej Radosz

Abstract

The supermassive black hole located in the galaxy M87 (BH M87) is four times larger than our solar system. If it is spherically symmetric, then a capsule free falling from a distance of 1 light year would cross BH M87's event horizon within some tens of years. Continuing that journey, any unfortunate astronomer traveling within the capsule would remain alive for a few further tens of hours; if the capsule were equipped with a powerful engine and could slow down, their lifetime inside the horizon beyond "the gates of Hell" would be slightly extended. How is this so? What are the other properties of the interior of BH M87? Maintaining the assumption of spherical symmetry of the exterior of BH M87, we briefly discuss some simple but intriguing properties of its interior, a region that turns out to be highly anisotropic, both expanding and contracting at the same time.

Keywords: Schwarzschild and Reissner-Nordström space-times, supermassive black hole M87, isotropic exterior, event horizon, anisotropic interior

1. Introduction

On April 10, 2019, the first ever image of a black hole was displayed. Due to the extensive efforts of very many teams of astronomers, working in parallel a picture of a supermassive black hole, of 6.5 billion solar masses, located in the galaxy M87, at a distance of 55 Mly, belonging to the Virgo supercluster, was produced. The size of that object, despite being four times the size of the solar system, is nonetheless still too small to be pictured by a single telescope, so worldwide cooperation through the Event Horizon Telescope project led to a synthesized Earth size-like device and the final vision (see **Figure 1**) [1].

The visible presence of such a supermassive black hole puts old questions in a new light. Traveling toward such an object, reaching its "edge"—the event horizon—crossing it, and entering the interior, what would be the experience of such a traveler, an unfortunate astronaut, who would be unable to share his/her views and experiences with colleagues who remain at the starting point in a "Mother Station?"

We will describe some particular features of such a trip focusing on the bizarre properties of the interior of the black holes.

In general there are four possible kinds of black holes (see, e.g., [2–4]). Isotropic, i.e., spherically symmetric and static, BHs are of the Schwarzschild type; rotating—hence axially symmetric—BHs are called Kerr BHs; both of these types could also be charged; then they are referred to as Reissner-Nordström and Kerr-Newman, respectively. The outer edge of the BHs, an event horizon, acts as a semitransparent membrane that might be crossed once and in one direction only. Apart from the Schwarzschild BH, the other three types of BH also possess an inner horizon referred to as a Cauchy horizon.

IntechOpen

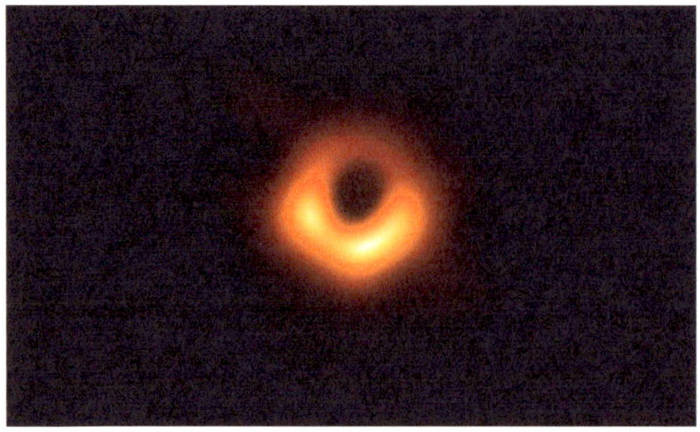

Figure 1.
First ever image of the black hole in galaxy Messier 87, here denoted as BH M87 (55 Mly from the Earth), April 10, 2019.

In our considerations we will limit our discussion to the case of spherically symmetric, static BHs: the Schwarzschild (S) and Reissner-Nordström (RN) types. In these two cases, the space-time metric tensor is diagonal in spherical polar coordinates t, r, θ, φ and is described by the line element:

$$ds^2 = g_{tt}c^2dt^2 - g_{tt}^{-1}dr^2 - r^2d\Omega^2 \tag{1}$$

where $d\Omega^2 = d\theta^2 + sin^2\theta d\varphi^2$ is a unit sphere element. The $\{tt\}$ element of the metric tensor takes the following form:

$$g_{tt}^{(i)} = \begin{cases} 1 - \dfrac{2GM}{c^2r} & i = S \\ 1 - \dfrac{2GM}{c^2r} + \dfrac{Q^2}{r^2} & i = RN \end{cases} \tag{2}$$

where M is the mass and Q is the charge of the BH. Hereafter we will use the notation $c = G = 1$. The zero value of $g_{tt}^{(S)}$ determines the location of the event horizon or gravitational radius, r_g:

$$r_g = 2M \tag{3}$$

There are two zeros of $g_{tt}^{(RN)}$,

$$r_{\pm} = M \pm \sqrt{M^2 - Q^2} \tag{4}$$

determining an outer, r_+, event horizon and an inner, r_- Cauchy horizon.

2. A capsule radially falling toward a black hole horizon

Consider the case of a test object, a capsule radially freely falling in a spherically symmetric and static space-time (1). We shall assume that capsule A (for Alice, see below) starts from rest at some initial position located at a Mother Station (MS) fixed at radial position r_{MS}. We will describe this radial infall answering some simple questions:

1. How long does it take, measured by an observer, termed A for Alice, within capsule A to reach the event horizon?

2. How much time does such a trip take from the point of view of another observer, termed static observer (SO) located at the Mother Station?

3. How does the speed of A change during this journey?

4. How can we verify these predictions?

Before doing this we will introduce some useful definitions. Firstly, every observer O whose history in the space-time is described by a world line, $\left\{x^{\mu}_{O}(\tau)\right\}$ such that

$$d\tau^2 = g_{\alpha\beta} dx^{\alpha} dx^{\beta} \tag{5}$$

is specified by a **unit velocity four-vector** $\left\{u^{\mu}_{O}(\tau) = \frac{dx^{\mu}}{d\tau}\right\}$,

$$u^2 \equiv g_{\alpha\beta} u^{\alpha} u^{\beta} = 1. \tag{6}$$

Light rays $\{x^{\mu}(\sigma)\}$ belong to light cones, and they are specified by a **null wave vector** $\left\{k^{\mu}(\sigma) = \frac{dx^{\mu}}{d\sigma}\right\}$,

$$k^2 \equiv g_{\alpha\beta} k^{\alpha} k^{\beta} = 0, \tag{7}$$

where σ is an affine parameter of the null geodesic. Due to the symmetry properties of the static and isotropic character of the S and RN space-times, there are two conservation laws: energy and angular momentum are conserved quantities (see, e.g., [5]). Energy conservation means that the t-component of the covariant velocity u_t/wave k_t vector is conserved:

$$u_t = g_{t\beta} u^{\beta} \equiv g_{tt} u^t = \varepsilon, \tag{8}$$

$$k_t = g_{t\beta} k^{\beta} \equiv g_{tt} k^t = \omega. \tag{9}$$

Conservation of the angular momentum results in the planar motion of both time-like geodesics (8) and light-like geodesics (9). Without loss of generality, one can consider then equatorial planar motion, $\theta = \frac{\pi}{2}$, where the corresponding velocity/wave vector component vanishes:

$$u^{\theta} = 0, \tag{10}$$

$$k^{\theta} = 0. \tag{11}$$

The value of the angular momentum is conserved, i.e.,

$$u_{\varphi} = g_{\varphi\varphi} u^{\varphi} \equiv L, \tag{12}$$

$$k_{\varphi} = g_{\varphi\varphi} k^{\varphi} \equiv l. \tag{13}$$

Therefore, geodesics determined by three nonvanishing components of the tangent vector, which is the velocity vector for the time-like world lines, Eq. (6), and the wave vector for the light-like world lines, Eq. (7), may be found from the two conservation laws and the normalization condition:

$$u^r = \pm\sqrt{\varepsilon^2 - g_{tt}\left(1 + \frac{L^2}{r^2}\right)}, \tag{14}$$

$$k^r = \pm\sqrt{\omega^2 - g_{tt}\frac{l^2}{r^2}}. \tag{15}$$

A special class of non-geodesic trajectories represents static observers (SO), whose position is fixed $(r_0, \theta_0, \varphi_0)$. Their only nonvanishing component of the velocity vector is a temporal one u^t_{SO}. It is determined by the normalization condition (Eq. (6)):

$$u^t_{SO} = \frac{1}{\sqrt{g_{tt}(r_0)}}. \tag{16}$$

Hence one can describe the trajectory of A, which is radially infalling, and the Mother Station (MS), which is static at r_0, by using their velocity vectors u_A and u_{MS}:

$$u_A = \left(u^t_A, u^r_A, 0, 0\right) = \left(\frac{\varepsilon}{g_{tt}}, -\sqrt{\varepsilon^2 - g_{tt}}, 0, 0\right), \tag{17}$$

$$u_{MS} = \left(u^t_{MS}, 0, 0, 0\right) = \left(\frac{1}{\sqrt{g_{tt}(r_{MS})}}, 0, 0, 0\right), \tag{18}$$

If A starts from the location of the Mother Station, being initially at rest, then

$$\varepsilon = \sqrt{g_{tt}(r_{MS})} \tag{19}$$

(see also below).

During the infall of A, one can measure its speed at some intermediate point r_1 (between r_{MS} and the event horizon) by arranging at r_1 an observer O determined by velocity vector u_O who measures an infinitesimal "distance of A" covered within an infinitesimal "time period." This results in a speed for A as measured by O, $v_A(O)$ expressed in terms of a *scalar product* $u_A u_O$:

$$u_A u_O = g_{\alpha\beta} u^\alpha_A u^\beta_O = \frac{1}{\sqrt{1 - v_A^2}}. \tag{20}$$

If O is a static observer located at r_1, then

$$v_A = \frac{\sqrt{\varepsilon^2 - g_{tt}(r_1)}}{\varepsilon} \tag{21}$$

as one can verify by using Eqs. (20) and (16).

2.1 How long does it take to Alice to reach the event horizon?

Now we can answer the questions concerning the duration associated with the infall of A. Applying the equations for the nonvanishing components of its velocity vector (Eq. (17))

$$\frac{dt}{d\tau} = \frac{\varepsilon}{g_{tt}}, \tag{22}$$

$$\frac{dr}{d\tau} = -\sqrt{\varepsilon^2 - g_{tt}}, \tag{23}$$

one obtains the equations for the *coordinate time t* and for the *proper time* τ:

$$t + C = -\int \frac{\varepsilon}{g_{tt}\sqrt{\varepsilon^2 - g_{tt}}} dr, \tag{24}$$

$$\tau + C' = -\int \frac{1}{\sqrt{\varepsilon^2 - g_{tt}}} dr. \tag{25}$$

The proper time is actually the time measured by Alice (A) traveling within the capsule. Hence the trip from MS to the event horizon of the BH is completed by Alice within the period:

$$\tau(r_{MS}; r_g) = -\int_{r_{MS}}^{r_g} \frac{1}{\sqrt{\varepsilon^2 - g_{tt}}} dr. \tag{26}$$

Specific examples of the free fall for both Schwarzschild and Reissner-Nordström space-times will be presented later. The important fact is that expression (26) leads to a finite value of the time $\tau(r_{MS}; r_g)$ recorded by Alice.

On the other hand, the coordinate time corresponding to the trip from MS to the event horizon is infinite (see also [6]):

$$t(r_{MS}; r) = -\int_{r_{MS}}^{r} \frac{\varepsilon}{g_{tt}\sqrt{\varepsilon^2 - g_{tt}}} dr \xrightarrow{r \to r_g} \infty \tag{27}$$

(see, however, below, Section 5). Coordinate time is associated with the time recorded by an observer(s) belonging to "our" part of the universe. It means that the perception of observers located outside the event horizon of a black hole is such that Alice would never complete her trip toward the horizon. In other words she could never reach the horizon in a finite time period.

This process of the asymptotic approach to the BH horizon as perceived by MS observers can be illustrated in a way presented in the following subsection.

2.2 Communication between the capsule and the Mother Station

Let us consider an exchange of electromagnetic signals, light rays between two observers: Alice, traveling within the capsule and Bob located at the Mother Station. Such signals are represented by radial rays (9) and (15) where $l = 0$ and

$$k = \left(\frac{\omega}{g_{tt}}, \pm\omega, 0, 0 \right). \tag{28}$$

The frequency of the signal recorded by an arbitrary observer O, ω_O, is given by the projection of the appropriate wave vector, k, on the unit time-like vector of O, i.e., on the four-velocity vector u_O. It is a scalar product $k \cdot u_O$, and

$$\omega_O = k \cdot u_O \equiv g_{\alpha\beta} k^\alpha u_O^\beta. \tag{29}$$

Hence, Alice sends back signals that are recorded by Bob (at MS), and the frequency ratio of the recorded, ω_B^r vs. emitted, ω_A^e signals, found from Eqs. (28), (29), (9), (15), (17) and (18) is (see also [7]):

$$\frac{\omega_B^r}{\omega_A^e}=\frac{g_{\alpha\beta}k^\alpha u_B^\beta}{g_{\alpha\beta}k^\alpha u_A^\beta}=\frac{\omega}{\sqrt{g_{tt}(r_{MS})}}\left[\frac{\omega\varepsilon}{g_{tt}(r_A)}\left(1+\frac{\sqrt{\varepsilon^2-g_{tt}(r_A)}}{\varepsilon}\right)\right]^{-1}=\frac{g_{tt}(r_A)}{g_{tt}(r_{MS})}\frac{1}{1+v_A}\equiv 1-v_A. \tag{30}$$

One can see that the signals are found to be **redshifted**: the frequency of the recorded signals is lower than the frequency of the emitted signals. But in this case it turns out to be of a special form: it may be referred to as a **critical redshift** as it tends to zero as A approaches the horizon. Indeed, the speed v_A of the capsule, once measured by the static observer, tends to the speed of light in a vacuum, $v_A \to 1$ (e.g., [6]) as the capsule approaches the horizon, $g_{tt}(r_g) = 0$ (see Eq. (21)). And it is the manifestation of the fact that from Bob's perspective, the capsule approaches event horizon asymptotically and will never reach the horizon (see however Section 5!): the frequency of the signals incoming from the capsule gradually decreases and eventually goes beyond the lower limits (however small!) of the sensitivity of recording devices.

Summarizing the findings of this section, one would like to point out some of intuitive and counterintuitive conclusions. Obviously the speed of the capsule freely falling toward the BH horizon increases as measured by static observers placed above the event horizon. Quite non-obvious is that this value tends to the speed of light as it approaches the horizon. And what is even more important is that this outcome is independent of the initial conditions: wherever the capsule starts from, the rest of the value of its speed asymptotically approaches the value of the speed of light. Moreover, there are no static observers residing on the horizon, so one cannot claim that a test object reaches the speed of light when crossing the BH horizon (see also Refs. [8, 9]). Accompanying this highly nonclassical behavior of the free fall speed is the duration of this trip toward the horizon—it turns out to be infinite for an observer located beyond the event horizon (see also Section 5). Nevertheless the trip is completed within a well-defined time period for a traveler, Alice, who is confined within the capsule. This may be regarded as a most dramatic illustration of time dilation where both kinematic and gravitational time dilations are combined. It is confirmed by the generalized Doppler frequency shift: signals emitted by Alice and recorded by Bob at MS are critically redshifted.

3. Approaching and crossing the event horizon

When Alice, confined to the capsule, approaches the event horizon and then if the BH is massive enough—greater than millions of solar masses—then tidal forces are not particularly large (see, e.g., [9]), and it is believed that she would not even notice the instant of crossing the horizon. But the further consequences would be quite dramatic: one may cross the event horizon only once and only in one direction toward the BH. One may ask the general question: in such a situation of free fall, would it be possible to identify the presence of the horizon?

On the one hand, there is an obvious outcome arising from the equivalence principle: in a freely falling frame, one cannot determine an external gravitational field. But this refers to possible experiments performed within a freely falling frame. It has recently been shown [7] that by using an appropriate communication channel Alice could identify the presence of the event horizon quite precisely, in order to stop the capsule, if it is equipped with a powerful enough engine, or to determine the instant of crossing the horizon. Indeed, by recording the electromagnetic signals coming from Bob (placed at MS), with $k^r = +\omega$ (see Eq. (28)), Alice finds the following frequency ratio of recorded ω_A^r and emitted ω_B^e signals:

$$\frac{\omega_A^r}{\omega_B^e} = \frac{1}{1 + v_A}. \tag{31}$$

This ratio tends to $\frac{1}{2}$ as the capsule approaches the horizon (see Eq. (21)). And this is the way to identify the presence of the horizon in general and to identify crossing instant in particular: the redshift of signals coming from MS equals ½.

One of the specific features of the event horizon relates to the singular character of time dilation described above for the trip toward the horizon: nobody residing in "our" part of universe could record the instant when the capsule (or any other test particle) reaches the edge of an (arbitrary) BH. This results in an effect referred to as "image collision" [10, 11] (also termed touching ghosts). If capsule A is followed by another capsule C (carrying Cindy), which started its free fall later than A, how would Cindy perceive capsule A crossing the horizon? This problem could be formulated in the following way. Let Alice release a signal "I'm crossing the Black Hole horizon!" at the particular instant (known perfectly well to her from the method described above) just as she passes the horizon. It does not need to be the message—it could be a specific, encoded light ray signal. How would such a signal be recorded by Cindy? One can answer this question in various ways, for instance by illustrating this using Kruskal-Szekeres coordinates (see Ref. [11]) or invoking an analytical description within a different singularity-free coordinate frame. But one also can give a reverse argument! Cindy must record Alice's signal only when she, herself, crosses the horizon. Otherwise, recording this signal before reaching the horizon, Cindy would be able to share this message with other residents in our part of the universe; she could even stop her capsule. But this would contradict the above paradigm, namely, that one cannot record in our part of the universe the event horizon crossing instant by capsule A (or any other test particle).

4. The interior of black holes: there is no black hole inside a black hole

There are two singularities in the expression for the line element (1). One is defined as the horizon of a BH—a horizon of a BH (1) is determined from the zero value of g_{tt} or as a singularity of $g_{rr} = g_{tt}^{-1}$. It is well known (see e.g., [12]) that there are different coordinate systems, other than that used in (1), that are free of this singular characteristic at the horizon. These include Gullstrand-Painleve, Kruskal-Szekeres, Eddington-Finkelstein coordinates, and many others [9, 12]. The other type of singularity corresponds to $r = 0$ and cannot be removed or avoided by applying a different system of coordinates. One uses then the phrase "coordinate singularity" to refer to the former type as a "horizon singularity as opposed to the "true singularity" representing the latter one. By applying an appropriate frame of reference, we no longer deal with singular behavior at the event horizon.

4.1 Cylindrical-like shape

Hence, the interior of a BH could be described within such a singularity-free frame of reference. It has been shown, however [13], that the interior of a Schwarzschild BH may also be described in the terms of above-the-horizon coordinates, t, r, θ, φ (see Eq. (1)). There are two important consequences of such an approach. The first is the singular character at the horizon, $r = r_g$. The second is even more important: inside the horizon one has to accept the interchange of the roles of coordinates t and r. The former takes on the role of a spatial coordinate, and the latter has to be regarded as a temporal coordinate. This means that within the BH's horizon, $g_{tt} < 0$ and r can only decrease, and $dr < 0$ representing the passage of time. This interchange leads to a new interpretation of the conservation law associated with the t-invariance in this case. Outside the horizon it is interpreted as energy conservation (Eqs. (8) and (9)); inside the horizon it is manifested as momentum, t-component, and conservation. One arrives then at the first rather counterintuitive property of a BH.

The interior of spherically symmetric black holes described by Eq. (1) turns out to be a **cylindrical-like shape**, homogeneous along its axis with spheres at the two ends.

Other counterintuitive properties are associated with the dynamical character of the interior. Indeed, inside the horizon of BHs $r < r_g$, where r plays the role of a temporal coordinate, one can see in expression (1) that all of the metric tensor elements are r, "*time*" dependent. Therefore, it is a dynamical space-time, or in other words, it may be regarded as a *cosmology*. What are the properties of such a cosmology, for instance compared to our homogenous and isotropic, expanding universe?

One can start with an extension of the case considered above of capsule A crossing the horizon and continuing its trip within the bounds of the horizon. As already mentioned we may apply the coordinates used outside the horizon remembering the important interchange of the roles of coordinates t and r. Hence, inside the horizon the velocity vector is still given by expression (17) where u^t_A and u^r_A refer now to spatial and temporal components, respectively. Alice, confined within the capsule, and being inside the horizon of the BH (and being aware of this!, see Section 3), still receives the electromagnetic signals released at the fixed location of MS by Bob. These are described by formula (28). Therefore, inside the horizon the frequency ratio

$$\frac{\omega^r_A}{\omega^e_C} = \frac{1}{1 + \frac{\sqrt{\varepsilon^2 - g_{tt}(r_A)}}{\varepsilon}} \overset{r \to 0}{\longrightarrow} 0 \tag{32}$$

decreases further below the horizon's value of ½ and tends to zero at the final singularity, $-g_{tt}(r) \overset{r \to 0}{\longrightarrow} \infty$.

This description may be deceptive when interpreted through the automatic application of formulae (31) naively leading to the (wrong!) conclusion that the speed of A, $v_A = \frac{\sqrt{\varepsilon^2 - g_{tt}(r_1)}}{\varepsilon} \overset{r \to 0}{\longrightarrow} \infty$. What is wrong with such an extension of the former interpretation?

One can ask for the speed of capsule A within the horizon measured in a way similar to the one applied outside the horizon. In order to do this, we need to introduce an analogue of a static observer, called an r-observer, *ro* (see below). This is one whose only velocity component is a "temporal" one, i.e.,

$$u_{ro} = \left(0, u^r_{ro}, 0, 0\right) = \left(0, -\sqrt{-g_{tt}(r)}, 0, 0\right). \tag{33}$$

The speed $\tilde{v}_A$ of capsule A within the horizon measured by *ro* (see Eq. (33)), by definition, is given as (see also Eq. (20))

$$u_A \cdot u_{ro} = \frac{1}{\sqrt{1 - \tilde{v}_A^2}} \tag{34}$$

which turns out to be:

$$\tilde{v}_A = \frac{\varepsilon}{\sqrt{\varepsilon^2 - g_{tt}(r_1)}}. \tag{35}$$

Hence inside the horizon, the speed of the capsule that has already crossed over the edge and entered that region is given by the expression that is formally inverse to the corresponding one above the horizon (c.f. Eqs. (18) and (35)). This outcome might be surprising only at first sight. Indeed, as the meaning of speed is the ratio of an (elementary) "distance"/(elementary) "time" and the numerator and denominator have already reversed their roles, then the ratio known as "speed" is expressed (formally) as the inverse of the one outside the horizon. That is why the speed outside the horizon, v_A Eq. (18), and the speed inside the horizon, $\tilde{v}_A$ Eq. (35), are expressed as mutually inverse quantities.

Another interesting feature of the speed inside the horizon $\tilde{v}_A$ (35) is that its value decreases from the asymptotic value 1 at the horizon to zero at the final singularity, $r = 0$. For different values of $\varepsilon = g_{tt}(r_{MS})$, i.e., different initial positions of the capsule, the speed changes differently (see **Figure 2**), but the asymptotic values at the horizon and at the ultimate singularity remain fixed.

The capsule's speed is plotted along the vertical axis (velocity) as a function of r for $M = 1$ and $r_g = 2$, and the horizontal axis represents distance for different initial conditions: the red line represents a fall from infinity, $\varepsilon = 1$.

This discussion throws new light on a BH's interior: the velocity of a freely falling test particle, which grows as it falls outside horizon, appears to decrease within the horizon (see also [7]).

To illustrate the behavior of the interior further, let us consider two r-observers placed along the axis of homogeneity t, exchanging electromagnetic signals. The frequency shift would in this case be a significant source of information about the dynamics of the BH's interior.

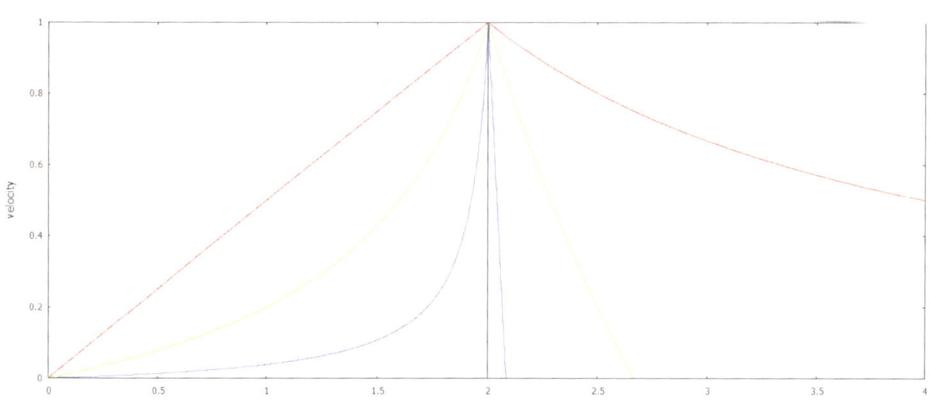

Figure 2.
The case of Schwarzschild space-time.

4.2 Expansion - exchange of electromagnetic signals along the t-axis

Let us consider then the exchange of signals between two observers located on the t-axis: Diana (D) receives signals sent by George (G), $r_D < r_G < r_g$. The frequency ratio is in this case expressed as follows:

$$\frac{\omega_D^r}{\omega_G^e} = \frac{\sqrt{-g_{tt}(r_G)}}{\sqrt{-g_{tt}(r_D)}} \tag{36}$$

It leads to distinct conclusions in S and RN space-times.

In the case of a Schwarzschild BH, $-g_{tt}(r) = \frac{2}{r} - 1$ is a monotonic function of r, and Eq. (36)

$$\frac{\omega_d^r}{\omega_G^e} < 1 \tag{37}$$

describes a Doppler-like *redshift* (see **Figure 3**). Hence, regarding this as a "cosmology," Eq. (36) represents a "cosmological redshift" due to expansion (along the t-axis!; see below).

In the case of a Reissner-Nordström BH, $-g_{tt}(r) = \frac{2}{r} - 1 - \frac{Q^2}{r^2}$ is a non-monotonic function of r, and Eq. (36) leads to:

a Doppler *redshift*,

$$\frac{\omega_D^r}{\omega_G^s} < 1, \tag{38}$$

for $r_m < r_D < r_G$, and a Doppler *blueshift*

$$\frac{\omega_D^r}{\omega_G^s} > 1 \tag{39}$$

for $r_D < r_G < r_m = Q^2$. This is illustrated in **Figure 4**, the ratio (36) in the RN case, M = 1, $Q = 0.6$, for fixed $r_G = 1.6$.

In this case Eq. (36) represents "cosmological redshift" due to expansion, followed by "cosmological blueshift" due to contraction (along homogeneity t-axis).

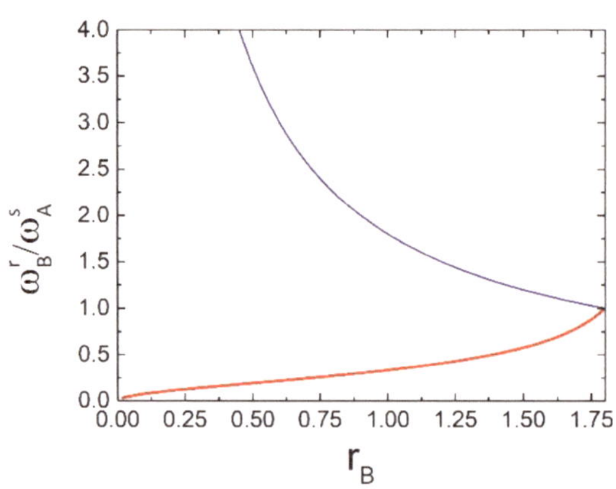

Figure 3.
The frequency shift inside the horizon of a Schwarzschild BH: signals propagating along the axis of homogeneity (t), Eq. (36) are redshifted (red), and signals propagating perpendicularly to the t-axis (57) are blueshifted (blue); (M = 1), $r_A = 1.75$.

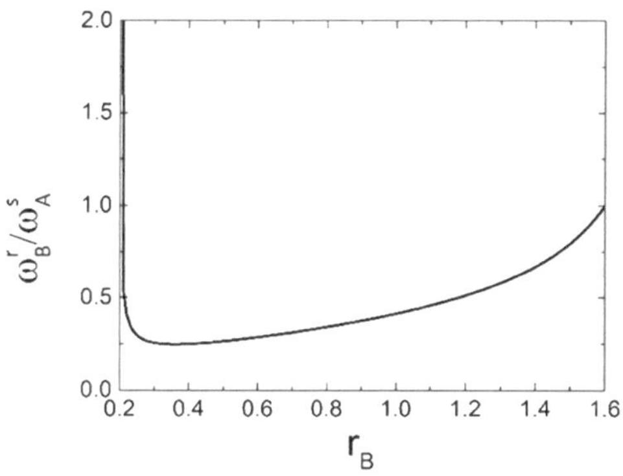

Figure 4.
Frequency shift inside the horizon of a RN BH (M = 1, Q = 0.6), for signals propagating along the t-axis $r_- = 0.2 < r_D < r_G = 1.6$: initial redshift Eq. (38) is followed by the final blueshift Eq. (39) (due to expansion followed by contraction).

4.3 Contraction – exchange of electromagnetic signals perpendicular to the *t*-axis

One may ask what happens if the exchanged signals travel perpendicularly to the t-direction? This means that the t-component of the position of Diana and George is the same. Assuming that the trajectory of the signal, the light ray, is confined within an equatorial plane, then it travels between φ_G and φ_D where D and G are placed at $\left(t_D = t_G, \frac{\pi}{2}, \varphi_D\right)$, $\left(t_G, \frac{\pi}{2}, \varphi_G\right)$. The signal is emitted at instant r_G and then recorded at instant r_D, so one will find (see [14]) both for S and RN BHs:

$$\frac{\omega^r_D}{\omega^s_G} = \frac{r_G}{r_D} > 1 \tag{40}$$

A Doppler blueshift is found in both S and RN space-times. This represents a contraction of this cosmology in a hyperplane perpendicular to the direction of homogeneity.

Therefore the cylindrically shaped interior of spherically symmetric, static (outside the horizon!), Schwarzschild and Reissner-Nordström black holes reveals anisotropic dynamics: they turn out to expand along the cylindrical axis of homogeneity and contract perpendicularly to this axis. In the case of Reissner-Nordström black holes, M = 1, the expansion stops at some instant, $r_m = Q^2$, and then contraction follows. It should be pointed out that the contraction perpendicular to the t-axis may simply be observed due to the form of the line element (1) inside the horizon, where the coefficient of its angular part, $r^2 d\Omega^2$, is an ever-decreasing coordinate r.

5. Traveling toward BH M87

The black hole in galaxy Messier 87, BH M87, is located at a distance of 55 Mly from the solar system. Its mass is estimated at 6.5 billion solar masses and its size, given as

$$R_{M87} \equiv \frac{2GM_{BHM87}}{c^2}, \tag{41}$$

appears to be 20 billion kilometers, four times the size of the solar system itself. It is probably rotating, so it cannot be regarded as spherically symmetric.

In our discussion we will assume, however, that this supermassive black hole, whose image was issued for the first time in history on April 10, 2019 (it looks like the gate to Hell) [1], is spherically symmetric and static—this implies that it is of Schwarzschild or Reissner-Nordström type. Having in mind our discussion above, we will try to indicate the specific features of such a trip, being of course absolutely fatal (as we will argue) once "the Gate" of the horizon of BH M87 has been crossed.

5.1 Free fall toward BH M87

Let us consider a spaceship starting its free fall from a Mother Station located at r_{MS} applying a coordinate system given by (1). We will consider various cases corresponding to different values of r_{MS}:

a. MS located at the Earth—$r_{MS} = 55\,Mly = 5.5 \cdot 10^{23} m$.

b. MS located within M 87—$r_{MS} = 1\,000\,ly = 10^{19} m$.

c. MS located (very) close to BH M87—$r_{MS} = 1\,ly = 10^{16} m$.

Our aim is to describe the trip itself and its perception by two specific observers: an astronaut, Archibald (A), located within a spaceship and a static observer, Barbara (B), located at MS. We will assume that A and B communicate simply by the exchange of electromagnetic signals and radial light rays of fixed frequency, characterized by a wave vector (28).

Firstly, free fall toward BH M87, the crossing of its horizon and eventually reaching the final outcome, will be considered within the two scenarios: fall from the rest (I) and fall with some nonzero initial speed simulating free fall from infinity (II).

5.1.1 How long does it take to reach position r_X?

The time to reach position r_X as measured by A is determined as follows (see Eq. (26)):

$$\tau(r_{MS}; r_X) = -\int_{r_{MS}}^{r_X} \frac{1}{\sqrt{\varepsilon^2 - g_{tt}}} dr = \begin{cases} -\int_{r_{MS}}^{r_X} \frac{1}{\sqrt{g_{tt}(r_{MS}) - g_{tt}}} dr \\ -\int_{r_{MS}}^{r_X} \frac{1}{\sqrt{1 - g_{tt}}} dr \end{cases} \tag{42}$$

for cases I and II, respectively. For Schwarzschild space-time, $Q = 0$, one finds in the scenarios (a)–(c) listed above the following results in cases I and II:

$$\tau_s(r_{MS}; r_X) = -\int_{r_{MS}}^{r_X} \frac{1}{\sqrt{\varepsilon^2 - g_{tt}}} dr = T_0 I_s(r_{MS}; r_X) \tag{43}$$

$$T_0 = \frac{R_g}{c} = 6.4 \cdot 10^4 s \tag{44}$$

$$I_s(r_2; r_1) = \begin{cases} -\int_{r_2}^{r_1} \frac{1}{\sqrt{g_{tt}(r_{MS}) - g_{tt}}} dr = I_{MS}((r_2; r_1)) = x_{MS}^{3/2} \left(arctgy + \frac{y}{1+y^2} \right)_{y_2}^{y_1} \\ -\int_{r_2}^{r_1} \frac{1}{\sqrt{1 - g_{tt}}} dr = I_\infty(r_2; r_1) = \frac{2}{3} \left(x_2^{3/2} - x_1^{3/2} \right) \end{cases} \tag{45}$$

$$x = \frac{r}{R_g} \qquad y = \sqrt{x_{MS}}\sqrt{\frac{1}{x} - \frac{1}{x_{MS}}} \tag{46}$$

a. $r_X = \frac{r_{MS}}{2}$

$$\tau_s(r_{MS}; r_X) = \begin{cases} \tau_{MS} = T_0 x_{MS}^{3/2}\left(arctgy + \frac{y}{1+y^2}\right)_{y_2}^{y_1} \sim 10^{21}s \sim 3 \cdot 10^{13}y \\ \tau_\infty = T_0 \frac{2}{3}\left(x_2^{3/2} - x_1^{3/2}\right) \sim 10^{20}s \sim 3 \cdot 10^{12}y \end{cases} \tag{47}$$

$$\tau_s(r_{MS}; r_X) = \begin{cases} \sim 10^6 y \\ \sim 5 \cdot 10^5 y \end{cases} \tag{48}$$

$$\tau_s(r_{MS}; r_X) = \begin{cases} \tau_{MS} \sim 30y \\ \tau_\infty \sim 10y \end{cases} \tag{49}$$

b. $r_X = 1.1 r_g$

$$\tau(r_{MS}; r_X) = \begin{cases} \tau_{MS} \sim\sim 3 \cdot 10^{13}y \\ \tau_\infty \sim\sim 3 \cdot 10^{12}y \end{cases} \tag{50}$$

$$\tau(r_{MS}; r_X) = \begin{cases} \tau_{MS} \sim 10^6 y \\ \tau_\infty \sim 5 \cdot 10^5 y \end{cases} \tag{51}$$

$$\tau(r_{MS}; r_X) = \begin{cases} \tau_{MS} \sim 30y \\ \tau_\infty \sim 13.5y \end{cases} \tag{52}$$

Barbara may make her own measurements of the time representing the instants indicated above in different ways: recording signals coming from A, communicating with A about his perception of time, etc. One may prefer to use a compromise based on this variety of approaches, namely, indicating the coordinate instant t_X corresponding to $\tau(r_{MS}; r_X)$. Indeed one or other method of measuring the instant when reaching coordinate position r_X by A applied by B refers to t_X and is determined by (see also Eq. (42)):

$$t_X \equiv t(r_{MS}; r_X) = -\int_{r_{MS}}^{r_X} \frac{\varepsilon}{g_{tt}\sqrt{\varepsilon^2 - g_{tt}}} dr = \begin{cases} -\int_{r_{MS}}^{r_X} \frac{\sqrt{g_{tt}(r_{MS})}}{g_{tt}\sqrt{g_{tt}(r_{MS}) - g_{tt}}} dr \\ -\int_{r_{MS}}^{r_X} \frac{1}{g_{tt}\sqrt{1 - g_{tt}}} dr \end{cases} \tag{53}$$

for cases I and II, respectively. As indicated in former sections, the coordinate time period becomes singular (goes to infinity) as A approaches the horizon, independently of the initial conditions:

$$r_X \to r_g, t_X \to \infty. \tag{54}$$

In analogy with the above results for A, one finds for B the following outcomes:

$$t_{\infty X} \equiv t(r_{MS}; r_X) = -\int_{r_{MS}}^{r_X} \frac{1}{g_{tt}\sqrt{1 - g_{tt}}} dr = T_0 I_{\infty t} \tag{55}$$

$$I_{\infty t} = \int_{x_{MS}}^{x_X} \frac{\sqrt{x}dx}{\left(1-\frac{1}{x}\right)} dr$$
$$= \frac{2}{3}\left(x_{MS}^{3/2} - x_X^{3/2}\right) + 2(\sqrt{x_{MS}} - \sqrt{x_X}) + ln\left|\frac{x_{MS}-1}{x_{MS}+1}\right| - ln\left|\frac{x_X-1}{x_X+1}\right| \quad (56)$$

$$t_{\infty X} = T_0\left\{\frac{2}{3}\left(x_{MS}^{3/2} - x_X^{3/2}\right) + 2(\sqrt{x_{MS}} - \sqrt{x_X}) + ln\left|\frac{x_{MS}-1}{x_{MS}+1}\right| - ln\left|\frac{x_X-1}{x_X+1}\right|\right\} \quad (57)$$

The dominant term in the coordinate time is the first one if the final position, r_X, is not too close to the horizon:

$$t_{\infty X} \approx T_0\left\{\frac{2}{3}\left(x_2^{\frac{3}{2}} - x_1^{\frac{3}{2}}\right)\right\} = \tau_{\infty X} \quad (58)$$

If the destination station, X gets close to the horizon, the duration of travel (45) becomes dominated by the last term which tends to infinity:

$$t_{\infty X} \approx T_0\left\{-ln\left|\frac{x_1-1}{x_1+1}\right|\right\} \xrightarrow{x_1\to 1} \infty \quad (59)$$

However, in practical terms, i.e., in all of the cases listed above

$$t_{\infty X} \approx T_0\left\{\frac{2}{3}\left(x_2^{\frac{3}{2}} - x_1^{\frac{3}{2}}\right)\right\} = \tau_{\infty X}. \quad (60)$$

The last term starts to dominate for

$$-ln\left|\frac{x_1-1}{x_1+1}\right| = x_2^{\frac{3}{2}} \quad (61)$$

i.e., it depends on the initial conditions. In case (c), the logarithmic term starts to dominate incrementally close to the horizon (on the horizon in fact, see below):

$$r_X = R_g\left(1 + e^{-6000}\right) \quad (62)$$

The meaning of this result is that the coordinate time is the corrected proper time, and the correction is moderate up to the vicinity of the horizon. In the close vicinity of the horizon, the singular term starts to dominate, and the coordinate time tends to infinity in this range. However, as shown above the "close vicinity of the horizon" means

$$\Delta r_X = R_g e^{-6000} \quad (63)$$

"effectively on the horizon!"

If the initial conditions are those described in (a) and (b), then that range is (formally) even smaller, i.e., it is a "stronger" zero.

Before entering the interior of the Schwarzschild BH, we will illustrate trip A via the Doppler shift.

5.1.2 Doppler shift

A and B are exchanging electromagnetic signals of fixed (emitter) frequency. Let us present the list of frequency ratios at various $r_X \equiv r_A$ as in the former subsection. Applying expressions (30) and (31), one finds:

$$v_A^2 = \frac{\varepsilon^2 - g_{tt}(r_A)}{\varepsilon^2} = \begin{cases} \dfrac{g_{tt}(r_{MS}) - g_{tt}(r_A)}{g_{tt}(r_{MS})} = \dfrac{\dfrac{R_g}{r_A} - \dfrac{1}{x_{MS}}}{1 - \dfrac{R_g}{r_{MS}}} = \dfrac{1}{x_{MS}} \dfrac{\dfrac{r_{MS}}{r_A} - 1}{1 - \dfrac{1}{x_{MS}}} & I \\ \dfrac{R_g}{r_A} = \dfrac{1}{x_{MS}} \dfrac{r_{MS}}{r_A} & II \end{cases} \tag{64}$$

a. $r_A = \dfrac{r_{MS}}{2}$

- $v_A^2 = \begin{cases} \dfrac{1}{x_{MS}} \dfrac{\dfrac{r_{MS}}{r_A} - 1}{1 - \dfrac{1}{x_{MS}}} = \dfrac{1}{x_{MS}} \dfrac{1}{1 - \dfrac{1}{x_{MS}}} \approx \dfrac{1}{x_{MS}} & I \\ 2\dfrac{1}{x_{MS}} & II \end{cases}$

- $\dfrac{\omega_A^r}{\omega_B^e} = \dfrac{1}{1 + v_A} \approx 1 - v_A = \begin{cases} 1 - \dfrac{1}{\sqrt{2.75}} \cdot 10^{-5} & I \\ 1 - 10^{-5} & II \end{cases}$

- $\dfrac{\omega_B^r}{\omega_A^e} = 1 - v_A$

- $\dfrac{\omega_A^r}{\omega_B^e} = \dfrac{1}{1 + v_A} \approx 1 - v_A = \begin{cases} 1 - \dfrac{1}{7} \cdot 10^{-2} & I \\ 1 - \dfrac{\sqrt{2}}{7} \cdot 10^{-2} & II \end{cases}$

- $\dfrac{\omega_B^r}{\omega_A^e} = 1 - v_A$

- $\dfrac{\omega_A^r}{\omega_B^e} = \dfrac{1}{1 + v_A} \approx 1 - v_A = \begin{cases} 1 - \dfrac{1}{22} & I \\ 1 - \dfrac{\sqrt{2}}{22} & II \end{cases}$

- $\dfrac{\omega_B^r}{\omega_A^e} = 1 - v_A$

b. $r_A = 1.1 R_{M87}$

- $v_A^2 = \begin{cases} \approx \dfrac{1}{1.1} & I \\ \dfrac{R_g}{r_A} = \dfrac{1}{1.1} & II \end{cases}$

- $\dfrac{\omega_A^r}{\omega_B^e} = \dfrac{1}{1 + v_A} = 0.512$

- $\dfrac{\omega_B^r}{\omega_A^e} = 1 - v_A \approx 0.046$

and (c) the same as (a)

c. $r_A = 1.01 R_{M87}$

- $\dfrac{\omega_A^r}{\omega_B^e} = \dfrac{1}{1 + v_A} = 0.5012$

- $\dfrac{\omega_B^r}{\omega_A^e} = 1 - v_A \approx 0.00496$

and (c) the same as (a)

d. $r_A = 1.001 R_{M87}$

- $v_A^2 = \dfrac{1}{1.001}$

- $\dfrac{\omega_A^r}{\omega_B^e} = \dfrac{1}{1 + v_A} = 0.50012$

- $\dfrac{\omega_B^r}{\omega_A^e} = 1 - v_A \approx 0.0004996.$

When A approaches the horizon of BH M87, $r_X \to R_{M87}$, the frequency of signals reaching B tends to zero

$$\frac{\omega_B^r}{\omega_A^e} \to 0 \tag{65}$$

and the signals themselves gradually disappear from the recording devices. Such a process becomes unboundedly extended in time. On the other hand, A receives the signals from B as redshifted toward a well-defined limit, and one finds in all cases (a–c)

$$\frac{\omega_A^r}{\omega_B^e}(r_X = R_{M87}) = 0.5 \tag{66}$$

as the indicator of the instant of crossing the horizon (see also Section 3).

5.2 Beyond "the gate of BH M87": how much time remains?

Archibald knows precisely the instant of his crossing of the horizon: whatever his starting point was, (a)–(c), he passes the horizon BH M87 when the frequency ratio hits ½. It is the irreversible instant in the whole trip: after this there is no way back. One may ask, however, the provocative question: why is there no way back?

Let us briefly discuss this point. During the radial fall toward BH M87, outside the horizon, $r > R_{M87}$, A can "see" both MS and BH M87, i.e., he can perceive the signals coming from B (located at MS) as well as the signals coming from regions located closer to BH M87 than his own current location. Radial light rays can obviously propagate along both increasing r and diminishing r. Upon crossing the horizon, the situation becomes quite different. The coordinate r changes its character—it becomes time-like, such that $dr < 0$. This means that the r coordinate only diminishes, reducing from R_{M87} to 0. Therefore, there is no way "back to the horizon" inside BH M87 because the horizon is "an instant in the past"—there is no way to "travel" to the past. It should be pointed out that this conclusion presented within this "pathological" (i.e., singular behavior at the horizon) system of coordinates remains valid as this also occurs in other, nonsingular coordinate systems.

Therefore, after crossing the horizon BH M87, Archibald no longer travels toward the center of black hole M87, but he travels along the t-axis of homogeneity until the final instant, $r = 0$.

How much time does this trip take? The answer is given by applying expression (42) to the interior of BH M87, $r < R_{M87} \equiv r_X, g_{tt} < 0$,

$$\tau(r_X;0) = -\int_{r_X}^{0} \frac{\varepsilon}{\sqrt{\varepsilon^2 - g_{tt}}} dr = \begin{cases} -\int_{r_X}^{0} \frac{1}{\sqrt{g_{tt}(r_{MS}) - g_{tt}}} dr \\ -\int_{r_X}^{0} \frac{1}{\sqrt{1 - g_{tt}}} dr \end{cases} \tag{67}$$

and it depends on the scenario, i.e., the boundary conditions, I or II.
Hence, in the case of free fall from infinity (or its simulation), one finds:

$$\tau_\infty(r_X;0) = 12\, hrs \tag{68}$$

In case I, a) – c) one obtains the same outcome:

Archibald, upon entering the interior of BH M87, would be left with only 12 hours in this fatal trip. Could this period be extended? Or what would be, if it were to exist, the maximal period, the maximal lifetime inside BH M87, hereafter termed lft BH M87?

As illustrated above lft BH M87 depends on the history (i.e., the initial conditions) of the trip, and it gets longer once MS gets closer to the horizon (much closer than 1 light year!). Actually as one finds from expression I (69), its maximal value corresponds to the case $g_{tt}(r_{MS}) = 0$. This cannot be achieved but it should be regarded as a limiting case. This limit represents the situation of Archibald's spaceship stopping just before reaching the horizon and then being released, maybe without Archibald who would prefer to avoid the particular experience of crossing the horizon. Then one finds the value of maximal lifetime of BH M87 as

$$\tau_{max}(r_X;0) = -\int_{r_X}^{0} \frac{1}{\sqrt{-g_{tt}}} dr = 28.4 hrs. \tag{69}$$

This is then the maximal extension of time, the maximal lifetime within the black hole M87.

So despite the fact that BH M87 is an enormously large object, you do not have much time left once you have crossed its border.

5.2.1 Tidal forces at the gate and beyond

Discussing even a hypothetical trip to the interior of BH M87, one should take into account aspects of human frailty. One of them concerns the forces applied to the human body during this particular journey. There are tidal forces applied to the body of the astronaut, in this case Archibald. They turn out to be quite moderate on the horizon in the case of a supermassive black hole as is a well-known fact. So at the horizon, $r \cong R_{M87}$, the differential force acting along Archibald's body, leads to a pressure of the order of (see, e.g., [9]) 10^{-15} *atm*. This effect increases, however, and at some stage, when $r \cong \frac{1}{1000000} R_{M87}$, it leads to a limiting value of the pressure, some 10^2 *atm*. And for Archibald who decided to undergo the unique experience of crossing the horizon of BH M87, that would be the ultimate *end*.

5.3 RN scenario

What changes if BH M87 is electrified with a charge Q? Then BH M87 is of the RN kind; it is a little smaller, but its radius cannot never be smaller than half of the

Schwarzschild value (see Eq. (4)), $\frac{GM_{M87}}{c^2} \equiv M_{M87} = 10^{13}m$. Moreover, for particular values of the electric charge, the estimations of this section remain to be verified, leading to different final outcomes. However, the qualitative character of the results remains unchanged: the frequency ratio at the horizon hits ½ for A, and signals coming to B are critically redshifted; there is the most dramatic manifestation of time dilation illustrating the "image collision" or "touching ghosts" effect, and there is a significant difference between the interiors of these two kinds of BHs. If BH M87 is electrically charged, then it possesses an inner horizon, and the process of expansion along the homogeneity axis, the t-axis, would stop at the instant

$$r_{min} \equiv \frac{Q^2}{M_{M87}} > r_- \tag{70}$$

and then contraction would follow. That process of contraction would continue up to the instant

$$r = r_- \tag{71}$$

During contraction along the homogeneity axis, it becomes of infinite length apart from the final instant (86) when its length suddenly becomes zero – the system reaches its inner horizon. However, the physical character of the inner horizon remains a questionable point (see [12]).

6. Concluding remarks

Supermassive BH M87 is a very large object with a size of some 20 light hours. Located at a distance 55 Mly, it does not seem to be reachable from the Earth. However, looking at its image (the very first of a black hole), it might be of interest to consider and present some issues representing and characterizing this kind of object. As a supermassive black hole, it exerts a very strong gravitational pull (see also: "strong gravitational fields" [15–17]). To illustrate this one could consider free fall due to the gravitational attraction of BH M87. The trip from the Earth would last 10,000 times longer than the age of the universe. But a test object falling from a distance of 1 light year would reach the BH M87 event horizon within some 30 years. On the other hand, traveling with a constant speed of 300,000 km/h (at the moment the greatest speed achieved by an object produced by humans), one could cover a distance of 1 light year within 3600 years, 120 times longer than the period given above.

Assuming it is spherical, we have presented a variety of features related to the hypothetical trip toward and within BH M87, emphasizing the dynamics of its anisotropic interior.

Finally we would like to comment on a remark on the image of BH M87 made by an anonymous columnists who said:

" ... it looks like the Gate to Hell".

Considering a hypothetical trip toward BH M87, one finds that the anonymous columnist was wrong: looking at the image of BH M87, one has to remember that in fact it functions in a way much worse than the Gate to Hell. After crossing such an "invisible, so apparently gentle gate," you are trapped: there is no way back and you are left with no more than 28 hours. By that time, your body would be stretched and compressed at the same time with no limits.

If BH M87 confines an electric charge, then it is possible that the process of stretching would be stopped, and contraction would follow. But this could hardly change your perspective: your lifetime within the horizon could never be substantially extended. And there is no way out.

Author details

Pawel Gusin[1], Andy T. Augousti[2] and Andrzej Radosz[1]*

1 Wroclaw University of Science and Technology, Wroclaw, Poland

2 Kingston University London, London, UK

*Address all correspondence to: andrzej.radosz@pwr.edu.pl

References

[1] The Event Horizon Telescope Collaboration. First M87 event horizon telescope results. I. The shadow of the supermassive black hole. The Astrophysical Journal Letters. 2019;**875**: L1. DOI: 10.3847/2041-8213/ab0ec7. (17p)

[2] Hawking SW, Penrose R. The singularities of gravitational collapse and cosmology. Proceedings of the Royal Society A. 1970;**314**(1519): 529-548. DOI: 10.1098/rspa.1970.0021

[3] Ellis GFR, Hawking SW. The Large Scale Structure of Space-Time. Cambridge: University Press; 1973. ISBN: 978-0-521-20016-5

[4] Penrose R. Gravitational collapse: The role of general relativity. General Relativity and Gravitation. 2003;**34**(7): 1141

[5] Hartle JB. Gravity: An Introduction to Einstein's General Relativity. San Francisco: Addison-Wesley; 2003

[6] Landau LD, Lifshitz EM. The Classical Theory of Fields. Oxford: Pergamon Press; 1971

[7] Augousti AT, Gusin P, Kuśmierz B, Masajada J, Radosz A. On the speed of a test particle inside the Schwarzschild event horizon and other kinds of black holes. General Relativity and Gravitation. 2018;**50**:131. DOI: 10.1007/s10714-018-2445-6

[8] Crawford P, Tereno I. Generalized observers and velocity measurements in general relativity. General Relativity and Gravitation. 2002;**34**:2075. DOI: 10.1023/A:1021131401034

[9] Misner CW, Thorne KS, Wheeler JA. Gravitation. San Francisco: W. H. Freeman. ISBN: 978-0-7167-0344-0

[10] Müller T. Falling into a Schwarzschild black hole—Geometric aspects. General Relativity and Gravitation. 2008;**40**:2185-2199. DOI: 10.1007/s10714-008-0623-7

[11] Augousti AT, Gawelczyk M, Siwek A, Radosz A. Touching ghosts: Observing free fall from an infalling frame of reference into a Schwarzschild black hole. European Journal of Physics. 2012;**33**:1. DOI: 10.1088/0143-0807/33/1/001

[12] Frolov VP, Novikov ID. Black Hole Physics: Basic Concepts and New Developments. Dordrecht: Kluwer Academic; 1998

[13] Doran R, Lobo FS, Crawford P. Interior of a Schwarzschild black hole revisited. Foundations of Physics. 2008; **38**:160-187

[14] Gusin P, Augousti AT, Formalik F, Radosz A. The (A) symmetry between the exterior and interior of a Schwarzschild black hole. Symmetry. 2018;**10**:366. DOI: 10.3390/sym10090366

[15] The LIGO Scientific Collaboration and The Virgo Collaboration. An improved analysis of GW150914 using a fully spin-precessing waveform model. Physical Review X. 2016;**6**(4):041014. DOI: 10.1103/PhysRevX.6.041014

[16] Abbott BP, et al. (LIGO Scientific Collaboration and Virgo Collaboration). Properties of the binary black hole merger GW150914. Physical Review Letters 2016;116(24):241102. DOI: 10.1103/PhysRevLett.116.241102

[17] Abbott BP, et al. (LIGO Scientific Collaboration and Virgo Collaboration). Observation of gravitational waves from a binary black hole merger. Physical Review Letters 2016;116(6):061102. DOI:10.1103/PhysRevLett.116.061102

Chapter 12

Dark Matter within the Milky Way

Aleksander Kaczmarek and Andrzej Radosz

Abstract

Dark matter is an invisible substance that seems to make almost 85% of all the mass and roughly 26% of mass-energy content of our Universe. We briefly present the history of its discovery, and we discuss the main attempts to resolve the problem of the origin of dark matter. Those attempts are as follows: dark matter particles (WIMPs), unseen astrophysical objects (MACHOs), or interactions of dark matter with ordinary (luminous) matter. We also introduce a different approach claiming no need for existence of the dark matter (MOND) and recent findings about the ultra-diffuse galaxies. Finally we present 21-cm line observations of neutral hydrogen in the Milky Way made by using 3 m in diameter radio telescope in the Astronomical Observatory of the Jagiellonian University. These studies yield rotational curve of our galaxy. Rotational curve we obtained is compared to those present in literature and constitutes a proof of presence of dark matter in the Milky Way.

Keywords: dark matter, WIMP, MACHO, MOND, rotational curve, ultra-diffuse galaxies, gravitational lensing, milky way

1. Introduction

In 1933 Fritz Zwicky [1] indicated a problem related to the galaxy cluster Coma. Galaxy cluster studied by Zwicky appeared to contain some 400 times more matter than an ordinary, visible, i.e., luminous matter. The content of the luminous matter was estimated form the amount of light emitted by the cluster. However, there was no response for that finding. Only 40 years later in 1970s the problem was rediscovered and concerned almost all of the galaxies. Research of Vera Rubin discovered that the galaxies rotate in a way that cannot be explained by taking into account visible, luminous matter. Today we know that most of the matter in the Universe is dark. Various attempts to resolve the problem of the existence of a mysterious form of matter, dark matter, have been taken ever since. One such idea is to find a particle to possibly complete the standard model. The most important property of such particle would be that it is not a subject to electromagnetic force; hence the dark matter is invisible in all electromagnetic wavelengths. In order to detect such particle, sensitive detectors are built, but still final conclusion has not been made. Another attempt of explaining the problem of missing matter was based on the assumption of existence of astrophysical objects such as black hole or dim brown dwarfs. This idea has rather been discredited as the abundance and masses of such objects are too small comparing to the amount of the matter that is missing. On different grounds stands the idea of modifying gravity in low acceleration regime. Modified Newtonian dynamics (MOND) proposed by M. Milgrom in 1983 is a phenomenological approach attempting to provide explanation of rotation of

IntechOpen

galaxies without invoking hidden matter at all. Yet such an approach seems to be in tension with recent findings of van Dokkum et al. about the ultra-diffuse galaxies. There appear to exist galaxies devoided of dark matter—then what about MOND predictions? This contribution is completed with the rotational curve of the Milky Way determined with 3 m in diameter radio telescope in the Astronomical Observatory of the Jagiellonian University. Obtained rotational curve is flat which indicates the presence of dark matter in the halo of our galaxy.

2. The dark matter problem

The term "dark matter" (DM) was introduced due to the contribution by Fritz Zwicky as early as in 1930s of the twentieth century. Studying the Coma cluster (of galaxies) located 320 million light-years away, Zwicky estimated [1] masses of the galaxies that make up this cluster based on the amount of light they emit. It turned out that such an amount of (*luminous*) matter wasn't large enough to explain the trajectories and velocities of those galaxies. Zwicky claimed then that the gravitational attraction exerted by the luminous matter was not enough to hold the cluster together and if there wasn't some kind of additional, nonluminous matter that provide extra gravity force, the galaxies would fly apart. These findings seemingly intriguing by themselves had not been taken seriously by scientific community. And only findings of Vera Rubin [2], some 40 years later, led to the formulation of the fundamental and still unresolved problem. Rubin studied rotational curves of galaxies. Rotational curve of a galaxy is a plot presenting how the orbital velocity of objects in this galaxy changes with increasing distance from the galaxy's center (see **Figure 1**). It turned out that the shapes of the curves did not comply with the theoretical predictions based on the mount of matter estimated due to the emitted light.

Figure 1 illustrates this discrepancy. When being close to the center of the galaxy, the plot agrees with what one would expect: the rotational curve increases

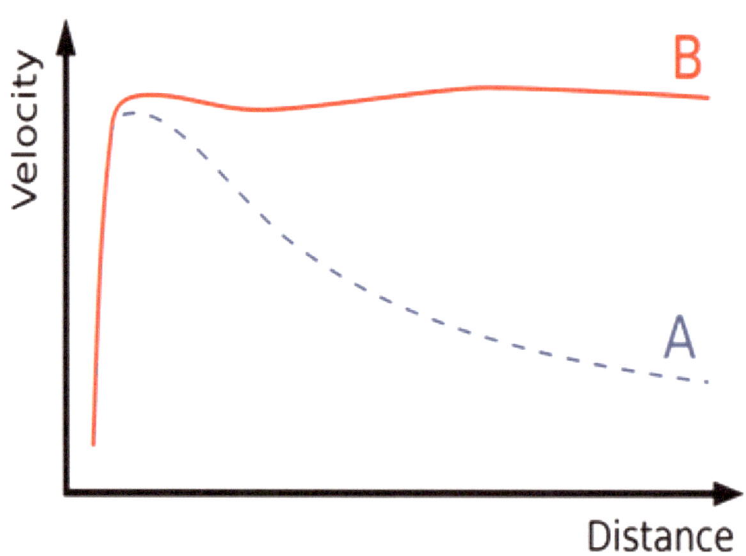

Figure 1.
Figure schematically representing discrepancy between observed (B) and predicted (A) rotational curves of galaxies that indicates presence of dark matter in halos of such galaxies. Credit: PhilHibbs, Wikipedia, https://pl.wikipedia.org/wiki/Krzywa_rotacji_galaktyki#/media/Plik:GalacticRotation2.svg, Creative Commons Attribution-Share Alike 3.0 Unported license.

rapidly that reflects an obvious fact that the velocity of a test object (a "star") increases as the effective gravitational force is growing (at a given radius, only the mass enclosed within a sphere of that radius is relevant in terms of excreting gravitational force—Newton's Shell Theorem). Past a certain distance though (when increasing a distance from the massive center of galaxy does not enclose adequately bigger amounts of mass), the effective force of gravity should decline (as R^2 will increase faster than the mass enclosed in a sphere of a radius being that distance from the center so the force of gravity will decline) which should result in lower orbital velocities.

Vera Rubin and Kent Ford published their first rotational curve in paper [2]. They presented there the rotation of Andromeda based on spectroscopic survey of emission regions applying neutral hydrogen, Hα, and [NII] λ6583 emission lines. Further works, see, e.g., [3], revealed that most of the galaxies have rather flat rotational curves like the one in **Figure 1**. The fact that more distant stars have almost constant velocity attracted the attention of scientists. The circular velocities of the stars are due to gravity which plays the role of centripetal force. Combining Newton's law of gravity with an expression for centripetal force yields the following relation:

$$\frac{GM}{R^2} = \frac{V^2}{R}, \tag{1}$$

where G is universal gravitational constant, M is mass exerting a gravitational force, V denotes velocity of a (test) object orbiting mass M, and R is the distance between them. One obtains from Eq. (1)

$$M = GV^2R. \tag{2}$$

Since G is constant and V appears to be constant as we can see in rotational curves (see **Figure 1**), it would mean that the mass of a galaxy increases linearly with the distance from its center:

$$M(R) \sim R. \tag{3}$$

As we know most of galaxies including the Milky Way have a bright massive center, a *bulge*, with majority of stars placed in that range and possibly a supermassive black hole in the middle. The farther away from the center, the fainter the regions are, i.e., less stars hence less matter is present, and linear dependence (3) is almost impossible to be obeyed. Computer simulations show that the galaxies move in a way we can observe them only if there is another than ordinary, luminous, form of matter, namely, dark matter. The amount of dark matter should be as large as almost five times more than the amount of ordinary matter. This is in agreement with calculations made within lambda-cold dark matter model (Λ-CDM) and the data from Wilkinson Microwave Anisotropy Probe (WMAP) [4] as well as Planck mission [5]. Λ-CDM model is a parametrization of the Big Bang cosmological model in which the Universe contains three major components: first, a cosmological constant denoted by lambda (Greek Λ) and associated with *dark energy*; second, the postulated *cold dark matter* (abbreviated CDM); and third, *ordinary matter*. It is often referred to as the standard model of Big Bang cosmology because it is the simplest model that provides a reasonably good description of the content of the Universe. WAMP was a satellite designed to map the cosmic microwave background (CMB) radiation over the entire sky in five frequency bands. The agreement between Λ-CDM model and the data from WAMP is good enough, which supports the validity of this model [4, 5]. The Λ-CDM model indicates that the matter the

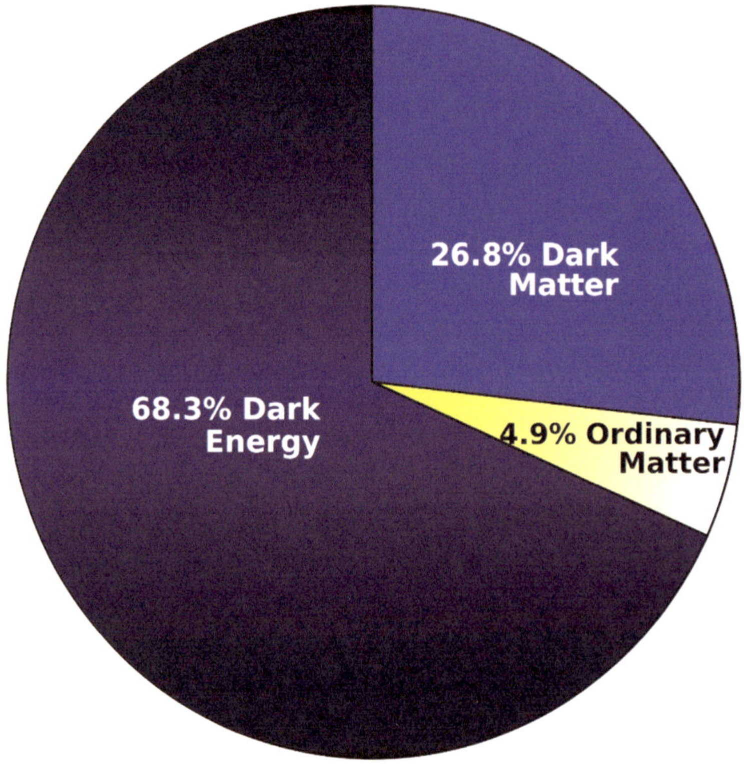

Figure 2.
Estimated distribution of matter and energy in the universe based on Planck data. Credit: ESA, Planck reveals an almost perfect Universe.

stars (and us) are made of is just a tiny part of the mass-energy content of the Universe (see **Figure 2**).

3. Possible solutions and even more problems

3.1 Wimps

Hypothetical particles that constitute the dark matter are called WIMPs which stands for weakly interacting massive particles. All the matter that we know (and us) is made of baryonic matter, i.e., the matter is made of baryons. WIMPS would be a new type of particles beyond the standard model. Those should be massive, subject to the gravitational force, and possibly other forces that are comparable to the weak force. One such candidate for WIMP could be a stable supersymmetric particle. Supersymmetric model has a particle of this property which was even called a "Wimp Miracle," but we have not yet observed any trace of supersymmetry, moreover, Wimp Miracle in any of the particle colliders. WIMPs also should not interact via electromagnetism; hence the DM is not visible in any wavelength. We only can "see" the DM due to its gravitational interactions, which are strong enough to cause a phenomenon known as gravitational lensing.

3.1.1 Gravitational lensing

This phenomenon is observed when the light rays passing near a very massive object are deflected (due to the curvature of space–time produced by this object) in such a way that a distant observer observes it lensed. **Figure 3** illustrates gravitational lensing: the stretched structures are distant galaxies, whose light was bent by the DM between them and the observer. This allows to calculate the mass required to cause such phenomenon [6]. Large aggregations of massive DM particles are able to produce such image letting us to know it's out there.

3.2 MACHOs

Massive astrophysical compact halo objects (MACHOs) was another hypothesis invoked to explain the presence of large amount of nonluminous matter in galactic halos. Those, contrary to the WIMPS, would have been regular astrophysical objects emitting little or no radiation such as black holes, neutron stars, as well as brown dwarfs and unassociated planets, which drift unseen through interstellar space providing extra gravity. Thorough investigations have shown that this concept rather fails to explain the expected amount of the DM. One way to detect MACHOS' influence, as described in [7], is to look for events of microlensing caused by them. Such microlensing would cause observable apparent amplification of star's flux. In [7] it was shown that the number of such events is far too less that would have been expected. That rules out MACHOS as the candidates for DM. Moreover, the studies of abundance of baryons created in the Big Bang show that baryon density is consistent with the mean cosmic density of matter visible optically and in X-rays. It implies that most of the baryons in the Universe are visible but not dark and that most of the matter in the Universe consists of nonbaryonic DM [7].

3.3 MOND

In the former sections, we have discussed the attempts of solving or explaining the problem of the missing matter. That is to find or to claim existence of unknown, invisible substance. Yet there is another idea based on a different assumption. In 1983 Milgrom [8] proposed an idea that maybe it is the theory that needs to be

Figure 3.
An image of gravitational lensing obtained with Hubble space telescope showing a distant image of galaxies which had been stretched due to the warping of space–time caused by a massive object between them and the observer. Credit: *ESA/Hubble https://www.spacetelescope.org/images/potw1506a/.*

modified rather than an invisible matter to be found. Modified Newtonian dynamics (MOND) is an empirically motivated modification of Newtonian dynamics at low accelerations, suggested as an alternative to dark matter concept [8, 9]. In Ref. [8] Milgrom considered the possibility that Newton's second law does not describe the motion of objects under the conditions which prevail in galaxies and systems of galaxies. Newton's laws have been tested in high-acceleration environment like the Earth or the solar system. The stars in the outer parts of the galaxies move in the circumstances of extremely low accelerations compared to what we know from everyday life. To illustrate how small such accelerations might be, let us calculate the acceleration of average star (the Sun) located on the suburbs of average galaxy (the Milky Way):

$$a = \frac{V^2}{R} = \frac{\left(220\ \frac{km}{s}\right)^2}{8.5\ kpc} \approx 1.845 \times 10^{-10} \frac{m}{s^2} \tag{4}$$

Milgrom proposed then a generalized form of Newton's second principle, claiming the inertia term not to be simply proportional to the acceleration of an object but being rather a more general function of it:

$$m \cdot \mu(a/a_0)\vec{\boldsymbol{a}} = \vec{\boldsymbol{F}}. \tag{5}$$

In expression (5) m is gravitational mass, a is acceleration, a_0 is some acceleration constant, and μ is a nonlinear function with the following properties:

$$\mu(x \gg 1) \approx 1, \mu(x \ll 1) \approx x \tag{6}$$

The acceleration constant is found to be $a_0 = 1.2 \pm 0.2 \times 10^{-10} \frac{m}{s^2}$ [8]. Phenomenological success of MOND is that applying it produces flat rotation curves of galaxies as observed and that this simple law is sufficient to make predictions for a broad range of galactic phenomena.

3.4 Ultra-diffuse galaxies

Recent studies of van Dokkum et al. [10, 11] have uncovered new class of object referred to as *ultra-diffuse galaxies*. NGC1052-DF2 and NGC1052-DF4 are large, faint galaxies with an excess of luminous globular clusters, and they have a very low-velocity dispersion. Velocity dispersion is the dispersion of radial velocities about the mean velocity for a group of objects. Low-velocity dispersion indicates that the galaxy has little or no dark matter. NGC1052-DF2 was studied with the Keck Cosmic Web Imager (KCWI), a new instrument on the Keck II telescope that was optimized for precision sky-limited spectroscopy of low surface brightness phenomena at relatively high spectral resolution. The spectroscopy data was used to describe kinematics of the galaxy. This result was based on the radial velocities of globular clusters that were assumed to be associated with the galaxies. It was claimed in Ref. [10] that taking observational uncertainties into account, the determined intrinsic velocity dispersion is consistent with the expected value found for the stars alone and lower than expected from DM halo (see **Figure 4**). The dynamical mass of NGC1052-DF2 determined in [10] was $1.3 \pm 0.8 \times 10^8 M_\odot$, and the stellar mass, i.e., luminous matter, was found to be $1 \pm 0.2 \times 10^8 \mathrm{M}_\odot$.

To give a reader some intuition and place this in some context, it is worth to notice that the stellar mass of the Milky Way found in [12] was $6.08 \pm 1.14 \times 10^{10} M_\odot$. It is broadly accepted in literature, and as will the following section

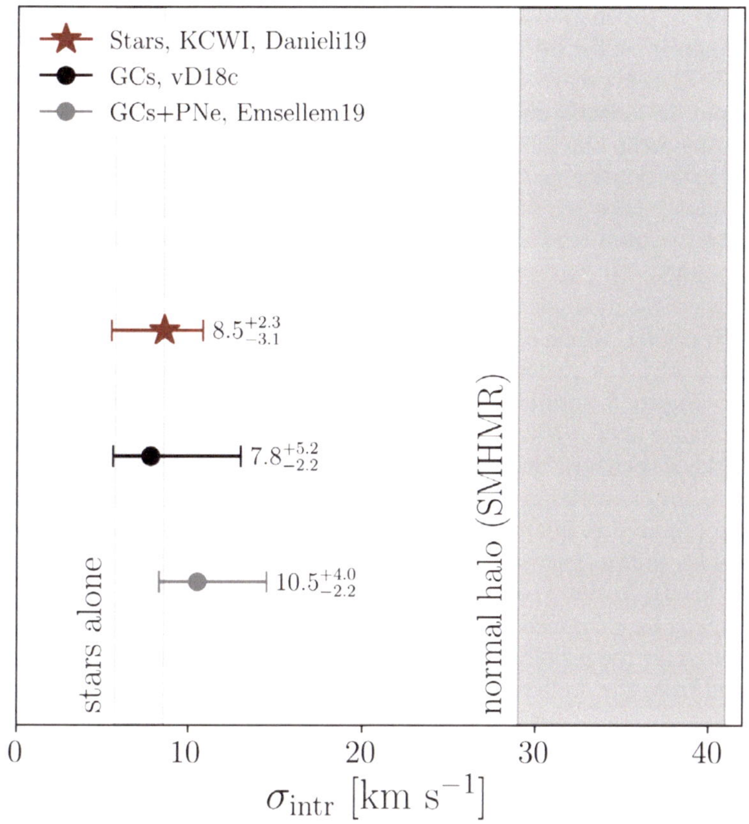

Figure 4.
Constraints on the intrinsic velocity dispersion of NGC1052-DF2. The result found in [8] (red dot star) is consistent with two other studies mentioned by authors and shows that such velocity dispersion indicates lack of the dark matter. Credit: *[10].*

present, the Milky Way contains big amount of the dark matter. The velocity dispersion of our galaxy is 75 km/s [13]. The NGC1052-DF2 is about 100 times lighter than the Milky Way; however, the velocity dispersion of NGC1052-DF2 was found to be only roughly 8.5 km/s [10]. If the galaxies can be formed and exist without the dark matter, i.e., the dark matter is not present in all existing galaxies, then the attempts to explain their dynamics by applying MOND might be at risk.

4. Detection of dark matter

4.1 Gravitational interaction with ordinary matter

In 2012 Moni Bidin et al. [14] published a paper in which they estimated surface mass density in the solar neighborhood. Results obtained match the expectations of visible matter alone without the need of adding the dark matter component. The difference between the measured mass of matter (derived in this study) and the mass of visible matter (i.e., mass of matter that is estimated in the independent way based on the amount of emitted) provides an estimate of the amount of DM in the volume under analysis, and constraints on the shape of the DM halo can be derived. The fundamental basis for this measurement is the application of the Poisson–Boltzmann and Jeans equations to a virialized system in steady state. This allows to estimate either the local density at the solar position or the surface density (mass per unit area) of the mass within a given volume. Authors in Ref. [14] derive analytical expression for surface density as a function of distance from the galactic disk plane Σ(Z) to estimate the surface mass density between 1.5 and 4.5 kpc distance from the galactic disk plane using data from of the kinematics studies of about 400 red giants kinematics. The authors in [14] claimed that the estimate of the surface mass density matches the expectation of visible mass alone and the degree of overlap between the two curves is striking. There is no need for any dark component to account for the results: the measured Σ(Z) implies a local DM density $\rho_{\odot DM} = 0 \pm 1\, M_{\odot} \cdot 10^{-3} pc^{-3}$ a. Further the authors in [14] compared this results with models of DM disk present in literature such as Ref. [15] hereafter OM; Ref. [16] hereafter SMH, which is standard DM halo model; or Ref. [17]—the model with minimal local DM density—hereafter MIN. Comparison of these findings is presented in **Figure** 5. Authors in Ref. [14] claim that the OM model is excluded at 8 sigma confidence level, SHM at 6 sigma, and even MIN model at 4.1 sigma. (Sigma confidence level says how many values lie within the number of standard deviation of the mean. For example, in particle physics there is a convention of a five-sigma effect being required to qualify as a discovery, that is to say that 99.99994% of the values must lay within 5 standard deviations of the mean; 8 sigma is even higher confidence level). Authors conclude that the measurement of the mass surface density at the solar galactocentric position between 1.5 and 4 kpc from the galactic plane accounts for the visible mass only. The DM density in the solar neighborhood, extrapolated from the observed curve of Σ(Z), is at variance with the general

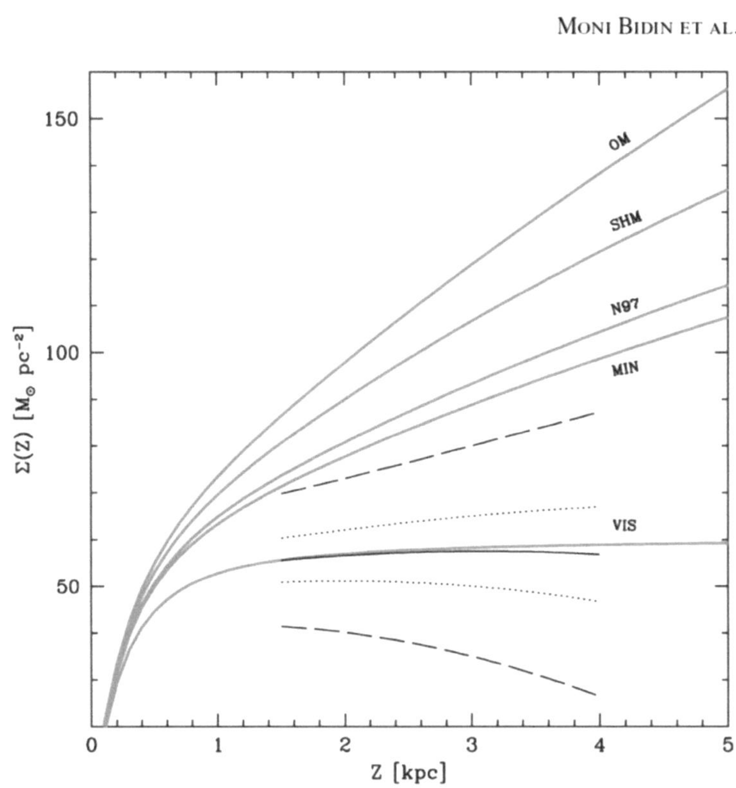

Figure 5.
Observational results for the surface mass density, as a function of distance from the galactic plane (black curve), compared to the expectations of the models discussed in the text (thick gray curves). The dotted and dashed lines indicate the observational 1σ and 3σ strip, respectively. Expectations for the known visible mass are indicated by the thick gray curve labeled as VIS. Credit: *[14].*

consensus that it must be in the range $5 - 13\,M_\odot \cdot 10^{-3} pc^{-3}$ (e.g., [18, 19]). Lack of DM is observed by using measurements of the thick disk kinematics and is independent of the choice of data, because very similar results were obtained by means of other kinematical results in the literature. It is clear that the local surface density as measured in Ref. [14], extrapolated to the rest of the galaxy, cannot retain the Sun in a circular orbit at a speed of $\sim$220 km s^{-1}. A deep missing mass problem is therefore confirmed by this study, and this finding tells us that indirect attempts to detect the dark matter by investigating its interactions with ordinary matter in that way have a little chance of success.

4.2 Direct detection

The experiments that aim at the direct detection due to scattering do not agree with each other yielding different constraints on the mass of the DM particles. The DAMA/LIBRA experiment [20] is the only one to claim positive result of detection which however has not been yet confirmed by the other groups (detectors). The aim of this experiment is detecting low-energy recoil photons from the scintillator crystals of thallium-doped sodium iodide NaI(Tl) placed in the detectors under the ground. Such photons would be emitted when the DM particle collides with one of the scintillators. If what we know about the DM is right, then since the Earth orbits the Sun, the DM particles should pass through the planet and hence have a chance to collide with those of the detectors. The idea of the experiment is that if one takes into account the revolution of the Earth around the Sun and the revolution of the Sun around the center of our galaxy, then the signals coming from the collisions should be modulated as in June the relative velocity of the Earth and the DM flux is the biggest hence yielding the biggest number of collisions. The data collected from the phase II of the experiment have all traits required to claim the presence of the DM in our part of the galaxy. The annual modulation is present only in the events concerning the photons with energies exactly within the energetic range theoretically predicted for the DM particles. Yet the DAMA/LIBRA is a singular case. Several groups have been working to develop experiments aiming at reproducing DAMA/LIBRA's results using the same target medium. To determine whether there is evidence for an excess of events above the expected background in sodium iodide and to look for evidence of an annual modulation, the COSINE-100 experiment [21] uses the same target medium to carry out a model-independent test of DAMA/LIBRA's claim. Their results from the initial operation of the COSINE-100 experiment were published in [21], and no excess of signal-like events above the expected background in the first 59.5 days of data from COSINE-100 has been observed. Assuming the so-called standard DM halo model, this result rules out spin-independent WIMP–nucleon interactions as the cause of the annual modulation observed by the DAMA/LIBRA collaboration. Another such experiment is the XENON100 experiment that searches for electronic recoil event rate modulation by measuring the scintillation light from a particle interacting in the liquid xenon. The results of this experiment published in [22] also exclude the DAMA/LIBRA results.

4.3 Others

We will present here very briefly the other two methods of detection of DM:

- *Production of DM particles in colliders*—If the DM particles were created, for instance, in LHC, they would escape through the detectors unnoticed (due to their non-electromagnetic nature). However, they would carry away energy and momentum, so one could infer their existence from the amount of energy and momentum “missing” after a collision. The LHC also search for existence of supersymmetric particles which are one of the candidates for DM particle.

- *Searching for products of annihilation of its particles*—Indirect detection. This experiments search for the products of the self-annihilation or decay of DM particles in outer space. For example, in regions of high DM density (e.g., the center of our galaxy), two DM particles could annihilate to produce gamma rays or standard model particle–antiparticle pairs. Alternatively if the DM particle is unstable, it could decay into standard model (or other) particles. These processes could be detected indirectly through an excess of gamma rays, antiprotons, or positrons emanating from high-density regions in the galaxy or others.

5. Milky way rotation curve

DM manifests its existence through the shape of rotational curves of galaxies, in particular, through the rotational curve of our own galaxy, the Milky Way. This is what motivated us to take a glimpse on that topic and to compare results to those present in literature [23]. We have studied the rotational curve of Milky Way with radio telescope located in the Astronomical Observatory of the Jagiellonian University provided by EU-HOU project (EU-HOU project was founded with support from the European Commission, grant 510,308-LLP-1-2010-FR-COMENIUS-CMP. https://www.astro.uni-bonn.de/hisurvey/euhou/LABprofile/).

5.1 The method

This 3 m in diameter telescope runs observations on 1420 MHz frequency which is the emission line of neutral hydrogen. When the hydrogen atom undergoes a transition from the state of higher energy when the spins of the proton and the electron are parallel to the state of lower energy that is when the spins are antiparallel, emitted photon is equivalent to radiation roughly 21 cm wavelength in vacuum (see **Figure 6**). Even though such process occurs very rarely, given the abundance of the hydrogen in the Universe (i.e., 74% of its baryonic mass), it is a common phenomenon. Hence the hydrogen is also present in the interstellar space around the stars, and radio observations yield information on how the matter is distributed inside the galaxy, and knowing the Doppler shift of the observed radiation, one can calculate the velocity of the hydrogen cloud from which it comes from. This in turn gives us an idea how the hydrogen and the nearby matter move within the galaxy, i.e., orbit around its center. Knowing the velocities and distance of such hydrogen clouds, one can plot the rotational curve of the galaxy. This is called tangent point method. Thus using the data obtained from the telescope, the Doppler equation:

$$V_r = \frac{f_0 - f}{f_0} \cdot c \tag{7}$$

one can calculate the source’s velocity (speed) relative to us (V_r).f_0 is the frequency emitted by the hydrogen atom, f is the frequency the radio telescope receives, and c denotes the speed of light. The frequencies registered by the radio telescope are of course slightly different than 1420 MHz which is the frequency of emitted photon as measured at the lab and as emitted by the hydrogen atom.

DOI: http://dx.doi.org/10.5772/intechopen.90267

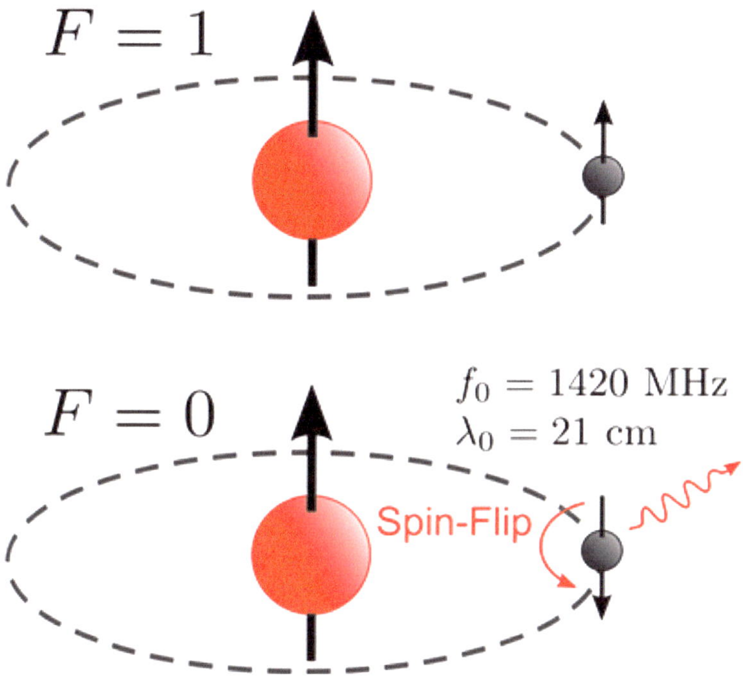

Figure 6.
Hydrogen 21-cm emission line.

The hydrogen atoms that we study are moving relatively to us so the signal coming from them is a subject to the same phenomenon as for the ambulance's siren applies. That is the change in frequency that enables us to calculate the radial velocity of such hydrogen cloud along the line of sight (which is defined by galactic longitude).

To find the speed of the hydrogen cloud, a simple fact is used, that is the radial velocity results in difference between the projection of ours (Sun's) velocity on the line of sight and the hydrogen cloud's velocity on the line of sight (see **Figure** 7). The line of sight is determined along the galactic longitude (see **Figure 8**) on which we set the radio telescope.

This results in the following equation for velocity of observed hydrogen cloud:

$$V_r = V\frac{R_0}{R} sin\,(l) - V_0\, sin\,(l) \tag{8}$$

Among the objects observed along the given line of sight, the one with the smallest distance will have the biggest velocity. The smallest possible distance between us and the source is when it lies in the tangent point; hence simple trigonometry allows us to determine the distance:

$$R = R_0\, sin\,(l) \tag{9}$$

which simplifies Eq. (8) to

$$V = V_r + V_0 \sin\,(l). \tag{10}$$

Eqs. (8) and (9) provide all required information to plot a rotational curve of the galaxy. This method works for objects in I and IV Quadrants of galactic longitude, that is for $0° < l < 90° and\ 270° < l < 360°$ and inside the galactocentric radius of the Sun.

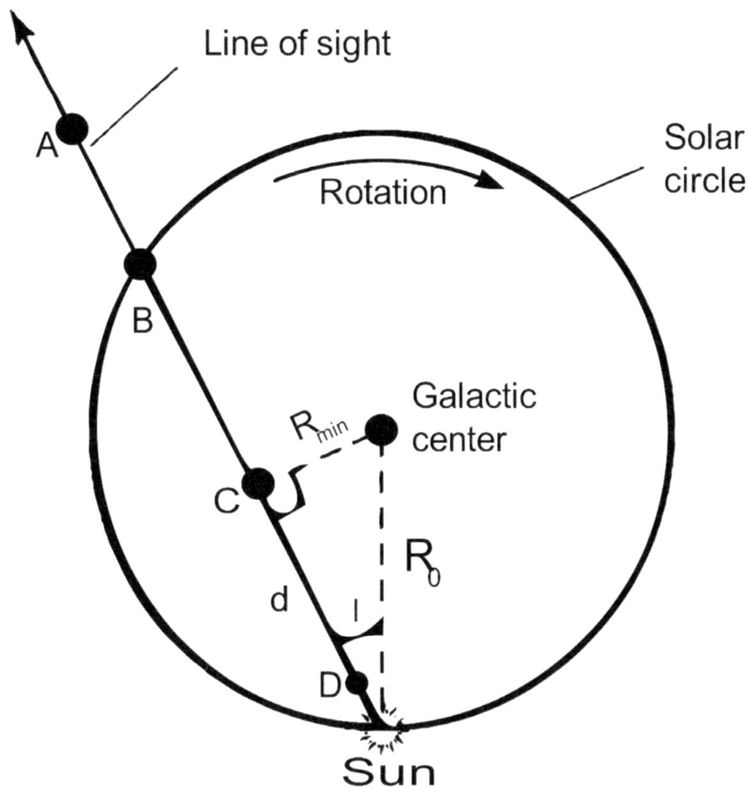

Figure 7.
Figure presenting two objects (A, B) along the line of sight. Hence object B lies in tangent point, i.e., its distance from the center of the galaxy is smaller, and its velocity is greater than the velocity of object A.

5.2 Results

Twenty-nine objects with galactic longitude $0° < l < 90°$ have been studied. Their positions on the map of the Milky Way are presented in **Figure 9**. Tangent point method applied to the data results in rotational curve presented in **Figure 10**.

Our rotational curve plot, **Figure 10**, is comparable to the plot obtained from data from LAB survey [24] and consistent with the ones that can be found in literature [23, 25]. We follow [25] in their choice of function to fit the data, namely

$$\frac{V}{V_0} = a\left(\frac{R}{R_0}\right)^b + c \tag{11}$$

where we put $V_0 = 220\frac{km}{s}$ and $R_0 = 8.5$ kpc and find the coefficients to be $a = -5.495e - 06$, $b = -21.28$, and $c = 0.9808$.

We conclude that the rotational curve reveals the existence of dark matter within the Milky Way. Taking (nonrelativistic) law of gravity, that is, the force of gravity is proportional to inverse squared distance, one would expect that the farther away the hydrogen clouds (constituting the distribution of matter) are from the massive center of the galaxy, the lower their velocities will be. As one see from the rotational curve, **Figure 10**, this is not the case; the velocities seem to be constant over a distance of roughly 3 kpc. Which means there is nonluminous matter distributed in such a way just to “keep up” with the increasing distance from

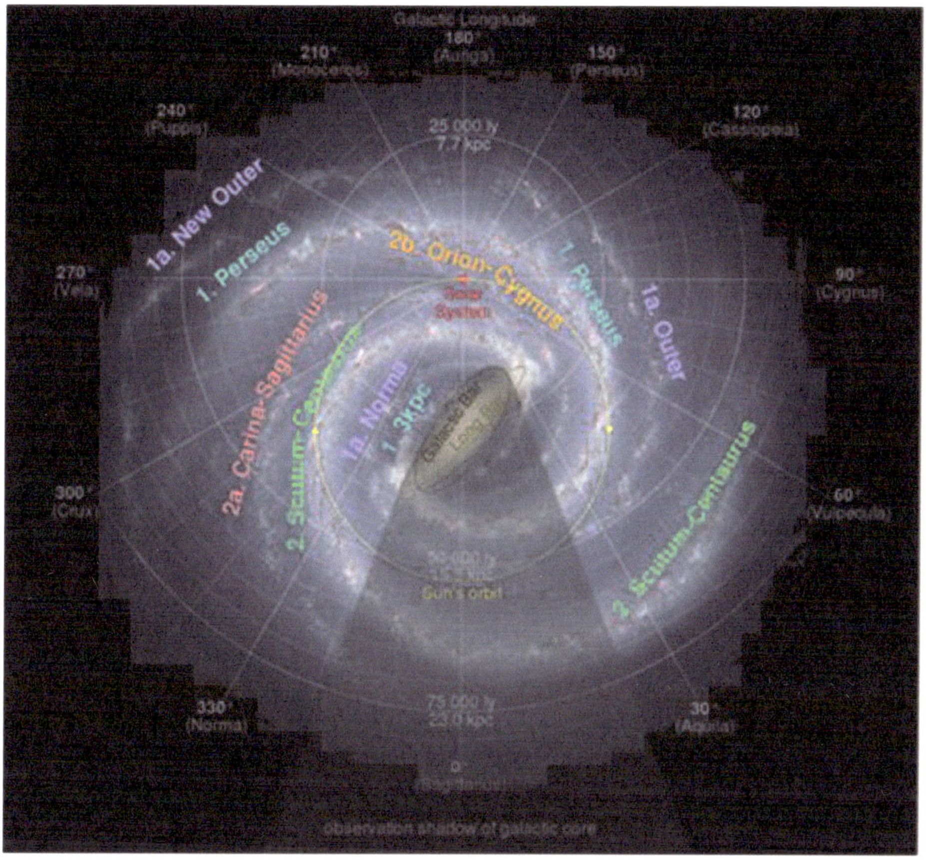

Figure 8.
Figure presenting galactic longitude. L = 0° is direction from the solar system to the center of galaxy. Credit: *File: Artist's_impression_of_the_Milky_Way.Jpg: NASA/JPL-Caltech/ESO/R.hurt.*

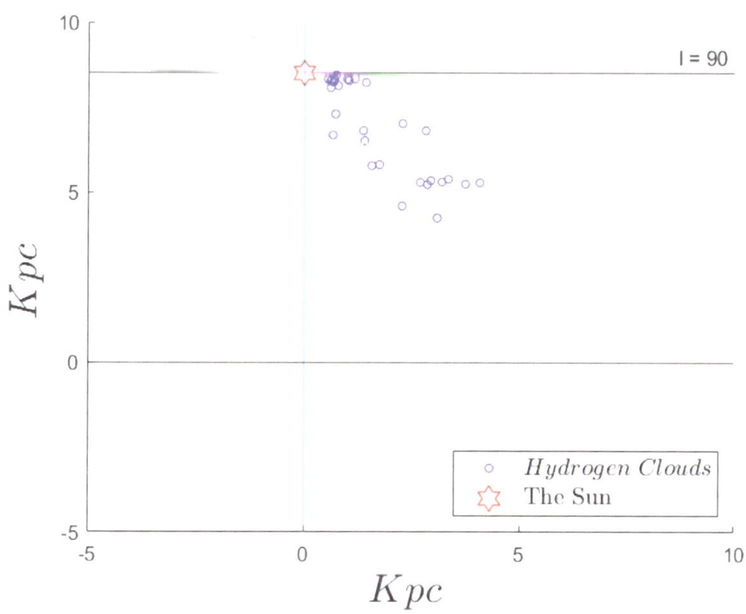

Figure 9.
Map of the hydrogen clouds used to determine the rotational curve of the Milky Way.

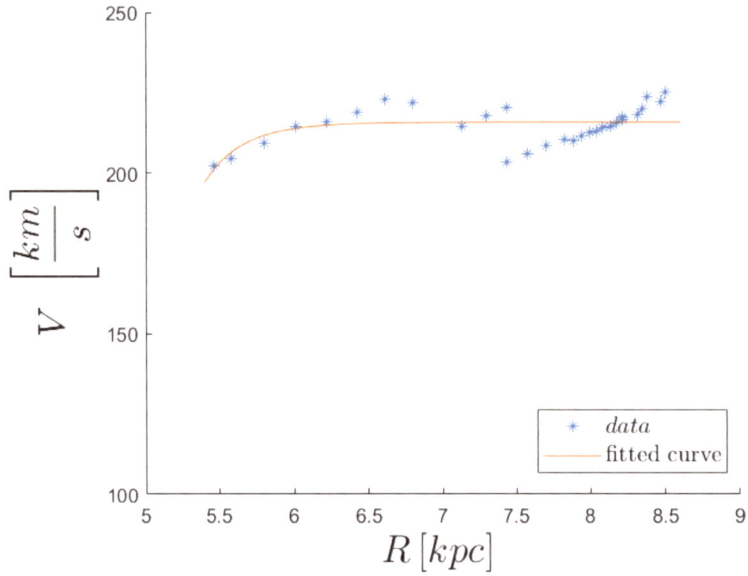

Figure 10.
Rotational curve obtained from 21-cm line observations of the Milky Way. Note that the velocity of the studied objects appears to be constant over roughly 3 kpc distance.

the center of galaxy and make it so that the velocities of hydrogen atoms are almost constant as the distance increases.

6. Conclusions

The problem of missing matter discovered by Fritz Zwicky in 1933 appears to be still an open question. The most important premise of existence of the dark matter is the shape of rotational curves of galaxies, introduced as a tool for studying galaxy rotation by Vera Rubin. With our current understanding of the Universe, the dark matter, still a mysterious substance, makes up 86% of all the matter in the Universe. Throughout the years various attempts have been made to explain its nature. Some of the ideas have been proven unlikely (MACHOs). Some of them contradict each other (DAMA/LIBRA, the COSINE-100 collaboration). Yet even simple Milky Way's observations as presented in Section 5 lead to the conclusion that the dark matter is present in the halo of our galaxy.

Author details

Aleksander Kaczmarek[1]* and Andrzej Radosz[2]

1 Faculty of Fundamental Problems of Technology, Wroclaw University of Science and Technology, Wroclaw, Poland

2 Department of Quantum Engineering, Wroclaw University of Science and Technology, Wroclaw, Poland

*Address all correspondence to: a.kaczmarek12@wp.pl

References

[1] Zwicky F. Die Rotverschiebung von extragalaktischen Nebeln. Helvetica Physica Acta. 1933;**6**:110-127

[2] Rubin VC, Kent Ford W. Rotation of the andromeda nebula from a spectroscopic survey of emission regions. The Astrophysical Journal. 1970;**159**. DOI: 10.1086/150317

[3] Rubin VC, Ford WK, Thonnard N. Extended rotation curves of high-luminosity spiral galaxies. Iv. Systematic dynamical properties. The Astrophysical Journal. 1978;**225**:L107-L111. DOI: 10.1086/182804

[4] Jarosik N, Bennett CL, Dunkley J, Gold B, Greason MR, Halpern M, et al. Seven-year Wilkinson microwave anisotropy probe (WMAP*) observations: Sky maps, systematic errors, and basic results. The Astrophysical Journal Supplement Series;**192**(2). DOI: 10.1088/0067-0049/192/2/14

[5] Ade PAR et al. Planck 2015 results XIII. Cosmological parameters. A&A. 2016;**594**(A13). DOI: 10.1051/0004-6361/201525830

[6] Amendola L, Appleby S, Avgoustidis A, Bacon D, Baker T, Baldi M, et al. Cosmology and fundamental physics with the Euclid satellite. Living Reviews in Relativity. 2018;**21**(1):2. DOI: 10.1007/s41114-017-0010-3

[7] Tisserand P, Le Guillou L, Afonso C, Albert JN, Andersen J, Ansari R, et al. Limits on the macho content of the galactic halo from the EROS-2 survey of the magellanic clouds. Astronomy and Astrophysics. 2006;**469**(2). DOI: 10.1051/0004-6361:20066017

[8] Milgrom M. A modification of the Newtonian dynamics as a possible alternative to the hidden mass hypothesis. The Astrophysical Journal. 1983;**270**:365-370. DOI: 10.1086/161130

[9] Milgrom M. A modification of the Newtonian dynamics: Implications for galaxies. The Astrophysical Journal. 1983;**270**:371-383. DOI: 10.1086/161131

[10] Danieli S, Pieter van Dokkum, Conroy C, Abraham R, Romanowsky AJ. Still missing dark matter: KCWI high-resolution stellar kinematics of NGC1052-DF2. The Astrophysical Journal Letters. 2019;**874**:L12 (8pp). DOI: 10.3847/2041-8213/ab0e8c

[11] Pieter van Dokkum, Danieli S, Abraham R, Conroy C, Romanowsky AJ. A second galaxy missing dark matter in the NGC1052 group. The Astrophysical Journal Letters. 2019;**874**:L5 (8pp). DOI: 10.3847/2041-8213/ab0d92

[12] Licquia TC, Newman JA. Improved estimates of the milky Way's stellar mass and star formation rate from hierarchical bayesian meta-analysis. The Astrophysical Journal. 2015;**806**:96 (20pp). DOI: 10.1088/0004-637X/806/1/96

[13] Gebhardt K, Bender R, Bower G, Dressler A, Faber SM, Filippenko AV, et al. A relationship between nuclear black hole mass and galaxy velocity dispersion. The Astrophysical Journal. 2000;**539**:L13-L16. DOI: 10.1086/312840

[14] Moni Bidin C, Carraro G, M'endez RA, Smith R. Kinematical and chemical vertical structure of the galactic thick disk. II. A lack of dark matter in the solar neighborhood. The Astrophysical Journal. 2012;**751**:30 (14pp). DOI: 10.1088/0004-637X/751/1/30

[15] Olling RP, Merrifield MR. Luminous and dark matter in the milky way. Monthly Notices of the Royal

Astronomical Society. 2001;**326**(1): 164-180. DOI: 10.1046/j.1365-8711.2001.04581.x

[16] Jungman G, Kamionkowski M, Griest K. Supersymmetric dark matter. Physics Reports. 1996;**267**(5–6):195-373. DOI: 10.1016/0370-1573(95)00058-5

[17] de Boer W, Sander C, Zhukov V, Gladyshev AV, Kazakov DI. EGRET excess of diffuse galactic gamma rays as tracer of dark matter. Astronomy and Astrophysics. 2005;**444**(1):51-67. DOI: 10.1051/0004-6361:20053726

[18] Weber M, de Boer W. Determination of the local dark matter density in our Galaxy. Astronomy and Astrophysics. 2010;**509**. DOI: 10.1051/0004-6361/200913381

[19] Garbari S, Read JI, Lake G. Limits on the local dark matter density. Monthly Notices of the Royal Astronomical Society. 2011;**416**(3):2318-2340. DOI: 10.1111/j.1365-2966.2011.19206.x

[20] Bernabeia R, Bellia P, Bussolotti A, Cappella F, Caracciolo V, Cerulli R, et al. First model independent results from DAMA/LIBRA–phase 2. Nuclear and Particle Physics Proceedings. 2018;**303–305**:74-79. DOI: 10.1016/j.nuclphysbps.2019.03.01

[21] The COSINE-100 Collaboration. An experiment to search for dark-matter interactions using sodium iodide detectors. Nature. 2018;**564**:83-86. DOI: 10.1038/s41586-018-0739-1

[22] Aprile E et al. Search for electronic recoil event rate modulation with 4 years of XENON100 data. Physical Review Letters. 2017;**118**:101101. DOI: 10.1103/PhysRevLett.118.101101

[23] Michael F, Blitz L, Stark Antony A. The rotation curve of the milky way to 2R0. The Astrophysical Journal. 1989; **342**:272-284. DOI: 10.1086/167591

[24] Kalberla PMW, Burton WB, Dap H, Arnal EM, Bajaja E, Morras R, et al. The Leiden/Argentine/Bonn (LAB) Survey of Galactic HI. Final data release of the combined LDS and IAR surveys with improved stray-radiation corrections. Astronomy and Astrophysics. 2005; **440**(2):775-782. DOI: 10.1051/0004-6361:20041864

[25] Brand J, Blitz L. The velocity field of the Outter galaxy. Astronomy and Astrophysics. 1993;**275**(1):67

Chapter 13

The Early Universe as a Source of Gravitational Waves

Vladimir Gladyshev and Igor Fomin

Abstract

In this chapter we consider the issues of the origin and evolution of relic gravitational waves (GW), which appear as a result of quantum fluctuations of the scalar field and the corresponding perturbations of the space-time metric at the early inflationary stage of the evolution of the universe. The main provisions of the inflationary paradigm and the methods of the construction of current cosmological models on its basis are considered. The influence of relic gravitational waves on the anisotropy and polarization of the relic radiation and the importance of estimating such an effect on the verification of cosmological models are discussed as well.

Keywords: universe, inflation, scalar field, cosmological perturbations, gravitational waves

1. Introduction

The general relativity (GR) theory proposed by A. Einstein more than a 100 years ago currently finds new confirmations. The possible existence of gravitational waves was predicted by A. Einstein on the basis of solving the equations of general relativity when calculating the power of gravitational radiation [1, 2].

Gravitational waves (GW) are space-time curvature disturbances, which propagate at the speed of light. They occur at any movements of material bodies, leading to inhomogeneous gravity force variation in the environment. Gravitational radiation was predicted by A. Einstein in the general relativity (GR) theory, but so far not detected by direct measurements.

According to general relativity, space-time is curved around the bodies due to the action of gravity and is represented by a symmetric tensor $g_{\mu\nu}$ with 10 independent components. However, far from the masses (the case of weak gravitational fields), the tensor can be divided into two terms $g_{\mu\nu} = \eta_{\mu\nu} + h_{\mu\nu}$ where the first term, i.e., tensor $\eta_{\mu\nu}$, corresponds to the flat space-time of the special theory of relativity and has only four components. The second tensor $h_{\mu\nu}$ contains information about the curvature caused by the gravitational field and makes small corrections. In the case of gravitational disturbances propagating far from their sources, the components of the tensor $h_{\mu\nu}$ can be calculated by the method proposed by Einstein [1], similar to that used in electrodynamics for delayed potentials.

The first evidence was received by experimental studies of Joseph H. Taylor and Joel M. Weisberg et al., who studied the effect of slowing down the period of the binary star system PSR 1913 + 16 due to energy losses on gravitational radiation [3].

Until recently, however, there has remained the main task: the direct recording of gravitational waves from space radiation sources by means of ground-based or space gravitational antennas.

Over the years, several methods have been proposed for recording gravitational radiation. Experimental work began in the 1960s of the twentieth century, but before the beginning of the twenty-first century, there was no reliable experimental proof of the ground-based recording of gravitational radiation [4].

This is due to the fact that gravitational waves have small amplitude; in addition, the proposed detection methods have insufficient sensitivity and are rather complicated in technical implementation.

These are broadband gravitational antennas, which offer a lot of opportunities as to the methods of recording gravitational waves and extracting signals, as well as the use of quantum non-perturbative measurements and the inclusion of gravitational antennas in the network.

The creation of new-generation gravitational antennas designed to reliably receive gravitational waves from remote space sources involves the use of high-power lasers, complex computer systems for processing large data arrays, the use of complex seismic protection systems, and the solution of other complex engineering and physical problems.

At present large international experience has been gained in the field of creating laser gravitational antennas, which ensured the ground-based recording of gravitational waves from black hole collision [5, 6] and neutron star merger [7]. Furthermore, the gravitational wave propagation velocity was estimated, which appeared to be equal to the speed of light in vacuum with an accuracy of 10^{-15} based on almost simultaneous recording of gravitational waves and a short gamma-ray burst from neutron star merger [8].

The modern theory of the early universe is based on the inclusion of the inflationary stage which precedes the stage of the hot universe. The theory of cosmological inflation [9] explains the origin of a large-scale structure and corresponds to observational data [10]. Inflationary expansion of the universe during very early times, once the universe emerged from the quantum gravity (Planck) era, has been proposed in the late 1970s and mainly in the beginning of the 1980s and is becoming more accepted as a necessary stage which modifies the standard Big Bang theory model. According to the theory of inflation, the primordial perturbations appear from quantum fluctuations. These fluctuations had essential amplitudes in scales of Planck length, and during inflation they generate the primordial perturbations which then lead nearer to scales of galaxies with almost the same amplitudes [11].

Thus, the theory of cosmological inflation connects large-scale structure of the universe with microscopic scales. The resultant range of inhomogeneities practically doesn't depend on scenarios of inflation and has a universal form. It leads to unambiguous predictions for a range of anisotropy of the background radiation.

The small quantum perturbations of the scalar field and the corresponding perturbations of the metric generate the relic gravitational waves. This type of gravitational waves was not directly observed; however, the possibility of such observations plays a key role to understand the physical processes in the early universe.

2. Inflationary stage of the early universe

The models of inflationary (accelerated) expansion of the universe at the early stage of its evolution, that is, at times close to the Planck time, are becoming increasingly convincing as a necessary step modifying the standard Big Bang theory, which is based on solutions of Einstein's equations for the universe filled with ordinary baryonic matter with a positive energy density obtained by Friedmann. However, the extrapolation of Friedmann solutions to early times leads to many insoluble problems when constructing on their basis the evolution scenarios of the universe [9].

The exponential (de Sitter) expansion, suggesting $p = -\rho$, or a close expansion of the early universe based on the evolution of a certain substance with the equation of state $p \approx -\rho$, i.e., with a negative pressure, is a feature of inflation models which allow to solve the problems of the standard model of the Big Bang theory, namely, the problems of the horizon, flatness, homogeneity, isotropy, low concentration of exotic states of matter (domain walls, monopoles, etc.), anisotropy of the background radiation, the initial singularity, and some other problems [9].

Thus, the cosmological models containing a combination of Friedmann solutions and (quasi) de Sitter solutions provide the basis for a current description of the evolution of the universe. In the context of the inflationary paradigm, the early universe expands for some time accelerated and, further, goes into a power-law expansion mode without acceleration corresponding to Friedmann solutions.

In most cosmological models, the geometric description of the universe is based on the Friedmann-Robertson-Walker (FRW) homogeneous isotropic space (space-time) model, which is associated with a high degree of isotropy of space, measured on the basis of the cosmic microwave background (CMB) radiation research. This identification also relies on a formal result known as the Ehlers-Geren-Sachs theorem, which refers to the universe filled with any ideal barotropic fluid [12].

The metric of Friedmann-Robertson-Walker (FRW) space-time is written as follows:

$$ds^2 = -dt^2 + a^2(t)\left(\frac{dr^2}{1-kr^2} + r^2\left(d\theta^2 + \sin^2\theta d\varphi^2\right)\right) \tag{1}$$

where $a(t)$ is a scale factor which characterizes the size of the universe, $\{r, \theta, \varphi\}$ are the spherical coordinates, and the values $k = 0$, $k = 1$, and $k = -1$ correspond to a spatially flat, closed, and open model of the universe.

The source of the accelerated expansion of the early universe with the equation of state $p = -\rho$ is a vacuum; the equation of the state $p \approx -\rho$ corresponds to a scalar (bosonic) field. The Bose-Einstein statistics for an ensemble of bosons, in contrast to an ensemble of fermions obeying the Pauli exclusion principle, implies that there can be several particles in one quantum state, which leads to the formation of boson condensate in which the increase in the concentration of massless bosons is associated with a decrease in the effective pressure corresponding to the equation of state $p \approx -\rho$. The initial (quasi) exponential expansion associated with a negative pressure, due to the exotic equation of state, is unstable, which leads to a phase transition, the termination of accelerated expansion, and the fragmentation of the original volume into many areas in which further evolution corresponds to the Friedmann solutions.

Also, the presence of a scalar field violates the symmetry of the system, which leads to the appearance of a mass of initially massless particles, for example, in the Higgs field [9].

Thus, the inclusion of the scalar field into cosmological models makes it possible to move from (quasi) de Sitter solutions to the Friedmann ones.

To prevent the rapid decay of the state of $p \approx -\rho$, it is necessary to assume the existence of some potential barrier, that is, the minimum potential energy of a scalar field. Consequently, in realistic inflation models, the scalar field evolves from a state of "false vacuum" with a non-zero potential energy to a state of "true vacuum," corresponding to the minimum of a potential $V(\phi)$. In other words, the scalar field rolls down (or tunnels) from some initial state to the minimum of $V(\phi)$, and the nature of this process is determined by the shape of the potential.

At the moment, there are many models of cosmological inflation with different potentials of a scalar field and different specifics of its evolution. A large number of current models of the early universe on the basis of the inflationary paradigm are considered in the review [13].

The physical justification for the inclusion of scalar fields in cosmological models is based on the experimental detection of the Higgs boson in the experiment at the Large Hadron Collider [14]. Thus, the scalar field corresponding to the Higgs bosons can be considered as the source of the gravitational field of the early universe. Moreover, the Higgs field can be considered as "inflation," leading to early accelerated expansion of the universe.

Now, in the system of units $8\pi G = c = 1$, we write the action that determines the dynamics of a scalar field ϕ based on Einstein's theory of gravity:

$$S_E = \int d^4x\sqrt{-g}\left[\frac{1}{2}R - g^{\mu\nu}\frac{1}{2}\partial_\mu\phi\partial_\nu\phi - V(\phi)\right] \tag{2}$$

where R is the Ricci scalar and $V(\phi)$ is the potential of a scalar field.

From the variation of this action with respect to the metric (1) and a field ϕ, for the case of the spatially flat universe, we obtain the equations defining the dynamics of a scalar field [9]:

$$3H^2 = \frac{1}{2}\dot{\phi}^2 + V(\phi) \tag{3}$$

$$-2\dot{H} - 3H^2 = \frac{1}{2}\dot{\phi}^2 - V(\phi) \tag{4}$$

$$\ddot{\phi} + 3H\dot{\phi} + \frac{dV(\phi)}{d\phi} = 0 \tag{5}$$

where $H = \dot{a}/a$ is the Hubble parameter, $X = \frac{1}{2}\dot{\phi}^2$ is the kinetic energy of a canonical scalar field ϕ, and the dot denotes the derivative with respect to cosmic time $\dot{a} = da/dt$.

Also, the state parameter w of a scalar field can be calculated as

$$w = \frac{p}{\rho} = \frac{X - V}{X + V} = -1 - \frac{2}{3}\frac{\dot{H}}{H^2} \tag{6}$$

To build a consistent model of cosmological inflation, the following conditions must be met:

a. The presence of the stage of accelerated expansion, which implies $-1 < w < -1/3$

b. The completion of the stage of accelerated expansion $w = -1/3$

c. The reheating of the scalar field with the subsequent formation of photons, i.e., the transition to the stage of predominance of radiation with $w = 1/3$

Currently, along with other models, several types of cosmological inflationary models are considered, which differ in both by the type of potential and the initial conditions under which an inflationary stage occurs: namely, a scalar field can be located at one of its potential minima, or accelerated expansion occurs for any conditions permitting the onset of inflation for scalar field energy density values comparable to the Planck mass [9].

The form of the scalar field's potential is determined from the physics of elementary particles and theories of the unification of fundamental interactions, such as supersymmetric theories and string theories in the context of the inflationary paradigm. Physical mechanisms corresponding to a large number of inflationary potentials were discussed in the review [14]. Due to the fact that the potential of the scalar field has a great importance for determining the physical processes at the stage of cosmological inflation, the potential $V(\phi)$ is given to build models of the early universe.

However, the finding of exact solutions to the system of Eqs. (3)–(5) for a given potential is impossible in most cases due to their nonlinearity. For this reason, a convenient tool for analyzing inflationary models based on a given scalar field potential is the "slow-roll approximation" which implies that $V(\phi) >> X$ and $\ddot{\phi} \approx 0$ and, therefore, simplifies the initial dynamic equations [9].

The dynamics of the expansion of the universe which determined by the scale factor $a(t)$ is no less important when analyzing cosmological models. By setting the expansion law $a(t)$, it is often possible to find the exact solutions of the system of Eqs. (3)–(5) and restore the evolution of the scalar field $\phi(t)$ and the potential $V(\phi)$. The different methods for constructing exact and approximate solutions of the equations of cosmological dynamics (3)–(5) can be found, for example, in the papers [15, 16]. We also note that the system of Eqs. (3)–(5) has many solutions that satisfy all the conditions for the inflationary stage that were outlined earlier.

Now, we consider the parameters that are necessary for the analysis of inflationary stage, namely, the e-fold number and the slow-roll parameters.

The e-fold number is usually noted as the natural logarithm of the ratio of the scale factor at the end of inflation to the scale factor at the beginning of inflation [9]:

$$N(t) = \ln \frac{a(t_{end})}{a(t_i)} = \int_{t_i}^{t_{end}} H dt \tag{7}$$

where t_i and t_{end} are the times of the beginning and ending of the inflation. The value of the number of e-folds at the end of the inflationary stage is estimated as $N = 50 - 60$ [9].

When analyzing inflationary models, the slow-roll parameters are important, and these parameters are defined as follows [13]:

$$\epsilon \equiv 2\left(\frac{H'_\phi}{H}\right)^2 = -\frac{\dot{H}}{H^2} \tag{8}$$

$$\delta \equiv 2\frac{H''_\phi}{H} = \epsilon - \frac{\dot{\epsilon}}{2H\epsilon} = -\frac{\ddot{H}}{2H\dot{H}} \tag{9}$$

$$\xi \equiv 4\frac{H'_\phi H'''_\phi}{H^2} = \frac{1}{H}(\dot{\epsilon} - \dot{\delta}) \tag{10}$$

Based on the relations (8)–(10), one can consider the slow-roll parameters as a function of time or field. During the inflationary stage, $\epsilon < 1$ and its completion are determined by the condition $\epsilon = 1$.

3. Cosmological perturbations

Cosmological perturbations are the source of the evolution of large-scale structure of the universe. An explanation of the distribution of galaxies and clusters of galaxies at large distances in the observable part of the universe on the basis of cosmological perturbations was originally proposed in the works of Harrison [17] and Zeldovich [18]. In the context of inflationary paradigm, the source of cosmological perturbations is quantum fluctuations of a scalar field and the corresponding fluctuations of the metric, which, in a linear order, correspond to three modes that evolve independently.

It is known from the classical theory of cosmological perturbations that the analysis of metric inhomogeneities can be simplified to the study of one perturbed quantity [11]. Thus, the quantum theory of cosmological perturbations can be reduced to the quantum description of the fluctuations of a certain scalar field.

Since the background in which the scalar field evolves depends on time, the field mass will also depend on time. This dependence of the field's mass on time will lead to the appearance of particles if the evolution begins with a certain vacuum state. Quantum particle production corresponds to the development and growth of cosmological perturbations.

In inflation models with one scalar field, at the crossing of the Hubble radius, cosmological perturbations "freeze," and their quantum state begins to change in such a way that the condition of constant amplitude is satisfied. The freezing of the vacuum state leads to the appearance of the classical properties [11]. Thus, the theory of cosmological perturbations provides a consistent approach for considering the generation and evolution of cosmological perturbations.

The influence of cosmological perturbations on the anisotropy and polarization of the background radiation is determined on the basis of spectral parameters and observational restrictions on the values of which form the basis of the experimental verification of theoretical models of the early universe. Also, within the framework of the cosmological perturbation theory, it is possible to calculate the spectra of initial density perturbations and relic gravitational waves depending on the values of the parameters of theoretical models [11].

After the end of the inflationary stage, the scalar field reheating and the formation of the first light particles of baryon matter begin. In the hot dense plasma, due to scattering on electrons, photons propagate much slower than the speed of light. When the universe expands so much that the plasma cools down to the recombination temperature, the electrons begin to connect with the protons, forming neutral hydrogen, and the photons begin to spread freely.

The points from which the photons reach the observer form the last scattering surface, whose temperature at the time of recombination is $\sim$3000 K and rapidly decreases with the expansion of the universe. The background radiation temperature is isotropic with an accuracy of 10^{-5}. The low anisotropy manifests itself as the temperature difference in different directions and its value is approximately equal to 3 mK [10].

The kinetic component of the anisotropy of the cosmic background radiation is due to the movement of the observer relative to the background radiation, which corresponds to the dipole harmonic.

In addition to the kinetic component in the anisotropy of the CMB, there are potential terms associated with effects in gravitational fields of very large scales that are comparable to the distance to the last scattering surface, namely:

a. Sachs-Wolfe effect, which corresponds to a change in the photon energy in a variable gravitational field of the universe

b. Silk effect due to adiabatic compression of radiation and baryon acoustic oscillations prior to the recombination epoch in high- and low-density zones

In the zero order of the cosmological perturbation theory, the universe is described by a single function of time, namely, by the scale factor $a(t)$. In the first (linear) order, the perturbations of the metric are the sum of three independent modes—scalar, vector, and tensor (relic gravitational waves), each of which is characterized by the spectral function of the wave number k [11].

For the inflationary stage in the linear approximation, one can write the Mukhanov-Sasaki equations for Fourier modes of the scalar v_k and tensor u_k perturbations [11]:

$$\frac{d^2 v_k}{d\eta^2} + \left(k^2 - \frac{1}{z}\frac{d^2 z}{d\eta^2}\right) v_k = 0 \tag{11}$$

$$\frac{d^2 u_k}{d\eta^2} + \left(k^2 - \frac{1}{a}\frac{d^2 a}{d\eta^2}\right) u_k = 0 \tag{12}$$

where $z = a\dot{\phi}/H$, k is a wave number, and η is the conformal time.

Eqs. (11) and (12) allow finding the power spectra $\mathcal{P}_S$ and $\mathcal{P}_T$ and spectral indices n_S and n_T of the scalar and tensor perturbations. The formulas for calculating the main cosmological parameters at crossing the Hubble radius $(k = aH)$ [19] are

$$\mathcal{P}_S(k) = \frac{1}{2\epsilon}\left(\frac{H}{2\pi}\right)^2 \tag{13}$$

$$\mathcal{P}_T(k) = 2\left(\frac{H}{2\pi}\right)^2 \tag{14}$$

$$n_S - 1 = 2\left(\frac{\delta - 2\epsilon}{1 - \epsilon}\right) \tag{15}$$

$$n_T = -\frac{2\epsilon}{1 - \epsilon} \tag{16}$$

$$r = \frac{\mathcal{P}_T}{\mathcal{P}_S} = 4\epsilon \tag{17}$$

The data on the effects of scalar and tensor modes can be obtained from observations of the anisotropy and polarization of the cosmic microwave background (CMB) radiation, which arose as a result of the joint effect on the photon distribution of the perturbation modes. Observational restrictions on the values of the parameters of cosmological perturbations according to the data of the PLANCK are [10]

$$10^9\mathcal{P}_S = 2.142 \pm 0.049 \tag{18}$$

$$n_S = 0.9667 \pm 0.0040 \tag{19}$$

$$r < 0.065 \tag{20}$$

In the context of such verification of cosmological models, let us pay attention to the tendency for the upper limit to decrease by the value of the tensor-scalar ratio for updated observational data [10].

Also, we note that the relic gravitational waves were not directly observed, which leads to a large number of theoretical models of cosmological inflation, which provide an explanation of the origin and evolution of the large-scale structure of the universe and correspond to the observational constraints.

4. Generalized exponential power-law inflation

The scheme for constructing models of the early universe based on the evolution of the scalar field in the context of the inflationary paradigm can be represented as follows:

a. The generating solutions of background dynamic equations (excluding quantum fluctuations of the scalar field) for a given potential, the law of accelerated expansion of the early universe, or the evolution of a scalar field.

b. Analysis of the quantum fluctuations of a scalar field and the corresponding metric perturbations on the basis of the theory of cosmological perturbation for the previously obtained background solutions. The result of this analysis is the values of the spectral parameters of cosmological perturbations which can be calculated from Eqs. (13)–(17).

c. Comparison of the obtained spectral parameters of cosmological perturbations with the corresponding observational data (18)–(20).

To build cosmological models corresponding to observational data, we propose the principle of constructing the inflationary models with generalized exponential power-law expansion. For this aim we consider any exact solutions $\{\phi, H, V\}$ of Eqs. (3)–(5) for which the substitution of the slow-roll parameters (8)–(10) into Eqs. (13)–(17) doesn't correspond to observational constraints (17)–(18).

After the following transformations

$$\overline{H} = nH + \lambda \tag{21}$$

$$\overline{a}(t) = Ca^n(t)e^{\lambda t},\ C = \overline{a}_0/a_0^n \tag{22}$$

$$\varphi = \sqrt{n}\phi \tag{23}$$

$$\overline{V}(\phi) = 3n^2H^2 + 6\lambda nH - nH'^2_{\phi} + 3\lambda^2,\ \overline{V}(\varphi) = \overline{V}(\phi(\varphi)) \tag{24}$$

one has new exact solutions $\{\varphi, \overline{H}, \overline{V}\}$ with new slow-roll parameters

$$\overline{\epsilon} = n\epsilon\left(n + \frac{\lambda}{H(\epsilon)}\right)^{-2} \tag{25}$$

$$\overline{\delta} = \delta\left(n + \frac{\lambda}{H(\epsilon)}\right)^{-1} \tag{26}$$

and with the conformity to observational constraints which can be achieved by choosing the values of free constant parameters n and λ.

The proposed approach has two limitations:

a. The original scale factor $a(t)$ doesn't violate the law of accelerated expansion.

b. The potential $\overline{V}(\varphi)$ corresponding to the scale factor (20) implies the evolution of the scalar field φ, according to the inflationary paradigm.

Transformations (21)–(24) define a class of models with the generalized exponential power-law dynamics, and the original scale factor $a(t)$ may not correspond to the condition of accelerated expansion $\ddot{a}>0$; however, the resulting scale factor $\overline{a}(t)$ implies a combination of the de Sitter solution (for $n=0$) and the power-law expansion (for $\lambda=0$), which corresponds to the basic feature of the inflationary paradigm implying a graceful exit from the stage of accelerated expansion to the power-law non-accelerated expansion.

5. Relic gravitational waves

As an additional verification tool for cosmological models, we consider the possibility of direct detection of the relic gravitational waves. The detection of relic gravitational waves is extremely important for determining the parameters of the models of early universe. Additionally, such a detection enhances the position of the inflationary paradigm compared to alternative scenarios, for example, the models with a rebound from singularity in which cosmological gravitational waves are absent [19].

As the main observational characteristic of relic gravitational waves, we consider the energy density, which is usually determined by the dimensionless quantity [20]:

$$\Omega_{GW}(f)=\frac{1}{\rho_c}\frac{d\rho_{GW}}{d\ln f} \tag{27}$$

where f is the linear frequency, $\rho_c=3H_0^2$ is the critical energy density, H_0 is the value of the Hubble parameter in the modern era, and ρ_{GW} is the energy density of gravitational waves.

Also, the energy density of relic gravitational waves can be represented in terms of the power spectrum:

$$\Omega_{GW}(k)=\frac{k^2}{12H_0^2}P_T(k) \tag{28}$$

The frequency and energy density of relic gravitational waves are limited by the following conditions [20]:

a. The energy density of relic gravitational waves should not exceed

$$\int_{f_0}^{\infty}\Omega_{GW}d\ln f<1.1\times10^{-5} \tag{29}$$

where $f_0\approx10^{-9}$ Hz.

a. The temperature of the scalar field T_* and the frequency of gravitational waves f at the end of the inflation stage are

$$T_*=5.85\times10^6\left(\frac{f}{Hz}\right)\left(\frac{g_*}{106.75}\right)^{1/6}GeV \tag{30}$$

$$f = 1.71 \times 10^{-7} \left(\frac{T_*}{GeV} \right) \left(\frac{g_*}{106.75} \right)^{-1/6} Hz \tag{31}$$

where g_* is the effective number of the degrees of freedom (in standard model of elementary particles $g_* = 106.75$).

Therefore, conditions (18) and (21)–(23) impose restrictions on the parameters of relic gravitational waves.

The application to the analysis of the models of the early universe only of the slow-roll approximation implies a low-frequency spectrum of relic gravitational waves in the range of $10^{-18} - 10^{-16}$ Hz [20]. However, the predominance of the kinetic energy of the scalar field during the evolution of the early universe provides a theoretical justification for the existence of high-frequency relic gravitational waves in models with one scalar field in the range of $10^2 - 10^4$ Hz [21] which can be used as affordable means of verification of models of the early universe in the presence of physical effects that increase the sensitivity of the detector to the required level.

Currently, the most productive method of direct detection of gravitational waves is the use of interferometers as detectors, which was proposed in the article by Gertsenshtein and Pustovoit [22]. This principle is widely used in modern laser interference gravitational antennas, the main element of which is the Fabry-Perot interferometer. These are broadband gravitational antennas, which offer a lot of opportunities as to the methods of recording of gravitational waves and extracting signals, as well as the use of quantum non-perturbative measurements and the inclusion of gravitational antennas in the network. The main element of laser interference gravitational antennas, as a rule, is Fabry-Perot multipath free-mass resonator, on whose properties the sensitivity and noise immunity of the entire gravitational antenna largely depend [4, 23, 24].

After creating the first laser interferometer for detecting gravitational waves, systematic work began on the creation and improvement of such devices in various laboratories around the world. The experience of gravitational antenna projects by VIRGO (Italy, France), LIGO (USA), TAMA (Japan), CLIO (Japan), GEO-600 (Germany), and OGRAN (Russia) will certainly be used to create more compact and highly sensitive antennas of new generation [4]. Also, as the most promising project for the direct detection of gravitational waves, work on the creation of a space interferometer in a helio-stationary orbit should be noted, in which the distance between the mirrors will be about 1 million kilometers. This project is called Laser Interferometer Space Antenna (LISA) [25]. The implementation of the LISA project is scheduled for 2029.

One of the promising methods for increasing the sensitivity of gravitational antennas in the high-frequency region of the spectrum is to use the phenomenon of low-frequency optical resonance, which distinguishes this approach from other projects on the detection of gravitational waves. The presence of this effect in Fabry-Perot interferometers was first considered in [23, 24]. At the moment, there is a high-frequency gravitational wave detector, which was built at the University of Birmingham, United Kingdom [26]. Also, it is planned to build the high-frequency gravitational wave detectors in Japan [27].

Thus, at the moment there are a large number of promising methods for direct observation of gravitational waves, which correspond to the ability to measure the characteristics of relic gravitational waves for a better understanding of the physical processes occurring in the early universe.

6. Conclusion

We considered the basis of building and verifying of the inflationary models of early universe. As the method for constructing the exact cosmological solutions corresponding to observational constraints, the models with generalized exponential power-law dynamics are proposed.

The verification of the relevance of such models is related to the estimation of the contribution of relic gravitational waves to the anisotropy and polarization of the cosmic microwave background radiation. Therefore, there are a lot of inflationary models with different scalar field potentials that will satisfy the observational constraints.

The most obvious way to significantly reduce the number of theoretical models of cosmological inflation is direct detection of relic gravitational waves.

The most promising methods in this area of experimental research are using the interferometers as detectors. The interesting direction of the observation is the detection of high-frequency relic gravitational waves using the effect of low-frequency optical resonance proposed in [23, 24].

Author details

Vladimir Gladyshev* and Igor Fomin
Bauman Moscow State Technical University, Moscow, Russia

*Address all correspondence to: vgladyshev@mail.ru

References

[1] Einstein A. Die Feldgleichungen der Gravitation. Berlin: Sitzungsberichte der Königlich Preussischen Akademie der Wissenschaften; 1915. pp. 844-847

[2] Einstein A. Näherungsweise Integration der Feldgleichungen der Gravitation. Berlin: Sitzungsberichte der Königlich Preussischen Akademie der Wissenschaften; 1916. p. 668

[3] Hulse R, Taylor J. Discovery of a pulsar in a binary system. Astrophysical Journal. 1975;**195**:L51-L53. DOI: 10.1086/181708

[4] Pustovoit V. On the direct detection of gravitational waves. Physics-Uspekhi. 2016;**59**:1034-1051. DOI: 10.3367/UFNe.2016.03.037900

[5] Abbott B et al. Observation of gravitational waves from a binary black hole merger. Physical Review Letters. 2016;**116**:061102. DOI: 10.1103/PhysRevLett.116.061102

[6] Abbott B et al. GW151226: Observation of gravitational waves from a 22-solar-mass binary black hole coalescence. Physical Review Letters. 2016;**116**:241103. DOI: 10.1103/PhysRevLett.116.241103

[7] Abbott B et al. GW170817: Observation of gravitational waves from a binary neutron star Inspiral. Physical Review Letters. 2017;**119**:161101. DOI: 10.1103/PhysRevLett.119.161101

[8] Abbott B et al. Gravitational waves and gamma-rays from a binary neutron star merger: GW170817 and GRB 170817 A. The Astrophysical Journal. 2017;**848**:L13. DOI: 10.3847/2041-8213/aa920c

[9] Linde A. Particle physics and inflationary cosmology. Contemporary Concepts in Physics. 1990;**5**:1-362

[10] Ade P et al. Planck 2015 results. XIII. Cosmological parameters. Astronomy and Astrophysics. 2016;**594**:A13. DOI: 10.1051/0004-6361/201525830

[11] Mukhanov V, Feldman H, Brandenberger R. Theory of cosmological perturbations. Physics Reports. 1992;**215**:203-333. DOI: 10.1016/0370-1573(92)90044-Z

[12] Ehlers J, Geren P, Sachs R. Isotropic solutions of the Einstein-Liouville equations. Journal of Mathematical Physics. 1968;**9**:1344-1349. DOI: 10.1063/1.1664720

[13] Martin J, Ringeval C, Vennin V. Encyclopædia Inflationaris. Physics of the Dark Universe. 2014;**5-6**:75-235. DOI: 10.1016/j.dark.2014.01.003

[14] Aad G et al. Measurements of the Higgs boson production and decay rates and coupling strengths using pp collision data at $\sqrt{s} = 7$ and 8 TeV in the ATLAS experiment. European Physical Journal C: Particles and Fields. 2016;**76**:6-70. DOI: 10.1140/epjc/s10052-015-3769-y

[15] Fomin I, Chervon S. Exact and approximate solutions in the Friedmann cosmology. Russian Physics Journal. 2017;**60**:427-440. DOI: 10.1007/s11182-017-1091-x

[16] Chervon S, Fomin I. The method of generating functions in exact scalar field inflationary cosmology. European Physical Journal C: Particles and Fields. 2018;**78**:301-323. DOI: 10.1140/epjc/s10052-018-5795-z

[17] Harrison E. Fluctuations at the threshold of classical cosmology. Physical Review D. 1970;**1**:2726-2730. DOI: 10.1103/PhysRevD.1.2726

[18] Ya Z. Gravitational instability: An approximate theory for large density

perturbations. Astronomy and Astrophysics. 1970;**5**:84-89

[19] Chervon S, Fomin I. On calculation of the cosmological parameters in exact models of inflation. Gravitation and Cosmology. 2008;**14**:163-167. DOI: 10.1134/S0202289308020060

[20] Maggiore M. Gravitational wave experiments and early universe cosmology. Physics Reports. 2000;**331**: 283-367. DOI: 10.1016/S0370-1573(99) 00102-7

[21] Sahni V, Sami M, Souradeep T. Relic gravity waves from brane world inflation. Physical Review D. 2002;**65**: 023518. DOI: 10.1103/PhysRevD.65. 023518

[22] Gertsenshtein M, Pustovoit V. On the detection of low frequency gravitational waves. Journal of Experimental and Theoretical Physics. 1963;**16**:433

[23] Gladyshev V, Morozov A. Low-frequency optical resonance in a multiple-wave Fabry-Perot interferometer. Technical Physics Letters. 1993;**19**:449-451

[24] Gladyshev V, Morozov A. The theory of a Fabry-Perot interferometer in a gravitational-wave experiment. Moscow Physical Society. 1996;**6**: 209-221

[25] Armano M et al. Sub-Femto- g free fall for space-based gravitational wave observatories: LISA pathfinder results. Physical Review Letters. 2016;**116**: 231101. DOI: 10.1103/ PhysRevLett.116.231101

[26] Cruise A, Ingley R. A prototype gravitational wave detector for 100-MHz. Classical and Quantum Gravity. 2006;**23**:6185-6193. DOI: 10.1088/0264-9381/23/22/007

[27] Nishizawa A et al. Laser-interferometric detectors for gravitational wave background at 100 MHz: Detector design and sensitivity. Physical Review D. 2008;**77**:022002. DOI: 10.1103/PhysRevD.77.022002

Chapter 14

Periodic Solution of Nonlinear Conservative Systems

Akuro Big-Alabo and Chinwuba Victor Ossia

Abstract

Conservative systems represent a large number of naturally occurring and artificially designed scientific and engineering systems. A key consideration in the theory and application of nonlinear conservative systems is the solution of the governing nonlinear ordinary differential equation. This chapter surveys the recent approximate analytical schemes for the periodic solution of nonlinear conservative systems and presents a recently proposed approximate analytical algorithm called continuous piecewise linearization method (CPLM). The advantage of the CPLM over other analytical schemes is that it combines simplicity and accuracy for strong nonlinear and large-amplitude oscillations irrespective of the complexity of the nonlinear restoring force. Hence, CPLM solutions for typical nonlinear Hamiltonian systems are presented and discussed. Also, the CPLM solution for an example of a non-Hamiltonian conservative oscillator was presented. The chapter is aimed at showcasing the potential and benefits of the CPLM as a reliable and easily implementable scheme for the periodic solution of conservative systems.

Keywords: Hamiltonian system, conservative system, nonlinear vibration, continuous piecewise linearization method, periodic solution, nonnatural system, perturbation method

1. Introduction

1.1 Hamiltonian and non-Hamiltonian conservative systems

Conservative systems can be defined as oscillating or vibrating systems in which the total energy content of the system remains constant. In order words, the total energy in the system is conserved. Ideally, such a system will continue to be in periodic oscillatory motion ad infinitum because the energy content of the system does not diminish due to the absence of dissipative force or increase due to additional energy input. However, for real cases where dissipative mechanisms such as friction or viscous damping cannot be completely eliminated, a conservative system can be thought of as one in which the energy dissipated is negligible during the time range under consideration. For example, the first few seconds of the oscillation of a simple pendulum may be considered conservative since the effect of air friction is negligible, but in the long run, the initial energy content is gradually dissipated until the pendulum comes to a halt. Other examples of practical conservative systems include mass-spring oscillator, structural elements (i.e., beams, plates, and shells), slider-crank mechanism [1], human eardrum [2], relativistic oscillator [3], planetary orbits around the sun [3], and current-carrying conductor in the electric field of an infinite rod [4]. Hence, a large number of oscillating physical systems can be studied as conservative systems.

At any point in time, the energy of a conservative system is composed of kinetic (T) and potential (V) energies except at critical points where the total energy may be only kinetic (T_{max}) or potential (V_{max}). Generally, it is expected that $T = T(q, \dot{q})$ and $V = V(q, \dot{q})$, where q is the generalized displacement. Naturally, q and $\dot{q}$ are not expected to form a product in the function $T(q, \dot{q})$, but in some cases, they do. Therefore, two types of conservative systems are distinguished namely: natural and nonnatural conservative systems. The natural conservative systems are those in which the kinetic energy can be expressed as a pure quadratic function of velocity, i.e., does not contain a product of the velocity and displacement. They are also known as *Hamiltonian systems* because they admit a Hamiltonian function ($H(q, \dot{q}) = T(q, \dot{q}) + V(q, \dot{q})$) that is always constant at any point in time. While this definition of Hamiltonian systems is a physical one, a mathematical definition has been discussed by Jordan and Smith [3]. Examples of Hamiltonian systems include mass-spring oscillator, simple pendulum, and a mass attached to the mid-point of an elastic spring. On the other hand, there are conservative systems in which the kinetic energy cannot be expressed as a pure quadratic function of the velocity because the kinetic energy expression contains a product of velocity and displacement. This second group of conservative systems is referred to as nonnatural because their kinetic energy is not a pure quadratic function of velocity. Although the total energy in such systems is conserved, their Hamiltonian function ($H(q, \dot{q})$) is not constant [4]. Hence, the nonnatural conservative systems may be referred to as *non-Hamiltonian conservative systems*. Examples of this category of conservative systems abound in artificial systems and include slider-crank mechanism [1], particle sliding on a vertical rotating parabola [4], pendulum attached to massless rolling wheel [4], rigid rod rocking on a circular surface without slip [4], and circular sector oscillator [5]. An important quality of the non-Hamiltonian conservative systems is that their vibration equation, which is normally derived by the Lagrangian approach, does not conform to the standard representation of conservative systems that clearly shows the restoring force. Rather, the derived vibration equation has a quadratic velocity term, which represents a coordinate-dependent parameter rather than a dissipative parameter.

1.2 Recent advances in solution schemes for nonlinear conservative oscillators

Exact analytical solutions for the nonlinear vibration models of conservative systems can be derived only in very few situations, and the solutions are usually derived in terms of special functions. Alternatively, highly accurate numerical solutions can be obtained for the nonlinear vibration model of any conservative system. However, as it is well recognized among the nonlinear science community, numerical solutions often have the limitations of lack of physical insight and convergence issues. Furthermore, there is the possibility of obtaining inaccurate convergent solutions for a nonlinear ordinary differential equation (ODE) [6], thus necessitating the independent verification of the convergent numerical solution by another numerical or analytical method. These limitations have driven the search for approximate analytical schemes capable of providing periodic solutions to nonlinear conservative oscillators. It can be rightly concluded that this search has been very fruitful considering the many approximate analytical schemes that now appear in the nonlinear science literature. The purpose of this section is to provide a brief survey of some of the notable achievements in the development of approximate analytical schemes for the periodic solution of nonlinear conservative oscillators. It should be noted that an approximate analytic method for nonlinear oscillators is considered adequate if it gives accurate predictions for the frequency-amplitude response and the oscillation history as well [7, 8].

Approximate analytical techniques to solve the nonlinear ODE governing the oscillations of a conservative system have been formulated for at least 100 years and can be classified as perturbation and nonperturbation methods. The first attempts were based on perturbation theory and are referred to as *classical perturbation methods*. The perturbation methods are formulated based on the concept that an unknown nonlinear system can be studied by introducing a small disturbance to a known linear system in equilibrium. For this reason, the classical perturbation methods (see Nayfeh [9] for a comprehensive treatment of classical perturbation methods) depend on the assumption of a small parameter. The problem with the small parameter assumption is that it has a small range of validity and only produces reliable solutions for cases of small-amplitude oscillations and weak nonlinearity. Nevertheless, the classical perturbation methods are still very relevant today for introducing and investigating various nonlinear concepts.

More recently, in the last four decades, a number of approximate analytical schemes have been proposed. Most of these recent schemes are nonperturbation methods, but some recent perturbation methods that attempt to improve on their classical counterparts have been formulated too. The recent perturbation methods include δ-method [10], Homotopy perturbation method [11] and its variants [12–17], modified Lindstedt-Poincare method [18–21], book-keeping parameter method [22], iteration perturbation method [23], parameterized perturbation method [24], perturbation incremental method [25], and linearized perturbation method [26]. A review article on some of the recent perturbation methods has been published by He [27]. The main point of the recent perturbation methods is to deal with the issue of the small parameter in order to formulate solutions that are applicable to small- and large-amplitude oscillations and also weak and strong nonlinear oscillations. Although the higher order approximations of the recent perturbation methods have been very successful in producing accurate estimates of the frequency-amplitude response, the same cannot be said of their estimation of the oscillation history. Studies [7, 28] have shown that the higher order approximations of the recent perturbation methods produce large unbounded errors in the oscillation history during large-amplitude oscillations and are, therefore, not better than the classical perturbation methods in this regard. A plausible explanation for this observation is that it occurs because perturbation methods are based on asymptotic series that are inherently divergent for amplitudes greater than unity [28]. Therefore, it may not be possible to formulate perturbation schemes that would correctly predict the oscillation history of large-amplitude vibrations.

In contrast to the perturbation methods, the nonperturbation methods do not use any small or artificial parameter. Examples include Adomian decomposition method [29], Homotopy analysis method [30], Variational iteration method [31], Energy balance method [2] and its modifications [32–34], He Chengtian's interpolation method [27] also called max-min approach, amplitude-frequency formulation [35], Hamiltonian approach [36], global error minimization method [37], Harmonic balance method [4] and its modifications [38–42], cubication methods [43–46], variational methods [47–49], differential transform method [50], and continuous piecewise linearization method [8]. Nonperturbation methods also have various limitations. For instance, a study [51] showed that the Adomian decomposition method does not converge to the correct solution in some cases, and the study proposed an optimal convergence acceleration parameter to deal with this issue.

Also, methods that rely on a simple harmonic approximation of the oscillation history, such as the energy balance method, amplitude-frequency formulation, Hamiltonian approach, max-min approach, and variational methods, can only give reliable estimate of the frequency-amplitude response. Sometimes these methods perform poorly in predicting the oscillation history during large-amplitude and/or strong nonlinear vibrations. Other methods that usually require high-order approximations, such as Adomian decomposition method, harmonic balance method, and variational iteration method, present algebraic complexities in their determination of higher order solutions and may be impractical for oscillators with highly complex nonlinearities such as the slider-crank mechanism [1] and the bifilar pendulum [52]. Furthermore, it has been observed that higher order estimates do not always improve the solution of the oscillation history [27]. Finally, some nonperturbation methods are heuristic in nature (e.g., energy balance method and variational methods) and require experience to choose the initial trial function and the condition for error minimization [8].

The continuous piecewise linearization method (CPLM) is an iterative analytic algorithm that was formulated to overcome most of the above challenges by providing simple and accurate solutions for the oscillation history and frequency-amplitude response of Duffing-type oscillators. In another study [53], the CPLM was modified in order to generalize it so that it can handle more complicated nonlinear conservative oscillators. Interestingly, the CPLM does not require higher order approximations or any small, artificial, or embedded parameter. Also, the algorithm is inherently stable, straightforward, and based on closed-form analytical approximations. This chapter is aimed at presenting the generalized CPLM algorithm as a veritable approach for accurate periodic solution of Hamiltonian and non-Hamiltonian conservative oscillators with complex nonlinearity. As is shown later, the CPLM retains the same simplicity in its implementation irrespective of the complexity of the nonlinear conservative system.

2. Continuous piecewise linearization method

2.1 Concept of the continuous piecewise linearization method

The main idea of the CPLM is based on the piecewise linearization of the nonlinear restoring force of a conservative oscillator. The linearization technique used by the CPLM was first applied in another algorithm [6, 54] for the solution of half-space impact models called force indentation linearization method (FILM). The FILM has been applied to formulate theoretical solutions for rigid body motions and local compliance response during nonlinear elastoplastic impact of dissimilar spheres [55]. However, because the FILM is limited to impact excitations that are nonoscillatory, it cannot be applied to nonlinear conservative oscillators. Hence, the CPLM applies the piecewise linearization technique of the FILM to provide a periodic solution for nonlinear conservative oscillators.

Essentially, the linearization technique of the CPLM involves n equal discretization of the nonlinear restoring force with respect to displacement (**Figure 1**) and formulating a linear restoring force for each discretization. Therefore, a linear ODE can be derived for each discretization. The solution of the linear ODE approximates the solution of the original nonlinear oscillator for a time-range that is automatically determined by the CPLM and updated continuously from one discretization to the next. Details on the discretization and linearization technique of the CPLM can be found in the following references [6, 8, 54], while the applications of the CPLM to nonlinear conservative systems are presented in [8, 56].

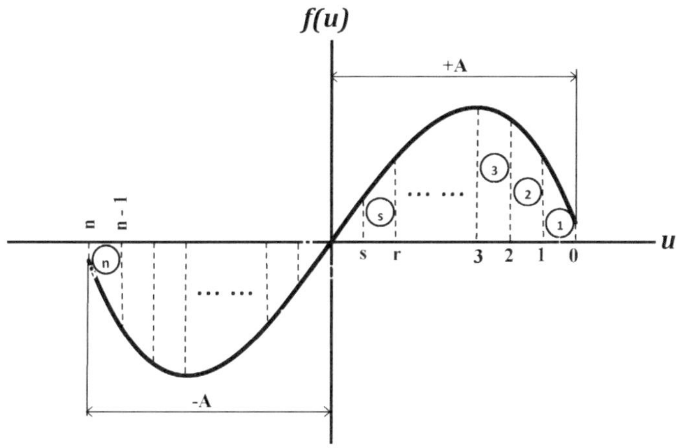

Figure 1.
Discretization of the restoring force of a typical nonlinear oscillator.

2.2 Mathematical formulation of the continuous piecewise linearization method

The standard form for representation of a nonlinear conservative oscillator moving in the u-direction is given as:

$$\ddot{u} + f(u) = 0, \tag{1}$$

where $f(u)$ is the nonlinear restoring force as shown in **Figure 1**. In **Figure 1**, the numbering on the horizontal axis represents the boundary points of the discretization. The sth discretization represents a general discretization with start point at r and endpoint at $s = r + 1$. **Figure 1** shows that for each discretization, the slope of the linear approximation of the restoring force can either be positive or negative. To account for this possibility, the linearized force for the sth discretization can be expressed as:

$$F_{rs}(u) = \pm|K_{rs}|(u - u_r) + F_r, \tag{2}$$

where $K_{rs} = [f(u_s) - f(u_r)]/(u_s - u_r)$ is the linear slope of $F_{rs}(u)$ and $F_r = f(u_r)$.

Since $F_{rs}(u)$ is an approximation of $f(u)$ for the sth discretization, then substituting Eq. (2) in (1) gives the approximate equation of motion for each discretization as follows:

$$\ddot{u} \pm |K_{rs}|u = \pm|K_{rs}|u_r - F_r. \tag{3}$$

Eq. (10) is a nonhomogeneous linear ODE and its solution depends on whether the sign is positive or negative.

2.2.1 Solution for positive linearized stiffness

If $K_{rs} > 0$, the solution for the displacement and velocity can be expressed as:

$$u(t) = R_{rs} \sin(\omega_{rs}t + \Phi_{rs}) + C_{rs} \tag{4a}$$

$$\dot{u}(t) = \omega_{rs}R_{rs} \cos(\omega_{rs}t + \Phi_{rs}), \tag{4b}$$

where $\omega_{rs} = \sqrt{K_{rs}}$, $C_{rs} = u_r - F_r/K_{rs}$, and $R_{rs} = \left[(u_r - C_{rs})^2 + (\dot{u}_r/\omega_{rs})^2\right]^{1/2}$. The initial conditions and other parameters are determined based on the oscillation stage. For the oscillation stage that moves from $+A$ to $-A$, the initial conditions for each discretization are $u_r = u_r(0) = A - r\Delta u$ and $\dot{u}_r = \dot{u}_r(0) = -\sqrt{\left|2\int_A^{u_r} -f(u)du\right|}$, where $\Delta u = A/n$, and the other parameters are calculated as:

$$\Phi_{rs} = \begin{cases} 0.5\pi & \dot{u}_r = 0 \\ \pi + \tan^{-1}[\omega_{rs}(u_r - C_{rs})/\dot{u}_r] & \dot{u}_r < 0 \end{cases} \tag{5a}$$

$$\Delta t = \begin{cases} (0.5\pi - \Phi_{rs})/\omega_{rs} & (u_s - C_{rs}) \geq R_{rs} \\ (0.5\pi + \cos^{-1}[(u_s - C_{rs})/R_{rs}] - \Phi_{rs})/\omega_{rs} & (u_s - C_{rs}) < R_{rs} \end{cases}. \tag{5b}$$

For the oscillation stage that moves from $-A$ to $+A$, the initial conditions are $u_r = u_r(0) = -A + r\Delta u$ and $\dot{u}_r = \dot{u}_r(0) = \sqrt{\left|2\int_A^{u_r} -f(u)du\right|}$; the other parameters are calculated as:

$$\Phi_{rs} = \begin{cases} -0.5\pi & \dot{u}_r = 0 \\ \tan^{-1}[\omega_{rs}(u_r - C_{rs})/\dot{u}_r] & \dot{u}_r < 0 \end{cases} \tag{6a}$$

$$\Delta t = \begin{cases} (0.5\pi - \Phi_{rs})/\omega_{rs} & (u_s - C_{rs}) \geq R_{rs} \\ (0.5\pi - \cos^{-1}[(u_s - C_{rs})/R_{rs}] - \Phi_{rs})/\omega_{rs} & (u_s - C_{rs}) < R_{rs} \end{cases}. \tag{6b}$$

The time at the end of each discretization is $t_s = t_r + \Delta t$, and the end conditions u_s and $\dot{u}_s$ are calculated by replacing r with s in the formulae for the initial conditions.

2.2.2 Solution for negative linearized stiffness

If $K_{rs} < 0$, the solution for the displacement and velocity can be expressed as follows:

$$u(t) = A_{rs}e^{\omega_{rs}t} + B_{rs}e^{-\omega_{rs}t} + C_{rs} \tag{7a}$$

$$\dot{u}(t) = \omega_{rs}(A_{rs}e^{\omega_{rs}t} - B_{rs}e^{-\omega_{rs}t}), \tag{7b}$$

where $\omega_{rs} = \sqrt{|K_{rs}|}$; $C_{rs} = u_r + F_r/|K_{rs}|$. Applying the initial conditions to Eqs. (7a) and (7b) gives: $A_{rs} = \frac{1}{2}(u_r + \dot{u}_r/\omega_{rs} - C_{rs})$; $B_{rs} = \frac{1}{2}(u_r - \dot{u}_r/\omega_{rs} - C_{rs})$. The initial and end conditions are determined in the same way as for $K_{rs} > 0$ above. The end conditions are applied in Eq. (7a) to get the time interval for each discretization as:

$$\Delta t = \begin{cases} \dfrac{1}{\omega_{rs}}\log_e\left[\dfrac{(u_s - C_{rs}) \pm \sqrt{(u_s - C_{rs})^2 - 4A_{rs}B_{rs}}}{2A_{rs}}\right] & (u_s - C_{rs}) > 2\sqrt{A_{rs}B_{rs}} \\ \dfrac{1}{\omega_{rs}}\log_e\left(\dfrac{u_s - C_{rs}}{2A_{rs}}\right) & (u_s - C_{rs}) \leq 2\sqrt{A_{rs}B_{rs}} \end{cases}. \tag{8}$$

The sign before the square root in Eq. (8) is negative for the oscillation stage that moves from $+A$ to $-A$ and vice versa. We note that if $\dot{u}_r = 0$, then $A_{rs} = B_{rs} = \frac{1}{2}(u_r - C_{rs})$ and

$$u(t) = (u_r - C_{rs})\cosh(\omega_{rs}t) + C_{rs}. \tag{9}$$

Therefore,

$$\Delta t = \frac{1}{\omega_{rs}}\cosh^{-1}\left(\frac{u_s - C_{rs}}{u_r - C_{rs}}\right). \tag{10}$$

2.2.3 Solution for zero linearized stiffness

In very rare situations, we may have $K_{rs} = 0$ for one or two discretization around the turning points or relatively flat regions of the restoring force. This is likely when Δu is very small, i.e., for very large n, and can be eliminated by increasing or decreasing n slightly. However, if we want to account for $K_{rs} = 0$, then we get [53]:

$$u(t) = H_{rs} + G_{rs}t - \frac{1}{2}F_r t^2, \tag{11}$$

where $G_{rs} = \dot{u}_r + F_r t_r$ and $H_{rs} = u_r - \dot{u}_r t_r - \frac{1}{2}F_r t_r^2$. Hence, the time interval is derived from Eq. (11) as:

$$\Delta t = \frac{G_{rs} + \sqrt{G_{rs}^2 + 2F_r(H_{rs} - u_s)}}{F_r}. \tag{12}$$

2.3 Remarks on the CPLM algorithm

1. From the above presentation of the CPLM formulation, it is obvious that the CPLM algorithm is simple and can be implemented by undergrads without difficulty.

2. The CPLM is inherently stable and does not have convergence issues [8].

3. For few discretization, say $n \leq 10$, the CPLM algorithm can be implemented with reasonable accuracy using a pocket calculator. However, the CPLM is better executed using a simple code in any programming language such as MATLAB and Mathematica or using a customized MS Excel spreadsheet.

4. When dealing with conservative oscillators with odd nonlinearity, which are symmetrical about the origin, discretization of the restoring force is only required for 0 to A. This means that there will be $2n$ discretizations from $-A$ to A.

5. The CPLM algorithm retains the same simplicity in implementation irrespective of the complexity of the restoring force. Only the K_{rs} constant and the integral of the restoring force are to be evaluated anew for any oscillator.

6. The CPLM relies on the explicit expression of restoring force, which means that the model for the oscillator must be expressed in the form of Eq. (1). For Hamiltonian systems, the oscillator model is formulated naturally in the form of Eq. (1). For non-Hamiltonian conservative systems, the oscillator model is not formulated naturally in the form of Eq. (1). Therefore, there is a need to transform the model of non-Hamiltonian conservative systems into the form of Eq. (1) before applying the CPLM algorithm. Fortunately, this

transformation only requires a simple algebraic manipulation as demonstrated in Section 4.

7. The phase equation gives the relationship between the state variables (displacement and velocity) and can be derived exactly in closed-form for all conservative systems. The CPLM extends this bilateral relationship into a tripartite one by finding the value of the corresponding independent variable (i.e., time) that matches the values of the state variables in each discretization.

8. For few discretization, say $n \leq 20$, it would be necessary to extract values within each discretization in order to obtain a smooth plot of the oscillation history. The values can be extracted using the approximate solution of the displacement. However, for many discretizations, say $n \geq 50$, there is no need to extract values from any discretization.

9. The usual initial conditions investigated for nonlinear conservative oscillators are nonzero displacement and zero velocity. However, the CPLM algorithm can comfortably handle nonzero initial conditions for displacement and velocity.

3. Periodic solution of typical Hamiltonian systems

3.1 Nonlinear simple pendulum

The simple pendulum is arguably the most investigated physical system and provides very interesting insights into nonlinear phenomena. Butikov [57] calls it "an antique but evergreen physical model." The undamped oscillation of a simple pendulum is a Hamiltonian system governed by the well-known nonlinear ODE as shown:

$$\ddot{u} + \Omega_0^2 \sin u = 0, \tag{13}$$

where u is the angular displacement, $\Omega_0 = \sqrt{g/l}$, l is the length of the pendulum, and $g = 9.8\ m/s^2$. The initial conditions are given as $u(0) = A$ and $\dot{u}(0) = 0$. The same initial conditions are applicable to all other oscillators discussed subsequently. The exact solution to Eq. (13) is expressed in terms of elliptic functions. The displacement and natural frequency are given as [58]:

$$u_{ex}(t) = 2\sin^{-1}\left[k\,\mathrm{sn}\left(\Omega_0 t + K\left(k^2\right); k\right)\right] \tag{14}$$

$$\omega_{ex} = \frac{\pi\Omega_0}{2K\left(k^2\right)}, \tag{15}$$

Where sn is the Jacobi elliptic sine function, $k = \sin(A/2)$, and $K\left(k^2\right)$ is the complete elliptic integral of the first kind given as:

$$K\left(k^2\right) = \int_0^{\pi/2} \frac{1}{\sqrt{1 - k^2 \sin^2\phi}} d\phi. \tag{16}$$

From Eq. (13), the restoring force for the pendulum is $f(u) = \Omega_0^2 \sin u$ and looks like the plot in **Figure 1**. This means that $K_{rs} = \Omega_0^2(\sin u_s - \sin u_r)/(u_s - u_r)$ and $\dot{u} = \pm\Omega_0\sqrt{2(\cos u - \cos A)}$. The initial and final velocities for each discretization

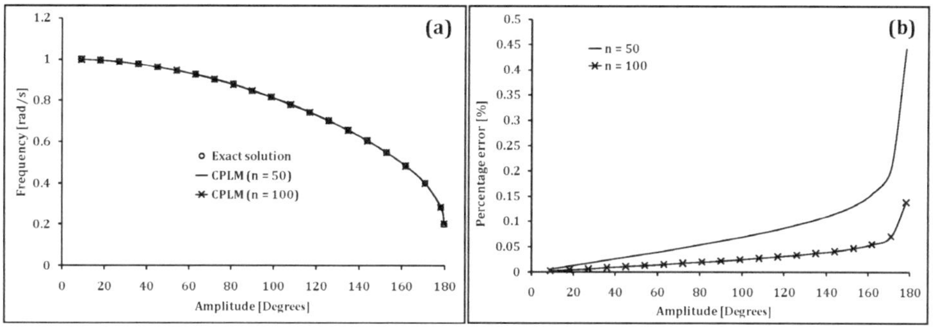

Figure 2.
(a) Frequency-amplitude response of the simple pendulum for 0° < A < 180°. (b) CPLM error analysis.

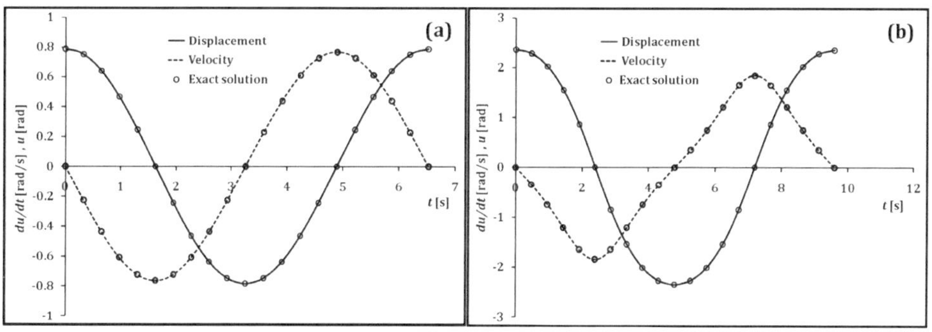

Figure 3.
Oscillation history of the simple pendulum for (a) A = 45° and (b) A = 135°. CPLM estimate—Lines; exact solution—Markers.

are determined using the expression for $\dot{u}$, and all the other constants for the solution of each discretization are determined based on K_{rs}. A plot of the frequency-amplitude response for the simple pendulum when $\Omega_0 = 1.0$ is given in **Figure 2a** and the corresponding error of the CPLM solution in comparison with the exact solution (Eq. (15)) is shown in **Figure 2b**. We see that for $A \leq 178°$, the maximum error in the CPLM estimate is less than 0.45% for $n = 50$ and 0.14% for $n = 100$. Also, the oscillation history for moderate-amplitude ($A = 45°$) and large-amplitude ($A = 135°$) is shown in **Figure 3** and an excellent agreement between the CPLM solution for $n = 100$ and the exact solution (Eq. (14)) is observed. We noted that trigonometric nonlinearity is usually difficult to deal with and that is why the CPLM shows a relatively slow convergence to accurate results. Hence, many discretizations (e.g., $n = 50 - 100$) are required to get an accuracy that is within 1.0% of the exact solution during large-amplitude oscillations ($90° < A < 180°$) of the simple pendulum.

3.2 Motion of satellite equidistant from twin stars

Consider the motion of a satellite along a path that is equidistant from two identical massive stars with mutually interacting gravitational fields. If the distance between the two stars is $2d$ and the coordinate of the satellite motion is u, then the equation of motion of the satellite is given as [3]:

$$\ddot{u} + \frac{2Mu}{\left(d^2 + u^2\right)^{3/2}} = 0, \tag{17}$$

where M is the mass of a star and the restoring force is $f(u) = 2Mu/(d^2 + u^2)^{3/2}$. Eq. (17) shows that the satellite-star interaction results in a conservative oscillation of the satellite. **Figure 4** shows the nonlinear restoring force, which is an irrational force because of the bottom square root. The restoring force spikes on both sides of the vertical axis close to the origin. The spikes indicate the point when the satellite is most influenced by the mutual gravitational field of the stars. Away from the origin, the restoring force decreases gradually and approaches the horizontal axis asymptotically. This means that the satellite is far away from the stars and experiences a much smaller gravitational force. This problem was discussed qualitatively by Jordan and Smith [3] and Arnold [59], but here, the periodic solution was investigated.

The main CPLM constant is $K_{rs} = 2M\left[\left(d^2 + u_s^2\right)^{-3/2} - \left(d^2 + u_r^2\right)^{-3/2}\right]$ and the velocity was derived as: $\dot{u} = \pm 2\sqrt{M\left[\left(d^2 + u^2\right)^{-1/2} - \left(d^2 + A^2\right)^{-1/2}\right]}$. The periodic solutions obtained by the CPLM and exact numerical solution are shown in **Figures 5** and **6**. The numerical solution was obtained by solving Eq. (17) using the NDSolve function in Mathematica™. The NDSolve function is a Mathematica subroutine for solving ordinary, partial, and algebraic differential equations numerically. In its basic form, it automatically selects the numerical method to use from a list of standard methods such as implicit Runge–Kutta, explicit Runge–Kutta, symplectic partitioned Runge–Kutta, predictor–corrector Adams, and backward difference methods. In some cases, the NDSolve function can combine two or

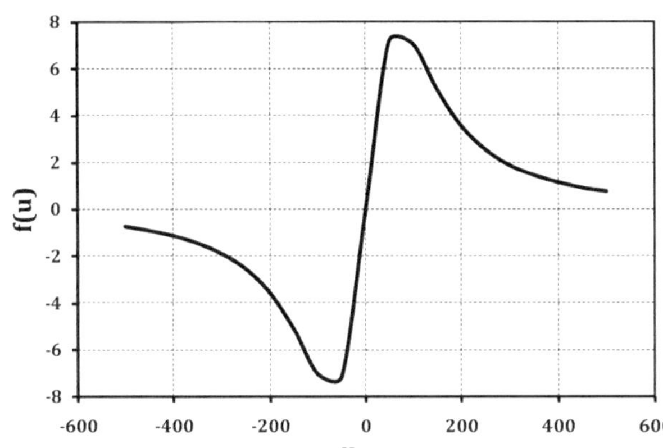

Figure 4.
Restoring force for Eq. (18): $M = 10^5$ [kg]; $d = 100$[m]; $A = 500$.

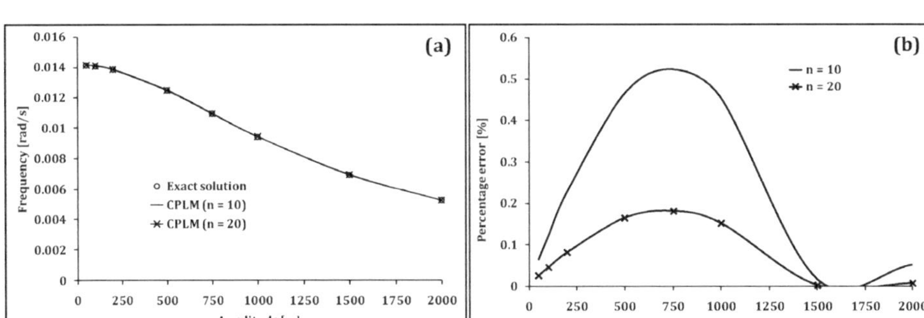

Figure 5.
(a) Frequency-amplitude response for satellite. (b) CPLM error analysis.

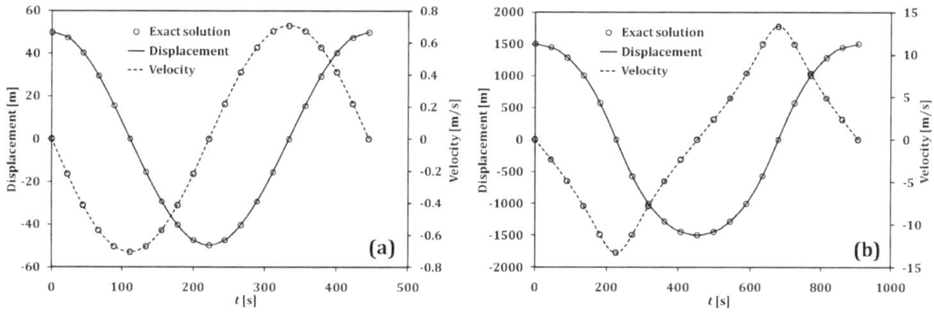

Figure 6.
Oscillation history of satellite for (a) A = 50 and (b) A = 1500. CPLM estimate—Lines; exact solution—Markers.

more methods to obtain the required solution. This basic form is preferable because the NDSolve function uses the method(s) that best solves the differential equation considering accuracy and solution time. Hence, the NDSolve function was used in its basic form for all numerical solutions obtained in this chapter.

The input values used for simulation are $M = 10^5$ $[kg]$ and $d = 1000[m]$. In contrast to the simple pendulum, the oscillation of the satellite requires less discretization for accurate results because there is no trigonometric nonlinearity. The maximum error of the CPLM estimate for the frequency-amplitude response is less than 0.55% for $n = 10$ and 0.20% for $n = 20$. Significantly higher accuracies can be achieved by increasing n, but the results show that $n = 10$ gives reasonably accurate estimates.

On the other hand, **Figure 6** shows the oscillation history of the satellite during small-amplitude ($A = 50$) and large-amplitude ($A = 1500$) oscillations. The former gives a simple harmonic response with a natural frequency that is independent of the amplitude and approximately equal to $\sqrt{2M/d^3}$, while the latter exhibits an *anharmonic* response with a natural frequency that depends strongly on the amplitude.

3.3 Mass-spring oscillator with fractional nonlinearity

An interesting oscillator that has been the subject of several studies [60–65] is the Hamiltonian oscillator with odd fractional nonlinearity. For the purpose of the present investigation, we consider an oscillator that is characterized by a general fractional nonlinearity as follows [65]:

$$\ddot{u} + u^{1/(2m+1)} = 0, \tag{18}$$

where the restoring force, $f(u) = u^{1/(2m+1)}$, has a fractional index for all $m \in \{\mathbb{Z}^+\}$. The main CPLM constant is evaluated as $K_{rs} = \left[u_s^{1/(2m+1)} - u_r^{1/(2m+1)}\right]/(u_s - u_r)$ and the velocity as $\dot{u} = \pm[(2m+1)/(m+1)]^{1/2}\sqrt{A^{(2m+2)/(2m+1)} - u^{(2m+2)/(2m+1)}}$. The periodic solution for the case of $m = 1$, i.e., $u^{1/3}$ oscillator, is shown in **Figures 7** and **8**. The exact frequency-amplitude response used for the verification of the CPLM solution is [63]:

$$\omega_{ex} = \frac{2\pi\Gamma(5/4)}{\sqrt{6}\Gamma(3/4)\Gamma(1/2)A^{1/3}} = \frac{1.070451}{A^{1/3}}. \tag{19}$$

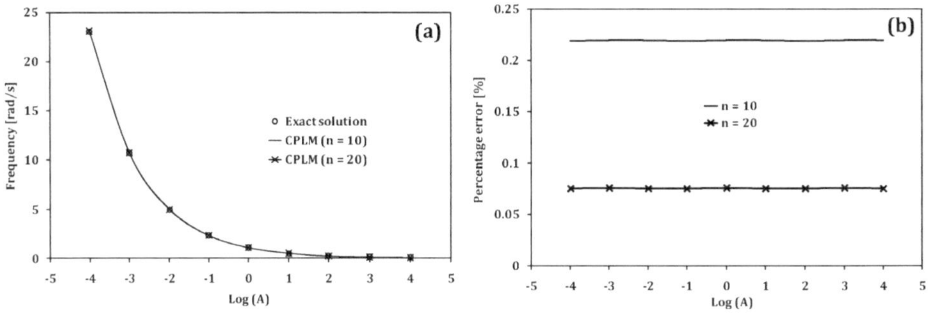

Figure 7.
(a) Frequency-amplitude response for $u^{1/3}$ oscillator. (b) CPLM error analysis.

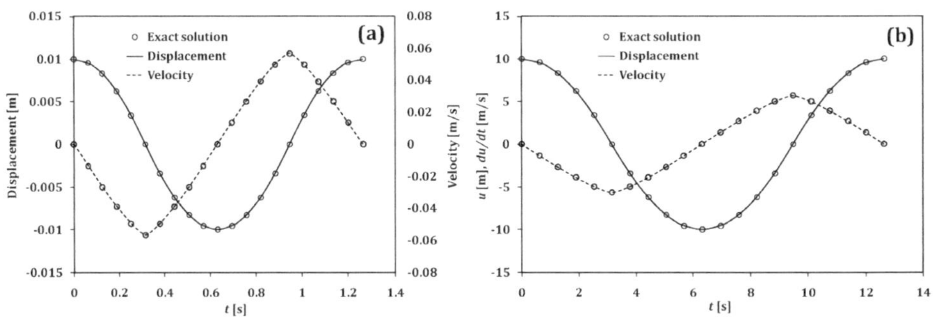

Figure 8.
Oscillation history of $u^{1/3}$ oscillator for (a) $A = 0.01$ and (b) $A = 10$. CPLM estimate—lines; exact solution—markers.

while the exact oscillation history was obtained by the numerical solution of Eq. (18) using the NDSolve function in Mathematica™. The CPLM solution demonstrates an excellent agreement with the exact solution.

Figure 7 compares the CPLM estimates of the frequency-amplitude response with Eq. (19), and the maximum error of the CPLM solution is 0.22% for $n = 10$ and 0.076% for $n = 20$. The error in the CPLM estimate is approximately constant for all amplitudes, and the maximum error is well below 1.0% for $n = 10$. The results also reveal that the frequency approaches zero as $A \to \infty$. In **Figure 8**, the oscillation history for small-amplitude ($A = 0.01$) and large-amplitude ($A = 10.0$) oscillations is shown to exhibit similar anharmonic response, which is an indication of strong nonlinearity. Therefore, it can be concluded that the $u^{1/3}$ oscillator is highly nonlinear for all amplitudes. This quality of possessing strong nonlinearity for all amplitudes is in contrast to most Hamiltonian oscillators that are linear for small amplitudes, e.g., the oscillators considered in Sections 3.1 and 3.2 above. Another Hamiltonian oscillator that possesses strong nonlinearity for all amplitudes is the geometrically nonlinear crank [1].

4. Periodic solution of non-Hamiltonian conservative systems

The non-Hamiltonian conservative systems are generally more complex and difficult to solve compared with the Hamiltonian systems. In order to demonstrate the application of the CPLM algorithm to non-Hamiltonian conservative systems, we consider the motion of a particle on a rotating parabola. This system consists of a frictionless mass sliding on a vertical parabolic wire described by $y = qu^2$ for $q > 0$

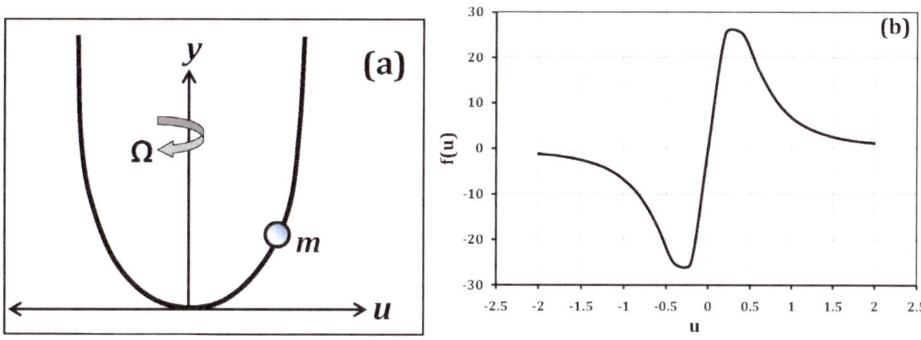

Figure 9.
(a) Schematic of particle on a rotating parabola. (b) Restoring force when $q = 1.0$; $\Lambda = 10.0$; $A = 2.0$.

and rotating at a constant speed, Ω, about the y-axis (**Figure 9a**). The u-axis represents the perpendicular displacement of the mass from the y-axis. The kinetic and potential energies of the system are given as [4]:

$$T = \frac{1}{2}m\left[\left(1 + 4q^2u^2\right)\dot{u}^2 + \Omega^2u^2\right]; V = mgqu^2. \tag{20}$$

Hence, the Lagrangian is:

$$L = T - V = \frac{1}{2}m\left[\left(1 + 4q^2u^2\right)\dot{u}^2 + \Omega^2u^2\right] - mgqu^2. \tag{21}$$

Next, we substitute Eq. (21) into the Euler–Lagrange equation to derive the equation of motion. The Euler–Lagrange equation can be written as:

$$\frac{d}{dt}\left(\frac{\partial L}{\partial \dot{u}}\right) - \frac{\partial L}{\partial u} = 0. \tag{22}$$

Therefore, using Eqs. (21) and (22), the motion of a particle on a rotating parabola is governed by:

$$\left(1 + 4q^2u^2\right)\ddot{u} + 4q^2u\dot{u}^2 + \Lambda u = 0, \tag{23}$$

where $\Lambda = 2gq - \Omega^2$ and the initial conditions are: $u(0) = A$ and $\dot{u}(0) = 0$.

To solve Eq. (23) using the CPLM, it must be recast in the form of Eq. (1). The conservation of energy for Eq. (23) is given as [4]:

$$\left(1 + 4q^2u^2\right)\dot{u}^2 + \Lambda u^2 = h, \tag{24}$$

where h is a constant representing the total energy in the system. Eq. (24) confirms that the Hamiltonian, $H = T + V = \frac{1}{2}m\left[\left(1 + 4q^2u^2\right)\dot{u}^2 + \left(\Lambda + 2\Omega^2\right)u^2\right] = \frac{1}{2}m\left(h + 2\Omega^2u^2\right)$, is not constant. Applying the initial conditions, we get $h = \Lambda A^2$ so that $\dot{u}^2 = \Lambda\left(A^2 - u^2\right)/\left(1 + 4q^2u^2\right)$. Substituting this expression for $\dot{u}^2$ in Eq. (24) and after algebraic simplification, we get:

$$\ddot{u} + \frac{\Lambda\left(1 + 4q^2A^2\right)u}{\left(1 + 4q^2u^2\right)^2} = 0. \tag{25}$$

Therefore, the restoring force is $f(u) = \Lambda(1+4q^2A^2)u/(1+4q^2u^2)^2$. **Figure 9b** shows that $f(u)$ is linear at small amplitudes and strongly nonlinear at large amplitudes. The main CPLM constant was calculated as $K_{rs} = \Lambda(1+4q^2A^2)\left[1/(1+4q^2u_s^2)^2 - 1/(1+4q^2u_r^2)^2\right]$, and the velocity was evaluated as $\dot{u} = \pm\sqrt{\Lambda(A^2-u^2)/(1+4q^2u^2)}$. The exact time period for this oscillator can be derived in terms of elliptic function as follows [4]:

$$T_{ex} = 4\left[(1+4q^2A^2)/\Lambda\right]^{1/2}E(k^2), \tag{26}$$

where $E(k^2) = \int_0^{\pi/2}(1-k^2\sin^2\phi)^{1/2}d\phi$ is the complete elliptic integral of the second kind and $k^2 = 4q^2A^2/(1+4q^2A^2)$. Then, the exact frequency was computed as $\omega_{ex} = 2\pi/T_{ex}$, while the exact oscillation history was obtained by solving Eq. (25) numerically using the NDSolve function in Mathematica™.

A comparison of CPLM frequency estimate and the exact frequency is shown in **Figure 10**, while the oscillation histories for $A = 0.50$ and $A = 2.0$ are shown in **Figure 11**. As demonstrated in [4], periodic solutions for this system exist only for $\Lambda > 0$. Hence, the simulations in **Figures 10** and **11** were conducted for $\Lambda = 10$ and $q = 1.0$. An excellent agreement is observed between the CPLM estimates and the exact results. For $0 < A \leq 20$, the maximum error in the CPLM estimate of the frequency-amplitude response is 0.642% for $n = 10$ and 0.101% for $n = 20$, both of which are well below 1.0%. Also, the CPLM solution gives an accurate prediction of the strong anharmonic response in the oscillation history as shown in **Figure 11**.

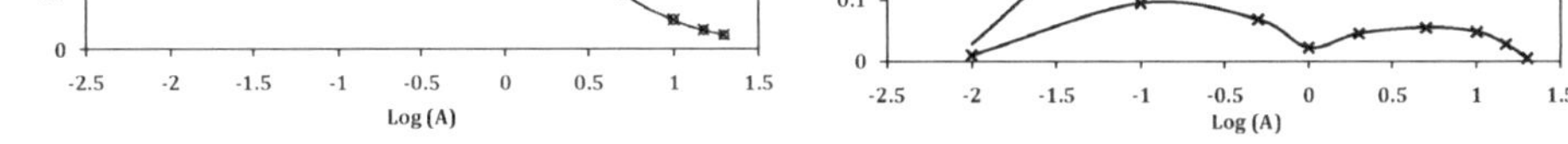

Figure 10.
(a) Frequency-amplitude response for particle on a rotating parabola. (b) CPLM error analysis.

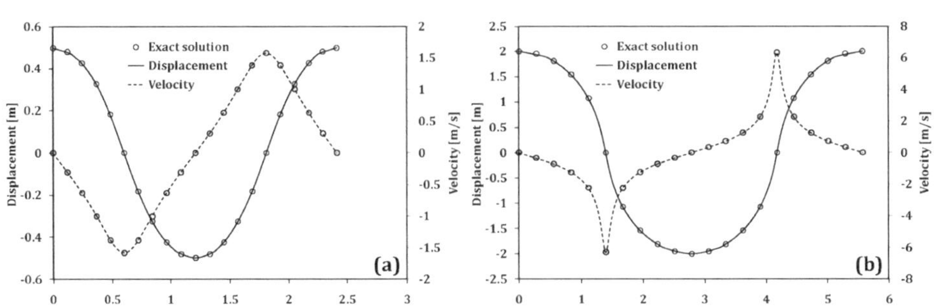

Figure 11.
Oscillation history of particle on a rotating parabola for (a) $A = 0.50$ and (b) $A = 2.0$. CPLM estimate—Lines; exact solution—Markers.

5. Concluding remarks

Conservative oscillators generally exhibit nonlinear response, and they form a large class of natural and artificially vibrating systems. Hence, the study of the dynamic response of nonlinear conservative systems is important for understanding many physical phenomena and the design of systems. The main challenge in the theoretical analysis of nonlinear conservative systems is that exact solutions are normally not available except for a few special cases where exact solutions are derived in terms of special functions.

To date, many approximate analytical methods have been formulated for the periodic solution of nonlinear conservative oscillators. This chapter provides a brief survey of the recent advances in the formulation of approximate analytical schemes and then introduced a recent approximate analytical algorithm called the continuous piecewise linearization method. The CPLM has been shown to overcome the challenges of solution accuracy and simplicity usually encountered in using most of the existing approximate analytical methods. The CPLM combines major desirable features of solution schemes such as inherent stability, accuracy, and simplicity. It is simple enough to be introduced at the undergraduate level and is capable of handling conservative oscillators with very complex nonlinearity. Conservative systems of broad interest were used to demonstrate the wide applicability of the CPLM algorithm. As demonstrated above, an accuracy of less than 1.0% relative error can be achieved for most oscillators using few discretizations, say $n \leq 20$, except for oscillators with trigonometric nonlinearity where such accuracy is achieved with many discretizations. This chapter has been designed to stimulate interest in the use of CPLM for analyzing various types of conservative systems, especially those with complex nonlinearity.

Acknowledgements

The authors are grateful to the Vice-Chancellor of the University of Port Harcourt, Prof. Ndowa E.S. Lale and the Dean of Engineering, Prof. Ogbonna Joel, for approving the publication grant for this chapter. The OAPF was funded by the African Centre of Excellence in Oilfield Chemicals Research (ACE-FOR), University of Port Harcourt.

Author details

Akuro Big-Alabo* and Chinwuba Victor Ossia
Department of Mechanical Engineering, Faculty of Engineering, University of Port Harcourt, Port Harcourt, Nigeria

*Address all correspondence to: akuro.big-alabo@uniport.edu.ng

References

[1] Fidlin A. Nonlinear Oscillations in Mechanical Engineering. New York: Springer; 2005

[2] He JH. Preliminary report on the energy balance for nonlinear oscillations. Mechanics Research Communications. 2002;**29**:107-111

[3] Jordan DW, Smith P. Nonlinear Ordinary Differential Equations: Problems and Solutions. Oxford: Oxford University Press; 2007

[4] Nayfeh AH, Mook DT. Nonlinear Oscillations. New York: John Wiley & Sons; 1995

[5] Shaban M, Ganji DD, Alipour MM. Nonlinear fluctuation, frequency and stability analyses in free vibration of circular sector oscillation systems. Current Applied Physics. 2010;**10**: 1267-1285

[6] Big-Alabo A, Cartmell MP, Harrison P. On the solution of asymptotic impact problems with significant localised indentation. Journal of Mechanical Engineering Sciences. 2017;**231**(5):807-822

[7] Sanchez NE. A view to the new perturbation technique valid for large parameters. Journal of Sound and Vibration. 2005;**282**:1309-1316

[8] Big-Alabo A. Periodic solutions of Duffing-type oscillators using continuous piecewise linearization method. Mechanical Engineering Research. 2018;**8**(1):41-52

[9] Nayfeh AH. Perturbation Methods. New York: John Wiley & Sons; 1973

[10] Bender CM, Milton KA, Pinsky SS, Simmons LM Jr. A new perturbative approach to nonlinear problems. Journal of Mathematical Physics. 1989;**30**(7): 1447-1455

[11] He JH. Homotopy perturbation technique. Computer Methods in Applied Mechanics and Engineering. 1999;**178**(3/4):257-262

[12] Odibat ZM. A new modification of the homotopy perturbation method for linear and nonlinear operators. Applied Mathematics and Computation. 2007; **189**(1):746-753

[13] Khan Y, Wu Q. Homotopy perturbation transform method for nonlinear equations using He's polynomials. Computer and Mathematics with Applications. 2019; **61**(8):1963-1967

[14] El-Dib YO, Moatimid GM. Stability configuration of a rocking rigid rod over a circular surface using the homotopy perturbation method and Laplace transform. Arabian Journal for Science and Engineering. 2019;**44**(7):6581-6591

[15] Kuang W, Wang J, Huang C, Lu L, Gao D, Wang Z, et al. Homotopy perturbation method with an auxiliary term for optimal design of a tangent nonlinear packaging system. Journal of Low Frequency, Noise, Vibration and Active Control. 2019. DOI: 10.1177/1461348418821204

[16] Adamu MY, Ogenyi P, Tahir AG. Analytical solutions of nonlinear oscillator with coordinate-dependent mass and Euler-Lagrange equation using the parameterized homotopy perturbation method. Journal of Low Frequency, Noise, Vibration and Active Control. 2019. DOI: 10.1177/1461348418821222

[17] Song H. A modification of homotopy perturbation method for a hyperbolic tangent oscillator arising in nonlinear packaging system. Journal of Low Frequency, Noise, Vibration and Active Control. 2019. DOI: 10.1177/1461348418822135

[18] Cheung YK, Chen SH, Lau SL. A modified Lindstedt-Poincare method for certain strongly non-linear oscillators. International Journal of Non-Linear Mechanics. 1991;**26**(3/4):367-378

[19] He JH. Modified Lindstedt-Poincare methods for some strongly nonlinear oscillations part III: Double series expansion. International Journal of Nonlinear Science and Numerical Simulation. 2001;**2**:317-320

[20] He JH. Modified Lindstedt-Poincare methods for some strongly nonlinear oscillations part I: Expansion of a constant. International Journal of Non-Linear Mechanics. 2002;**37**:309-314

[21] He JH. Modified Lindstedt-Poincare methods for some strongly nonlinear oscillations part II: A new transformation. International Journal of Non-Linear Mechanics. 2002;**37**:315-320

[22] He JH. Bookkeeping parameter in perturbation methods. International Journal of Nonlinear Science and Numerical Simulation. 2001;**2**:257-264

[23] He JH. Iteration perturbation method for strongly nonlinear oscillations. Journal of Vibration and Control. 2001;**7**(5):631-642

[24] He JH. Some new approaches to Duffing equation with strongly and high order nonlinearity (II) parameterized perturbation technique. Communications in Nonlinear Science and Numerical Simulation. 1999;**4**(1):81-83

[25] Chan HSY, Chung KW, Xu Z. A perturbation incremental method for strongly nonlinear oscillators. International Journal of Non-Linear Mechanics. 1996;**31**(1):59-72

[26] He JH. Linearized perturbation technique and its applications to strongly nonlinear oscillators. Computers & Mathematics with Applications. 2003;**45**(1-3):1-8

[27] He JH. Some asymptotic methods for strongly nonlinear equations. International Journal of Modern Physics B. 2006;**20**(10):1141-1199

[28] He JH. Author's reply to 'A view to the new perturbation technique valid for large parameters' by N.E. Sanchez. Journal of Sound and Vibration. 2005; **282**:1317-1320

[29] Adomian GA. Review of the decomposition method in applied mathematics. Journal of Mathematical Analysis and Applications. 1988;**135**: 501-544

[30] Laio SK. On the homotopy analysis method for nonlinear problems. Applied Mathematics and Computation. 1994; **147**(2):499-513

[31] He JH. Variational iteration method: A kind of nonlinear analytical technique: Some examples. International Journal of Nonlinear Mechanics. 1999;**34**(4): 699-708

[32] Durmaz S, Kaya MO. High-order energy balance method to nonlinear oscillators. Journal of Applied Mathematics. 2012;**2012**. Article ID 518684, 7p

[33] Khan Y, Mirzabeigy A. Improved accuracy of He's energy balance method for analysis of conservative nonlinear oscillator. Neural Computing and Application. 2014;**25** (3-4):889-895

[34] Molla MHU, Razzak MA, Alam MS. An analytical technique for solving quadratic nonlinear oscillator. Multidiscipline Modeling in Materials and Structures. 2017;**13**(3):424-433

[35] He JH. An improved amplitude-frequency formulation for nonlinear oscillators. International Journal of Nonlinear Science and Numerical Simulation. 2008;**9**(2):211-212

[36] He JH. Hamiltonian approach to nonlinear oscillators. Physics Letters A. 2010;**374**(23):2312-2314

[37] Farzaneh Y, Tootoonchi AA. Global error minimization method for solving strongly nonlinear oscillator differential equations. Computers and Mathematics with Applications. 2010;**59**:288-295

[38] Hosen MA, Chowdhury MSH, Ali MY, Ismail AF. A new analytical technique for solving nonlinear non-smooth oscillators based on the rational harmonic balance method. In: Saian R, Abbas MA, editors. Proceedings of the Second International Conference on the Future of ASEAN (ICoFA) 2017. Vol. 2. 2018

[39] Qian YH, Pan JL, Chen SP, Yao MH. The spreading residue harmonic balance method for strongly nonlinear vibrations of a restrained cantilever beam. Advances in Mathematical Physics. 2017;**2017**. Article ID 5214616, 8p

[40] Lai SK, Lim CW. Nonlinear vibration of a two-mass system with nonlinear stiffnesses. Nonlinear Dynamics. 2007;**49**:233-249

[41] Gimeno E, Belendez A. Rational-harmonic balancing approach to nonlinear phenomena governed by pendulum-like differential equations. Zeitschrift für Naturforschung. 2009; **64a**:819-826

[42] Lim CW, Lai SK, Wu BS. Accurate higher-order analytical approximate solution to large-amplitude oscillating systems with a general non-rational restoring force. Nonlinear Dynamics. 2005;**42**:267-281

[43] Bravo YS, Sanchez MA. A weighted mean-square method of "cubication" for nonlinear oscillators. Journal of Sound and Vibration. 1989; **134**(3):423-433

[44] Belendez A, Alvarez ML, Fernandez E, Pascual I. Cubication of conservative nonlinear oscillators. European Journal of Physics. 2009; **30**(5):973-981

[45] Elıas-Zuniga A, Martınez-Romero O. Accurate solutions of conservative nonlinear oscillators by the enhanced cubication method. Mathematical Problems in Engineering. 2013;**2013**:9. Article ID 842423

[46] Big-Alabo A. A simple cubication method for approximate solution of nonlinear Hamiltonian oscillators. International Journal of Mechanical Engineering Education. 2019. DOI: 10.1177/0306419018822489

[47] He JH. Variational approach for nonlinear oscillators. Chaos, Solitons and Fractals. 2007;**34**:1430

[48] Ismail GM. An analytical coupled homotopy-variational approach for solving strongly nonlinear differential equation. Journal of the Egyptian Mathematical Society. 2017;**25**(4):434-437

[49] Yazdi MK. Approximate solutions to nonlinear oscillations via an improved He's variational approach. Karbala International Journal of Modern Science. 2016;**2**:289-297

[50] Ebaid AE, Ali E. On a new aftertreatment technique for differential transformation method and its application to non-linear oscillatory systems. International Journal of Nonlinear Science. 2009;**8**(4):488-497

[51] Zhang X, Zou L, Liang S, Lui C. A novel analytic approximate method with a convergence acceleration parameter for solving nonlinear problems. Communications in Nonlinear Science and Numerical Simulation. 2018;**56**:354-364

[52] Uhler HS. Period of the bifilar pendulum for large amplitudes. Journal of the Optical Society of America. 1923; **7**(3):263-269

[53] Big-Alabo A. Approximate periodic solution to the large-amplitude oscillations of a simple pendulum. International Journal of Mechanical Engineering Education. 2019. DOI: 10.1177/0306419019842298

[54] Big-Alabo A, Harrison P, Cartmell MP. Algorithm for the solution of elastoplastic half-space impact: Force-indentation linearisation method. Journal of Mechanical Engineering Sciences. 2015;**229**(5):850-858

[55] Big-Alabo A. Equivalent impact system approach for elastoplastic impact analysis of dissimilar spheres. International Journal of Impact Engineering. 2018;**113**:168-179

[56] Big-Alabo A. Continuous piecewise linearization method for approximate periodic solution of the relativistic oscillator. International Journal of Mechanical Engineering Education. 2018. DOI: 10.1177/0306419018812861

[57] Butikov EI. The rigid pendulum–An antique but evergreen physical model. European Journal of Physics. 1999;**20**: 429-441

[58] Lima FMS. Simple but accurate periodic solutions for the nonlinear pendulum equation. Revista Brasileira de Ensino de Física. 2019;**41**(1): e20180202-1-6

[59] Arnold VI. Mathematical Methods of Classical Mechanics. 2nd ed. New York: Springer; 1989. pp. 261-264

[60] Mickens RE. Oscillations in an $x^{4/3}$ potential. Journal of Sound and Vibration. 2001;**246**:375-378

[61] Gottlieb H. Frequencies of oscillators with fractional-power nonlinearities. Journal of Sound and Vibration. 2003;**261**(3):557-566

[62] Hu H, Xiong ZG. Oscillations in an $x^{(2m+2)/(2n+1)}$ potential. Journal of Sound and Vibration. 2003;**259**:977-980

[63] Belendez A. Homotopy perturbation method for a conservative $x^{1/3}$ force nonlinear oscillator. Computers and Mathematics with Applications. 2009; **58**:2267-2273

[64] Cveticanin L, Kovacic I, Rakaric Z. Asymptotic methods for vibrations of the pure non-integer order oscillator. Computers and Mathematics with Applications. 2010;**60**:2616-2628

[65] Cveticanin L. Strong Nonlinear Oscillators: Analytical Solutions. 2nd ed. Switzerland: Springer; 2018